Dynamic Analysis of Machines

McGraw-Hill Series in
MECHANICAL ENGINEERING
Robert M. Drake, Jr.
Stephen J. Kline, *Consulting Editors*

Beggs, *Mechanism*
Cambel and Jennings, *Gas Dynamics*
Durelli, Phillips, and Tsao, *Introduction to the Theoretical and Experimental Analysis of Stress and Strain*
Eckert, *Introduction to Heat and Mass Transfer* (Translated by Joseph F. Gross)
Eckert and Drake, *Heat and Mass Transfer*
Gröber, Erk, and Grigull, *Fundamentals of Heat Transfer*
Ham, Crane, and Rogers, *Mechanics of Machinery*
Hartman, *Dynamics of Machinery*
Hinze, *Turbulence*
Jacobsen and Ayre, *Engineering Vibrations*
Phelan, *Fundamentals of Mechanical Design*
Raven, *Automatic Control Engineering*
Sabersky, *Elements of Engineering Thermodynamics*
Schenck, *Theories of Engineering Experimentation*
Schlichting, *Boundary Layer Theory*
Shigley, *Dynamic Analysis of Machines*
Shigley, *Kinematic Analysis of Mechanisms*
Shigley, *Mechanical Engineering Design*
Shigley, *Theory of Machines*
Spalding and Cole, *Engineering Thermodynamics*
Stoecker, *Refrigeration and Air Conditioning*
Wilcock and Booser, *Bearing Design and Application*

DYNAMIC ANALYSIS
OF MACHINES

Joseph Edward Shigley
PROFESSOR OF MECHANICAL ENGINEERING
THE UNIVERSITY OF MICHIGAN

McGRAW-HILL BOOK COMPANY

New York   Toronto   London

1961

DYNAMIC ANALYSIS OF MACHINES

Copyright © 1961 by the McGraw-Hill Book Company, Inc. Printed in the United States of America. All rights reserved. This book, or parts thereof, may not be reproduced in any form without permission of the publishers. *Library of Congress Catalog Card Number* 60-16465

9 10 11 12 – MAMM – 7 5 4 3 2

ISBN 07-056858-8

PREFACE

This book is written for the student of engineering, yet it has many features which should make it a useful addition to the practicing engineer's library.

When the plan for this book was originally prepared, the objectives were expressed to fulfill a group of needs which seemed to be gently insisting on satisfaction, but only during the writing of the manuscript these needs have grown from a somewhat muffled requirement to a loud and lusty demand. An attempt has thus been made in this book to satisfy the following requirements:

1. Utilize the unit vector approach in solving dynamics problems
2. Treat mechanical vibrations as an integral part of dynamics of machinery and utilize it in the development of the subject matter
3. Include a study of transient forces as well as steady-state forces in the analysis of machines
4. Include an introduction to the automatic control systems of mechanical engineering

The introduction of the unit vector approach in mathematics, mechanics, physics, and other areas of undergraduate engineering instruction makes it imperative that these same problem-solving tools be utilized in the professional subjects which follow. In fact, the power and versatility of vector analysis are so great and the advantages to be gained from using this method are so many, that its use has become quite widespread for the solution of problems in dynamics of machinery even without the impetus furnished by its introduction into the fundamental instructional areas. Furthermore, the use of digital computers to solve engineering problems requires that problems be expressed in analytical or numerical form before they are solved.

Similarly, now that differential equations occupy a firm position in undergraduate engineering education, mechanical vibrations must be placed in its proper position and used to develop the subject matter of dynamics of machinery and to strengthen the design sequence which usually follows. This arrangement has many advantages. It utilizes more of the student's mathematical preparation, and it does this while this preparation is still fresh in his memory. In dynamics it makes it possible to study the effect of transient as well as harmonic forcing on the operation of machines. Furthermore, the study of balance, or rather of unbalance, is a complete one when the subject of mechanical vibrations is considered first. Mechanical vibrations is treated as a separate subject in many colleges. This study can be strengthened when the student comes into it with advance preparation; perhaps it can even be extended to include nonlinear systems. And finally, it should be observed that the electronic differential analyzer, as a method of solving dynamics problems by analogy, is a tool which has been neglected too long in the education of a mechanical engineer. The addition of vibrations in dynamics courses makes this tool immediately useful. Also the analog computer can be used to build up a very worthwhile series of laboratory experiences for students of dynamics.

Machines do not run at constant speeds, they do not deliver constant outputs, and they are not driven by constant input sources. Machines do have to be started up, and when running, they may be subjected to unexpected disturbances and load fluctuations. It is wrong to reveal to the student only the steady-state conditions or a few instantaneous conditions. He should, at least, have an appreciation and an awareness of variable operating conditions, mechanical transients, and load disturbances. The nonlinearities in mechanical engineering—springs that do not obey Hooke's law, followers that jump off the cams, and gears that have backlash—should be known to the student even though he is not ready to analyze them in detail. Observation and awareness, after all, are the first steps toward understanding. This book provides the reader with the first tools for exploring the subject of mechanical transients and a means for showing him the nonlinearities. I am sure that we shall be able to do much more for the student than this after we have learned the use of digital and analog computers and are generally using them in undergraduate engineering education.

A study of automatic control systems is included in this book because it ought to be included somewhere in the education of a mechanical engineer and because I think the most logical place for it is in the study of the dynamics of machines. After the student has learned to analyze the forces in static and dynamic systems, after he has studied vibrations and utilized this material in the analysis of machines, and after a base of understanding in transient force analysis has been established, an investigation of automatic control systems follows easily and naturally. Thus the study of control systems terminates the book because it deserves this position.

The use of the unit vector approach will make possible some economies in instructional time because it is no longer necessary to work through the special cases up to the general case. Once the general approach is presented by vector methods the special cases drop out almost automatically. However, if vibration studies have not previously been offered, then this is not a sufficient saving to permit such an offering now. The additional time will have to be found elsewhere (but not from the humanities, I hope) because today's mechanical engineer cannot successfully compete with others unless he has a thorough understanding of dynamics.

The reader of this book should have a mathematical background through calculus including some instruction in linear differential equations, or he should be engaged in the study of differential equations. It is also expected that he will have completed the usual mechanics course in dynamics and a study of the kinematics of mechanisms. ("Kinematic Analysis of Mechanisms" and "Dynamic Analysis of Machines" are also available in a single combined volume under the title "Theory of Machines.")

Chapters 15 through 18 are fundamental and are necessary for an understanding of the remaining portion of the book. Certain sections of Chapter 17 may be omitted, however, without destroying the continuity. Chapter 20 on engine dynamics is the only one that I would be willing to omit in order to save instructional time. The remaining chapters are dependent and should be offered in the same order in which they are presented.

Robert F. Timm, Teaching Fellow at the University of Michigan, formulated the problems on feedback control systems. I am, indeed, appreciative of his help.

I am grateful, too, to the authors of three outstanding books for pointing out the direction in which education in dynamics of machines must proceed. These are James B. Hartman, "Dynamics of Machinery"; Lydik S. Jacobsen and Robert S. Ayre, "Engineering Vibrations"; and Harold A. Rothbart, "Cams."

I wish I could adequately express my thanks to Prof. Richard M. Phelan of Cornell University for his penetrating and thoughtful reviews of the manuscript. Let it at least be known that his comments, caustic but effective, rescued me from numerous pitfalls and that the final form of the book owes much to his wise counsel.

Joseph E. Shigley

CONTENTS

Preface v

14. Introduction to Dynamic Analysis 341

14-1. Principles of Dynamics. 14-2. Force Analysis. 14-3. Balancing. 14-4. Mechanical Vibrations. 14-5. Transient Effects. 14-6. Automatic Control Systems. 14-7. Notation.

15. Static-force Analysis 347

15-1. Forces. 15-2. Couples. 15-3. Conditions of Equilibrium. 15-4. Free-body Diagrams. 15-5. Analysis of a Four-bar Linkage. 15-6. Spur Gears. 15-7. Helical Gears. 15-8. Straight Bevel Gears. 15-9. Resisting Forces. 15-10. Force Analysis Using Coulomb Friction. 15-11. Forces in the Slider-crank Mechanism with Friction. 15-12. Efficiency. 15-13. Friction in Gear Teeth. 15-14. Worm Gears.

16. Dynamic-force Analysis—Plane-motion Mechanisms 384

16-1. D'Alembert's Principle. 16-2. Inertia Forces. 16-3. Analysis of a Floating Link. 16-4. Rotation. 16-5. An Example of Dynamic-force Analysis. 16-6. Shaking Forces. 16-7. The Method of Virtual Work.

17. Dynamic-force Analysis—Space Mechanisms 410

17-1. Linear Impulse and Momentum. 17-2. Moment of Momentum. 17-3. The Components of Moment of Momentum. 17-4. Motion of a Rigid Body. 17-5. Moments and Products of Inertia. 17-6. Translation of Axes. 17-7. Rotation of Axes. 17-8. Measuring Moment of Inertia. 17-9. Euler's Equations of Motion. 17-10. Rotation about a Fixed Axis. 17-11. Gyroscopes.

18. Vibration Analysis 438

18-1. Introduction. 18-2. Free Vibration without Damping. 18-3. Step-input Transient Forcing. 18-4. Phase-plane Representation. 18-5. Transient Disturbances. 18-6. Free Vibration with Damping. 18-7. Phase-plane Representation of Damped Vibration. 18-8. Harmonic Forcing. 18-9. Forcing Due to Unbalance. 18-10. Vibration Isolation. 18-11. Natural Frequencies of Beams and Rotating Shafts. 18-12. Torsional Systems. 18-13. Holzer's Tabulation Method. 18-14. Equivalent Systems.

19. Balancing 500

19-1. Static Unbalance. 19-2. Static Balancing Machines. 19-3. Dynamic Unbalance. 19-4. Graphical Analysis of Unbalance. 19-5. Dynamic Balancing. 19-6. Pivoted-cradle Balancing. 19-7. Nodal-point Balancing. 19-8. Mechanical Compensation. 19-9. Balancing Flexible Rotors. 19-10. Field Balancing.

20. Dynamics of the Reciprocating Engine 526

20-1. Engine Types. 20-2. Gas Forces. 20-3. Equivalent Masses. 20-4. Inertia Forces in the Single-cylinder Engine. 20-5. Bearing Loads in the Single-cylinder Engine. 20-6. Crankshaft Torque Delivered by the Single-cylinder Engine. 20-7. Balancing Single-cylinder Engines. 20-8. Balance of Multicylinder Engines. 20-9. Flywheels. 20-10. Torsional Models.

21. Cam Dynamics 569

21-1. Forces in Rigid Systems. 21-2. Mathematical Models. 21-3. Response of a Uniform-motion Undamped Cam Mechanism—Analytical Method. 21-4. Position Error. 21-5. Follower Response by Phase-plane Method. 21-6. Jump and Crossover Shock. 21-7. Johnson's Numerical Analysis. 21-8. Unbalance, Spring Surge, and Windup.

22. Dynamics of Feedback Control Systems 599

22-1. Examples of Automatic Control Systems. 22-2. Writing the Equations. 22-3. Standard Input Functions. 22-4. Solution of Linear Differential Equations. 22-5. Analysis of the Proportional-error Feedback System. 22-6. Response of the Proportional-error System to Harmonic Input. 22-7. Transfer Functions. 22-8. Stability. 22-9. Types of Controls. 22-10. Nonlinear Systems.

Appendix III. Moments of Inertia 633

Appendix IV. Indicator Diagrams 635

Index 641

CHAPTER 14

INTRODUCTION TO DYNAMIC ANALYSIS

In the study of kinematic analysis of mechanisms we were concerned only with the geometry of the motions and with the relationships between displacement and time. The forces which produced that motion or the motions which would result from the application of a given force system were completely neglected.

Consideration of a problem in the design of a machine in which only the units of *length* and *time* are involved is a tremendous simplification. It frees the mind of the complicating influence of many other factors which ultimately enter into the problem and permits one's attention to be focused on the primary problem—that of designing a mechanism to obtain a desired motion.

A suitable motion for a machine having been determined, the next step in the design is to give the members of the mechanism size and shape. When this is started, the designer is faced with the fact that the machine has an input and an output requiring the receiving, transformation, and delivery of energy. Sometimes forces are required to do work, and so forces must be applied to the input members and must be exerted by the output members. The size, shape, and material of these machine members must be such that they are strong enough to resist the forces which are applied to them. This is another way of saying that the internal forces must have a reasonable distribution in the members and that the intensity of these internal forces should not be too large. In addition to designing machine members with force intensities of reasonable magnitude, the designer is also expected to specify members which are stiff enough to carry the forces without bending or buckling excessively.

Forces are transmitted into machine members through mating surfaces —from a gear to a shaft or from one gear through meshing teeth to another gear, from a connecting rod through a bearing to a lever, from a V belt to a pulley, from a cam to a follower, or from a brake drum to a brake shoe. The distribution of the forces at these boundaries or mating surfaces must be reasonable, and their intensities must be within the working limits of the materials composing the surface. If the force operating on a sleeve bearing becomes too high, it will squeeze out the oil film and cause metal-to-metal contact, overheating, and rapid failure of the bearing. If the forces between gear teeth are too large, the oil film may be squeezed out from between them. If the unit force exceeds the surface strength of the material, flaking and spalling of the metal may occur, resulting in noise, rough motion, and eventual failure.

Determination of the size of machine members requires a knowledge of the magnitude of the forces involved. If these forces are underestimated, then the parts may be designed with insufficient strength; then failure will occur rapidly. If the forces are overestimated, the machine members may be designed with more strength than is necessary. This is almost as bad because the resulting machine is overly expensive to manufacture, is heavy, and may not be competitive with others. Even if only a single member is designed too heavy, its excess weight results in an increased inertia force which is carried over into other members. Thus the size of the other members will have to be increased to carry the additional inertia force.

In studying the dynamics of a machine we are still interested in the motion of the elements, but we are also interested in the forces which result because of the motion and the forces which are required to produce the motion. The designer's responsibility is to define the size, shape, and material of a machine member, but this means that he has also defined the weight and stiffness of the member. When members have acceleration, then their weight adds inertia forces to the system for which we must account. As an example, a rotating system requires the application of torque to bring it up to speed; the magnitude of this torque depends upon the shape and weight of the mechanism and how fast it is to be brought up to speed.

The fundamental units in kinematic analysis are *length* and *time;* in dynamic analysis they are *length, time,* and *force.*

14-1. Principles of Dynamics. The fundamental principles of dynamics are *Newton's laws of motion.* They are not new to most readers of this book and so are repeated here for reference purposes only.

1. *A body will remain at rest or in a state of uniform motion in a straight line unless it is acted upon by an external force.*

2. *The rate of change of momentum of a body acted upon by an external force or forces is proportional to the resultant external force and in the direction of that force. If the mass is invariant, then the magnitude of the acceleration of the body is proportional to the force acting upon it and inversely proportional to the mass of the body. The direction of the acceleration is the same as the direction of the resultant external force.*

3. *For every action of a force there is an equal and opposite reaction.*

The first law is called the *law of inertia.* It states that the velocity of a body changes when that body is acted upon by a force; it does *not* quantitatively relate the velocity change to the force. Note that *both* force and velocity are vector quantities and have direction as well as magnitude.

The second law quantitatively relates the velocity change to the force and is the law with which we are most concerned in the study of dynamic analysis. Note that the second law states nothing about the velocity before or after the application of the force. It is concerned only with the change in velocity, that is, the acceleration. $a = \frac{F}{m}$

We shall find the third law quite useful in studying the forces which act upon bodies at rest or upon bodies in motion when they have no acceleration.

14-2. Force Analysis. In analyzing the forces and motions of machine parts we shall find a great variety of force-time relationships. The piston of an automotive engine delivers energy to the crankshaft only once in two revolutions of the crankshaft; during a part of these two revolutions the crankshaft is delivering energy to the piston. The force exerted by a cam against its follower is greatest when it is lifting the follower and load. The quick-return linkage of a shaper has a certain force-time relationship during the cutting stroke and a different one during the return stroke when the cutter is doing no work. It is a rare machine indeed in which the relation between force and time is a constant.

The power stroke of an automotive engine begins with the piston near the top dead position. As the piston moves downward and delivers energy to the crank, the forces existing at any moment depend upon many factors. The geometry of the slider-crank mechanism and the phase or position of the mechanism at any instant in time, the relationship between the gas pressure and the piston displacement, and the speed of the engine, all affect the forces existing in the parts of the mechanism. Inertia forces, too, will exist. The energy which the piston delivers to the crankshaft through the connecting rod is used up in many ways. Some of it is stored momentarily in the flywheel, some is delivered to other pistons for their compression stroke and to overcome friction, and some of it goes directly to the rear wheels to propel the car. The relationship between force and time is, indeed, not a simple one for an automotive engine.

In some cases force-time relationships are quite simple and can be expressed in the form of a sine wave, a square wave, or a triangular wave. More complicated relationships can often be approximated as a series of sine waves which are summed to obtain the total relationship. A series of trigonometric terms, each term having its own amplitude and frequency, can be used to represent many complex force-time relationships. In this book we shall develop analytical as well as graphical methods for determining the force-time relations in machines.

14-3. Balancing. When a body is rotated about an axis which is not its axis of symmetry, inertia forces and couples (rather loosely termed centrifugal forces and couples) are produced. The addition or subtraction of masses to reduce or eliminate the effect of these forces and couples is termed *balancing*. Thus a rotating system is said to be *balanced* if these couples and forces do not exist and *unbalanced* if they do exist. Large unbalanced forces may produce objectionable vibrations in the foundation or supporting structure of the machine. When the machine designer is aware of the nature and source of unbalance, there is a great deal that he can do to minimize this problem.

Many machines have parts which move with a motion other than rotation about a fixed axis. For example, the piston of an engine reciprocates, and other parts oscillate on circular arcs or curved paths. The designer is in a unique position, since he has the power to control, reduce, and eliminate the forces and couples which come about because of the masses of these oscillating machine parts.

The problem of obtaining balanced moving parts is emphasized by the present tendency toward increasing the operating speeds. For example, an eccentric mass rotating at 3,000 rpm will create 100 times more unbalanced force than the same mass rotating at 300 rpm. For these reasons we shall investigate the subject of balancing in considerable detail.

14-4. Mechanical Vibrations. A spring acted upon by a force deforms; the amount of deformation depends upon the stiffness or elasticity of the spring and the magnitude of the force. The deflection given to a spring by any action, say a cam, may cause the spring to oscillate. The amplitude of the resulting oscillation depends upon the mass and stiffness of the spring, upon the amount of friction or damping which spring oscillation must overcome, and upon the frequency and magnitude of the cam-imposed deflection. For particular relationships of these parameters it is possible to destroy the spring, using a low-amplitude force, simply by the violence of the oscillation.

All machine parts are elastic and, like springs, deform when a force is applied to them. The crankshaft of an automotive engine twists and bends when the forces of the exploding gas are transmitted to the cranks. The combination of a certain crankshaft stiffness and mass together with a certain speed may result in violent torsional vibrations of the crankshaft and lead to rapid destruction. If automotive valve springs are operated near their natural frequency, violent surges may begin. Sometimes these surges have an amplitude which is sufficient to permit the valve to lift from its seat. High-amplitude surges such as these produce stresses in the springs which often greatly exceed the design limits.

Much has been written on the subject of mechanical vibrations, and space requirements in this book will permit only an introduction to this important subject. However, machine speeds are continually increasing and the volume, size, and weight of the moving parts are decreasing because improved materials and more accurate methods of analysis are available to the engineer. This means that it is increasingly important that every mechanical engineer have basic knowledge of mechanical vibrations.

14-5. Transient Effects. The velocity of a steam turbine in an electric generating plant may vary only a few revolutions per minute from a constant speed of 3,600 rpm; the turbine runs for hour after hour and day after day at the same speed and perhaps at the same load. This is termed a *steady-state* condition of operation. The forces due to speed (neglecting load changes) are the same today as they were yesterday.

Now contrast the operation of the steam turbine to that of an automotive engine which seldom runs under steady-state speed conditions for more than 15 min at a time. When the operating conditions are not steady-state, they are described as *transient*. A transient condition is any operating condition of relatively short duration. Transient conditions will sometimes exist in all machines because they must be started as well as stopped. A great many machines are operated under transient conditions during their entire lifetime.

Expected or unexpected disturbances of a transient character may operate on the input of a machine, on the output, or on one or more intermediate points. An engine lathe running into a heavy cut, a power failure on a motor, the sudden application of brakes to an automobile, tripping of the main output power switches in a power plant by a lightning storm, the sudden decision of a jet pilot not to land after he is over the runway with wheels down are all examples of disturbances to mechanical systems. It is just as important to study the force situation under transient conditions as it is to study it under steady-state conditions. Consequently, this aspect of dynamic analysis will be considered in some detail in this book. It turns out that, though a cam mechanism is usually considered as a steady-state mechanism, it can be analyzed excellently using transient techniques. Our study of cam dynamics, therefore, presents a particularly favorable opportunity to investigate transient phenomena.

14-6. Automatic Control Systems. An automatically controlled machine is one which continuously inspects its own output, compares the output with what it should be, and, if a difference is found, makes its own correction. Some have labeled automatic control systems as "thinking machines" because they do replace human operators. Our interest in automatic control systems here results because of the fact that they are used to control physical processes and to do this means that their elements must themselves be at least partly mechanical.

Therefore, quantities like force, torque, displacement, velocity, and acceleration are involved. The very fact that a machine requires control indicates that something is changing. For this reason the problem does involve an application of the principles of dynamics. This, together with the knowledge of the tremendous growth of automatic control in recent years, means that the subject of automatic control is a necessary and proper part of dynamics of machinery.

14-7. Notation. In Part 1 of this book boldface capital letters were employed as vectors and lightface lower-case letters as scalars. This notation cannot be used in the study of dynamics of machinery without deviating sharply from the established notation of the subject. For example, if we were to designate **T** as a torque vector, then its magnitude, in the notation of Part 1, would be represented by the scalar t. But t is used universally to represent time. Both torque and time appear frequently in the same equation in dynamic analysis, and so the former notation cannot be employed.

It is assumed here that the reader possesses a good background in both two- and three-dimensional vector analysis. Our notation shall be simply to employ boldface characters for vectors and lightface for scalars without regard to whether they are upper- or lower-case characters. If the reader is familiar with vector analysis, he will have no difficulty with this notation. Of course, when kinematic relations must be established, the equations of Part 1 will be used without any change in notation. Thus the notation of everything which has been developed in Part 1 will be preserved in Part 2.

CHAPTER 15

STATIC-FORCE ANALYSIS

It is appropriate to begin the study of dynamics with a brief review of statics, but with particular attention paid to the statics of machines. This also serves to introduce fundamental concepts, procedures, and notation to be employed throughout the book.

15-1. Forces. Force has *magnitude* and *direction*, and so it is a vector quantity and can be manipulated in exactly the same manner as other vector quantities. Force also has a *point of application*.

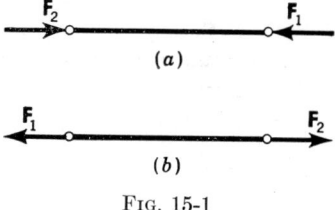

Fig. 15-1

When we are considering only the equilibrium of a rigid body, the point of application is not important. In Fig. 15-1a the link is in equilibrium under the action of forces F_1 and F_2. These forces place the link in compression. If we interchange F_1 and F_2 as in Fig. 15-1b, the link is still in equilibrium but now it is tension. The same forces are used in (a) as in (b) because, for example, F_1 has exactly the same magnitude and direction in (a) as it has in (b).

The point of application of a force is important when the stresses created by the force are to be found.

In this book we shall employ the boldface capital letter **F** to represent a force vector and the lightface capital letter F to represent the magnitude of the force vector. We can then use the triad of unit vectors **I**, **J**, **K**, as defined in Sec. 12-1, to represent a force vector in space. Thus

$$\mathbf{F} = F^x\mathbf{I} + F^y\mathbf{J} + F^z\mathbf{K} \qquad (a)$$

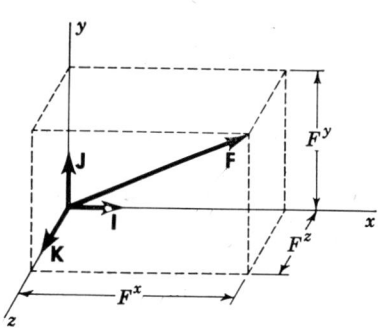

Fig. 15-2. The components of a force vector.

is a force vector in space. The force magnitudes F^x, F^y, and F^z are the

347

components of the force vector parallel to the x, y, and z axes, respectively, as shown in Fig. 15-2.

15-2. Couples. Two equal and opposite forces acting along two non-coincident parallel straight lines in a body cannot be combined to obtain a single resultant force. Any two such forces acting on a body constitute a *couple*. The *arm of the couple* is the perpendicular distance between their lines of action, and the *plane of the couple* is the plane containing the two lines of action.

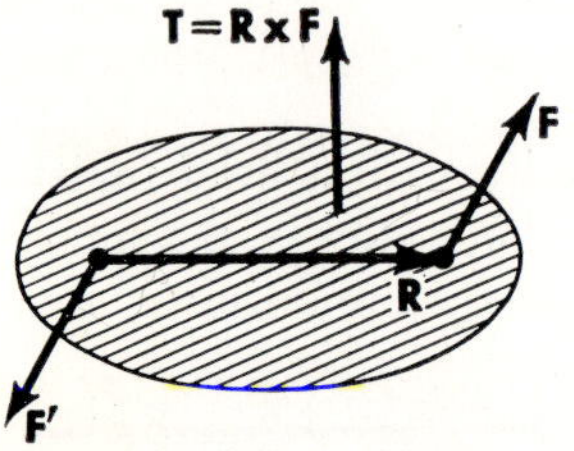

The *moment of a couple* is another vector **T** directed normal to the plane of the couple; the sense of **T** is in accordance with the right-hand rule for rotation. The magnitude of the moment is the product of the arm of the couple and the magnitude of one of the forces. If we designate the magnitude of the moment by T, then

$$T = hF \tag{15-1}$$

where h is the moment arm.

Fig. 15-3. Note that **R** is a position vector but **F** and **F'** are force vectors; the free vector **T** is the moment of the couple formed by **F** and **F'**.

As shown in Fig. 15-3 the moment vector is the cross product[1] of the relative position vector **R** and the force vector **F** and is defined by the equation

$$\mathbf{T} = \mathbf{R} \times \mathbf{F} \tag{15-2}$$

Designating

$$\mathbf{R} = x\mathbf{I} + y\mathbf{J} + z\mathbf{K} \tag{a}$$
$$\mathbf{F} = F^x\mathbf{I} + F^y\mathbf{J} + F^z\mathbf{K} \tag{b}$$

then the moment vector is

$$\mathbf{T} = \mathbf{R} \times \mathbf{F} = \begin{vmatrix} \mathbf{I} & \mathbf{J} & \mathbf{K} \\ x & y & z \\ F^x & F^y & F^z \end{vmatrix} \tag{c}$$

and its components are

$$\begin{aligned} T^x &= yF^z - zF^y \\ T^y &= zF^x - xF^z \\ T^z &= xF^y - yF^x \end{aligned} \tag{d}$$

where

$$\mathbf{T} = T^x\mathbf{I} + T^y\mathbf{J} + T^z\mathbf{K} \tag{e}$$

Some of the rather interesting properties of couples can be determined by an examination of Fig. 15-4. Here $\mathbf{F}_1$ and $\mathbf{F}_2$ are two equal and opposite forces. Choose any point on each line of action, and define these points by the position vectors $\mathbf{R}_1$ and $\mathbf{R}_2$. Then the relative position

---

[1] See Sec. 12-3 for the definition of vector products.

STATIC-FORCE ANALYSIS

vector is
$$R_{21} = R_2 - R_1 \tag{f}$$

The moment of the couple is the sum of the moments of each force and is
$$T = R_1 \times F_1 + R_2 \times F_2 \tag{g}$$

But $F_1 = -F_2$ and so Eq. (g) can be written
$$T = (R_2 - R_1) \times F_2 = R_{21} \times F_2 \tag{h}$$

Equation (h) shows:

1. The value of the moment of the couple is independent of the choice of the center about which the moments are taken because the relative position vector R_{21} is the same for all positions of the origin.

2. Since R_1 and R_2 define any set of points on the lines of action, the relative position vector R_{21} is not restricted to perpendicularity with F_1 and F_2. This is a very important result of the vector product because it means that the value of the moment is independent of how R_{21} is chosen. The magnitude of the moment can be obtained as follows: Resolve R_{21} into two components $R_{21}{}^t$ and $R_{21}{}^n$, parallel and perpendicular, respectively, to F_2. Then we have

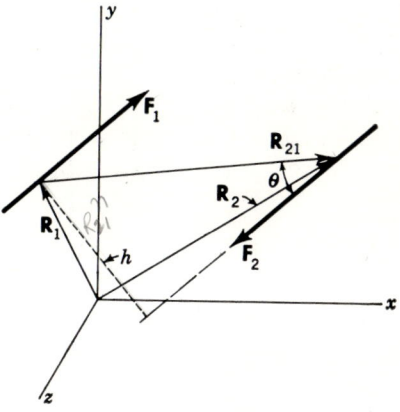

FIG. 15-4

$$T = R_{21}{}^t \times F_2 + R_{21}{}^n \times F_2 \tag{i}$$

But $R_{21}{}^n$ is the perpendicular distance between the lines of action and $R_{21}{}^t$ is parallel to F_2. Therefore $R_{21}{}^t \times F_2 = 0$ and
$$T = R_{21}{}^n \times F_2$$

is the moment of the couple. Since $R_{21}{}^n = R_{21} \sin \theta$, where θ is the angle between R_{21} and F_2, the magnitude of the moment is
$$T = (R_{21} \sin \theta) F_2 \tag{j}$$

3. The moment vector T is independent of any particular origin and is thus a *free vector*.

4. The forces of a couple may be rotated together within their plane, keeping their magnitudes and the distance between their lines of action constant, or they may be translated to any parallel plane without chang-

ing the magnitude or sense of the moment vector. Also, two couples are equal if they have the same moment vectors, regardless of the values of the forces or moment arms. In other words, it is the vector product of the two that is significant rather than their individual values.

15-3. Conditions of Equilibrium. A rigid body is in static equilibrium if:

1. *The vector sum of all the forces acting upon it is zero.*
2. *The sum of the moments of all the forces acting about any single axis is zero.*

This requires that
$$\Sigma F = 0 \quad \text{and} \quad \Sigma T = 0 \tag{15-3}$$

Since the forces and their moments are often defined in terms of their components relative to the three coordinate axes, another form of Eq. (15-3) is

$$\begin{aligned} \Sigma F^x &= 0 & \Sigma T^x &= 0 \\ \Sigma F^y &= 0 \quad \text{and} \quad \Sigma T^y &= 0 \\ \Sigma F^z &= 0 & \Sigma T^z &= 0 \end{aligned} \tag{15-4}$$

A great many problems in dynamic analysis have all the forces acting in a single plane. When this is true, it is appropriate to choose the plane of the forces as the xy plane. Under these conditions Eqs. (15-4) become

$$\Sigma F^x = 0 \quad \Sigma F^y = 0 \quad \Sigma T^z = 0 \tag{15-5}$$

and the remaining terms of Eqs. (15-4) are identically zero. Sometimes T^z will be written simply as T with the z direction implied by the fact that the forces exist only in the xy plane.

15-4. Free-body Diagrams. The term "body" as used here may be an entire machine, several parts of a machine, a single machine part, or a portion of a machine part. A *free-body diagram* is a sketch of the body, isolated from the machine, on which the forces and moments are shown in action. It is usually desirable to include on the diagram the known magnitudes and directions as well as other pertinent information.

The diagram so obtained is called "free" because the part or portion of a machine part has been freed from the remaining machine elements and their effects replaced by forces and moments. If the free-body diagram is of an entire machine part, then the forces shown on it are the *external* forces and moments exerted by adjoining parts. If the diagram is a portion of a machine part, then the forces and moments acting on the cut portion are the *internal* forces and moments exerted by the part which has been cut away.

The construction and presentation of clear and neatly drawn free-body diagrams represent the heart of engineering communication. This is so because they represent a part of the thinking process whether they are actually placed on paper or not and because the construction of these diagrams is the *only* manner in which the results of thinking can be communicated to others.[1] The reader should acquire the habit of drawing free-body diagrams no matter how simple the problem may appear to be. They are a means of storing one thought while proceeding toward the solution of the next step in a problem. Construction of the diagrams speeds up the problem-solving process and greatly decreases the chances of making mistakes.

The advantages of drawing free-body diagrams is excellently summarized in the following words by an unknown engineering educator:

Students do not generally recognize the importance of free-body diagrams (1) for translation of words or thoughts into physical models, (2) for general understanding of all facets of the problem, (3) for planning the method of attack on the problem, (4) for aid in setting up the mathematical relations which will lead to a solution, (5) for recording progress in solution so that judgment can be exercised when required, such as in making simplifying assumptions, (6) for illustration of the methods used in the solution, and (7) for presentation of the results in a clear manner. Some of these are more important to those reviewing the work than to the student making the solution; but practically all an engineer's work is done for others; therefore, a student should learn this form of communication.

15-5. Analysis of a Four-bar Linkage. *Graphical Solution.* As an example of the static-force analysis of a linkage we select the four-bar linkage of Fig. 15-5a. The force $\mathbf{P} = 120$ lb acting at Q on link 4 is given. It is required to determine the forces and couples acting on each link as well as those which act upon the frame.

The notation to be employed is illustrated in the figure. $\mathbf{F}_{34}$ is the force exerted by link 3 on link 4. $\mathbf{F}_{43}$ is the force exerted by link 4 on link 3. $\mathbf{F}_{34}$ and $\mathbf{F}_{43}$ have the same magnitudes but are opposite in sense. $\mathbf{T}_2$ is an external couple acting on link 2. The graphical solution is obtained as follows:

1. (Fig. 15-5a.) Select a space scale and a force scale and draw the mechanism. Draw the given force or forces to scale and in their proper direction.

2. (Fig. 15-5b.) Select a link or links from which the analysis can be started and draw a free-body diagram of the link. In this case we have selected link 4 with which to begin. The magnitude and direction of $\mathbf{P}$ are given. Link 3 is pivoted at the ends, and so it must be either in

---

[1] It is important for the reader to realize that he can advance in his profession only if another person understands how outstanding the reader really is! This requires some form of communication—written, oral, or graphical. Successful people are those who *are* really outstanding and, in addition, have successfully communicated this information to others. Advancement in the engineering profession requires that one become proficient in all three modes of communication.

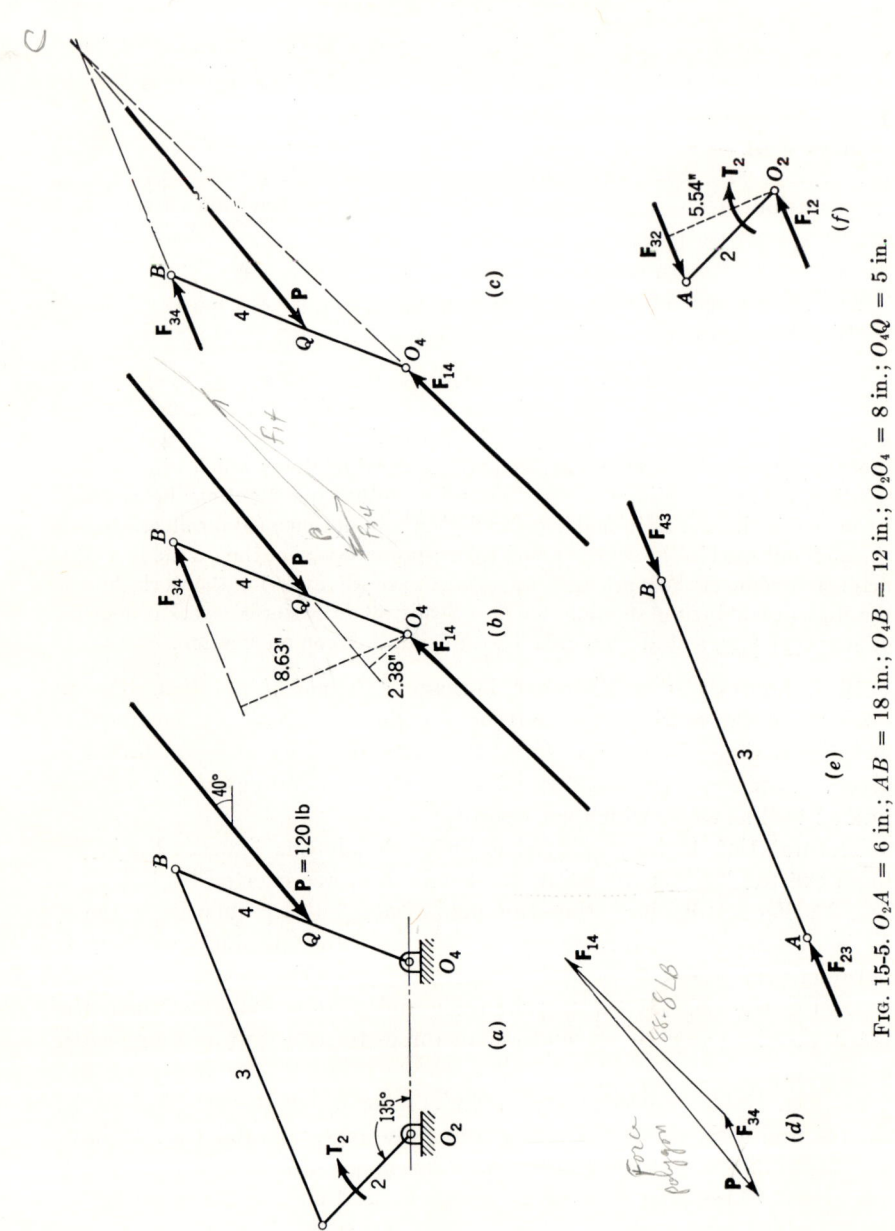

Fig. 15-5. $O_2A = 6$ in.; $AB = 18$ in.; $O_4B = 12$ in.; $O_2O_4 = 8$ in.; $O_4Q = 5$ in.

tension or in compression. Therefore the line of action of $\mathbf{F}_{34}$ is known, though not its magnitude. Neither the direction nor the magnitude of $\mathbf{F}_{14}$, the reaction of the frame at O_4, is known.

Draw and measure the moment arms of $\mathbf{P}$ and $\mathbf{F}_{34}$ about O_4. These are found to be 2.38 and 8.63 in., respectively. Summing the moments of these two forces about O_4 gives the magnitude of $\mathbf{F}_{34}$.

$$\Sigma T_{O_4} = 0 \qquad (2.38)(120) + 8.63 F_{34} = 0$$

$$F_{34} = -\frac{(2.38)(120)}{8.63} = -33.1 \text{ lb}$$

Here F_{34} is the magnitude of the vector $\mathbf{F}_{34}$. The negative sign indicates that the moment of $\mathbf{F}_{34}$ about O_4 is clockwise as shown in the figure.

3. (Fig. 15-5c.) Since link 4 is a three-force member, it may be desirable to determine the direction of $\mathbf{F}_{14}$, using the point of concurrency, instead of finding the magnitude of $\mathbf{F}_{34}$ as was done in step 2. If the lines of action of $\mathbf{P}$ and $\mathbf{F}_{34}$ are extended, they will intersect at C. The line of action of $\mathbf{F}_{14}$ is now determined because it must pass through both O_4 and C.

4. (Fig. 15-5d.) Using the results either of step 2 or of step 3 construct the force polygon for link 4. In either event the polygon is the solution of the vector equation

$$\mathbf{P} + \mathbf{F}_{34} + \mathbf{F}_{14} = 0$$

If step 2 is used, both $\mathbf{P}$ and $\mathbf{F}_{34}$ are given and the magnitude and direction of $\mathbf{F}_{14}$ are found as the closing vector. If step 3 is used, $\mathbf{P}$ is given, together with the direction of $\mathbf{F}_{34}$ and $\mathbf{F}_{14}$. The magnitudes of the latter two vectors are found when the polygon is constructed. The magnitude of $\mathbf{F}_{14}$ is

$$F_{14} = 88.8 \text{ lb}$$

5. (Fig. 15-5e.) Construct the free-body diagram of link 3. Here $\mathbf{F}_{23} = -\mathbf{F}_{43} = \mathbf{F}_{34}$.

6. (Fig. 15-5f.) Construct the free-body diagram of link 2. Here $\mathbf{F}_{32} = -\mathbf{F}_{23}$ and $\mathbf{F}_{12} = -\mathbf{F}_{32}$. If the moment arm of $\mathbf{F}_{32}$ about O_2 is measured, it is found to be 5.54 in. The external moment is, therefore,

$$T_2 = -(33.1)(5.54) = -183 \text{ in.-lb}$$

The negative sign indicates that the external moment must be applied in a clockwise direction.

Analytical Solution. To apply this method, the angular location of each link must first be found. This is done using the methods of Appen-

354 DYNAMIC ANALYSIS OF MACHINES

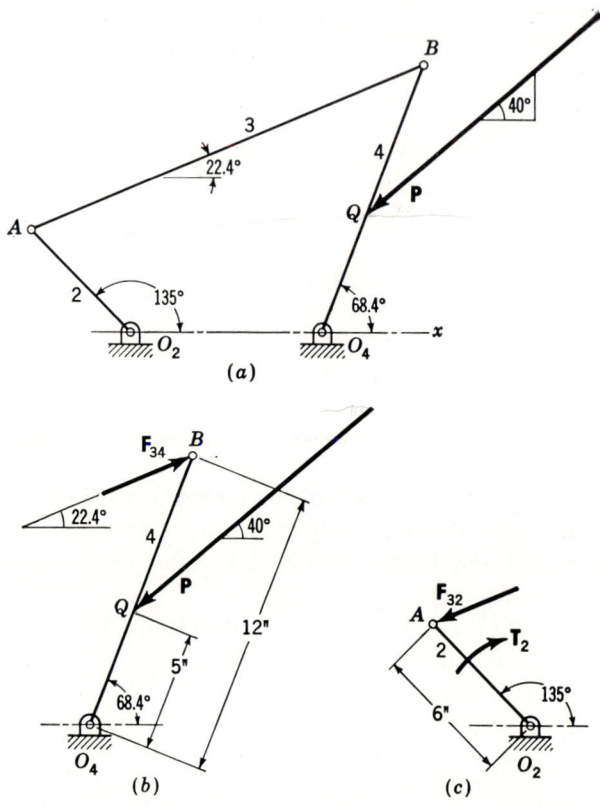

Fig. 15-6

dix I, and the results are shown in Fig. 15-6a. Referring now to Fig. 15-6b we shall apply the condition that $\Sigma T^z = 0$ about an axis through O_4. The values of the forces and position vectors are[1]

$$P = 120\underline{/-140°} = -91.9I - 77.1J \qquad (a)$$
$$F_{34} = F_{34}\underline{/22.4°} = F_{34}{}^x I + F_{34}{}^y J \qquad (b)$$
$$R_Q = 5\underline{/68.4°} = 1.84I + 4.65J \qquad (c)$$
$$R_B = 12\underline{/68.4°} = 4.41I + 11.17J \qquad (d)$$

The moments of these forces about O_4 are

$$R_Q \times P = \begin{vmatrix} I & J & K \\ 1.84 & 4.65 & 0 \\ -91.9 & -77.1 & 0 \end{vmatrix} = 285K \qquad (e)$$

[1] The transformation of plane vectors from the polar form to rectangular coordinates is very easily accomplished on the slide rule. See Sec. 2-12, Example 2-1.

$$R_B \times F_{34} = \begin{vmatrix} I & J & K \\ 4.41 & 11.17 & 0 \\ F_{34}{}^x & F_{34}{}^y & 0 \end{vmatrix} = (4.41F_{34}{}^y - 11.17F_{34}{}^x)K \qquad (f)$$

Applying $\Sigma T = 0$ to Eqs. (e) and (f) gives

$$4.41F_{34}{}^y - 11.17F_{34}{}^x + 285 = 0 \qquad (g)$$

But, from Eq. (b), we can write

$$\frac{F_{34}{}^y}{F_{34}{}^x} = \tan 22.4° = 0.412 \qquad (h)$$

Solving Eqs. (g) and (h) simultaneously yields the components of F_{34}. The result is

$$F_{34} = 30.6I + 12.6J = 33.1\underline{/22.4°} \text{ lb}$$

Reference again to Fig. 15-5b indicates that link 4 is a three-force member having the reaction of link 3, F_{34} acting at B, the applied force P acting at Q, and the frame reaction F_{14} acting at O_4. Equations (15-5) are written

$$P + F_{34} + F_{14} = 0 \qquad (i)$$

Substituting the values of P and F_{34} produces

$$(-91.9I - 77.1J) + (30.6I + 12.6J) + F_{14} = 0 \qquad (j)$$
or
$$F_{14} = 61.3I + 64.5J = 88.8\underline{/46.4°} \text{ lb} \qquad (k)$$

Next, in Fig. 15-6c we write $\Sigma T = 0$ for link 2. Since $F_{32} = -F_{34}$, its moment about O_2 is $R_A \times F_{32}$, where

$$R_A = 6\underline{/135°} = -4.25I + 4.25J \qquad (l)$$
$$F_{32} = -F_{34} = -33.1\underline{/22.4°} = -30.6I - 12.6J \qquad (m)$$

Therefore

$$R_A \times F_{32} = \begin{vmatrix} I & J & K \\ -4.25 & 4.25 & 0 \\ -30.6 & -12.6 & 0 \end{vmatrix} = 183K \qquad (n)$$

Applying Eqs. (15-5) gives

$$T_2 + R_A \times F_{32} = 0 \qquad \text{or} \qquad T_2 = -183K = 183 \text{ in.-lb}$$

In both of the preceding solutions it has been assumed that the forces all act in the same plane. For the connecting link 3 it was also assumed that the line of action of the forces and the center line of the link were coincident. A careful machine designer will sometimes go to extreme measures in order to approach these conditions as nearly as possible.

356 DYNAMIC ANALYSIS OF MACHINES

Note that if the pin connections are arranged as shown in Fig. 15-7a, then such conditions are theoretically obtained. On the other hand if the connection is like the one of Fig. 15-7b, then the pin itself as well as each link will have turning couples acting upon them. If the forces are not in the same plane, then couples exist whose moments are proportional to the distance between the force planes.

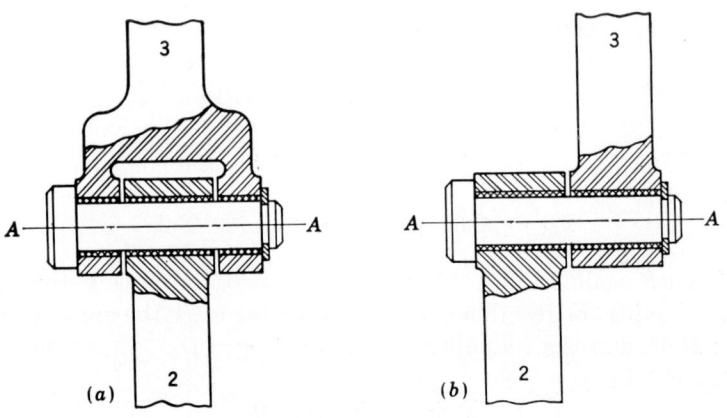

FIG. 15-7. (a) A balanced connection. (b) This connection produces a turning moment on the pin and on each link.

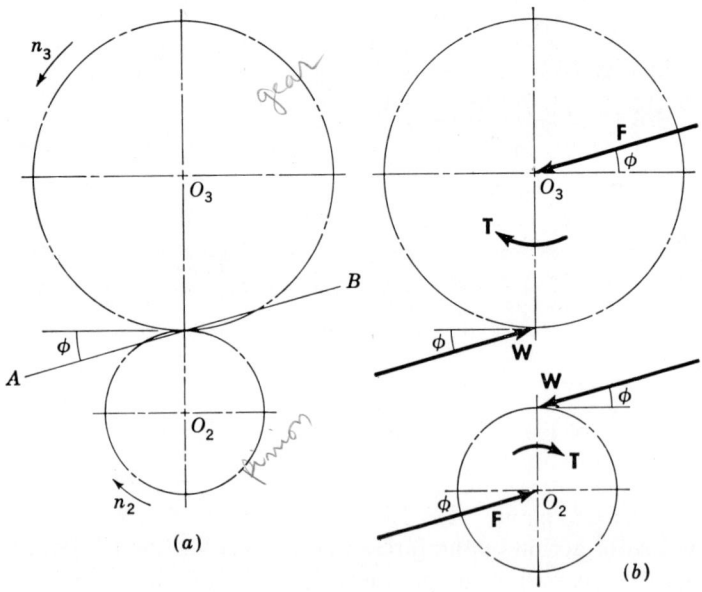

FIG. 15-8

15-6. Spur Gears. Figure 15-8a shows a pinion with center at O_2 rotating clockwise at n_2 rpm and driving a gear with center at O_3 at n_3 rpm. The reactions between the teeth occur along the pressure line AB. Free-body diagrams of the pinion and gear are shown in Fig. 15-8b. The action of the pinion on the gear has been replaced by the force **W** acting at the pitch point in the direction of the pressure line. Since the gear is supported by its shaft, an equal and opposite force **F** must act at the center line of the shaft. A similar analysis of the pinion shows that the same observations are true. In each case the forces are equal in magnitude, opposite in direction, parallel, and in the same plane. Therefore, they constitute a couple.

In Fig. 15-9 the free-body diagram of the pinion has been redrawn and the forces resolved into components. Here we employ the superscripts r and t to indicate the radial and tangential directions with respect to the pitch circle. It is expedient to utilize the same superscripts for the components of the force **F** which the shaft exerts on the gear. The moment of the couple $\mathbf{W}^t$ and $\mathbf{F}^t$ is the torque which must be applied to drive the gearset. Designating the pitch radius of the pinion as r_2, the torque is

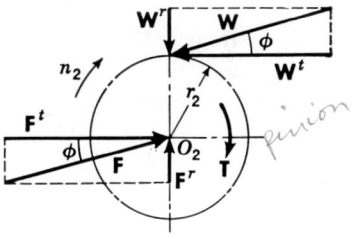

Fig. 15-9

$$T_2 = r_2 W^t \qquad (15\text{-}6)$$

where T_2 is the applied torque, positive for the counterclockwise direction, and W^t is the magnitude of the force vector $\mathbf{W}^t$.

Note that the radial force W^r serves no useful purpose as far as the transmission of power is concerned. For this reason $\mathbf{W}^t$ is frequently called the *transmitted* force.

If the horsepower and speed of the pinion are given, then the tangential force W^t can be obtained from the equation

$$W^t = \frac{(33{,}000)(12)\ \text{hp}}{2\pi r_2 n_2} \qquad (15\text{-}7)$$

where r_2 is the pitch radius in inches and n_2 is the rpm. The following relations are then evident from Fig. 15-9:

$$W^r = W^t \tan \phi \qquad W = \frac{W^t}{\cos \phi} \qquad (15\text{-}8)$$

where ϕ is the pressure angle.

15-7. Helical Gears. In the treatment of the forces on helical gears it is convenient to determine the axial force, work with it independently, and treat the remaining force components the same as for straight spur gears. Figure 15-10 is a drawing of a helical gear with half of the face removed to show the forces acting at the pitch point. The gear is imagined to be rotating clockwise. The driving gear has been removed, and its effect replaced by the forces shown acting on the teeth. The resultant force **W** is divided into three components $\mathbf{W}^a$, $\mathbf{W}^r$, and $\mathbf{W}^t$, which are the axial, radial, and tangential forces, respectively. The

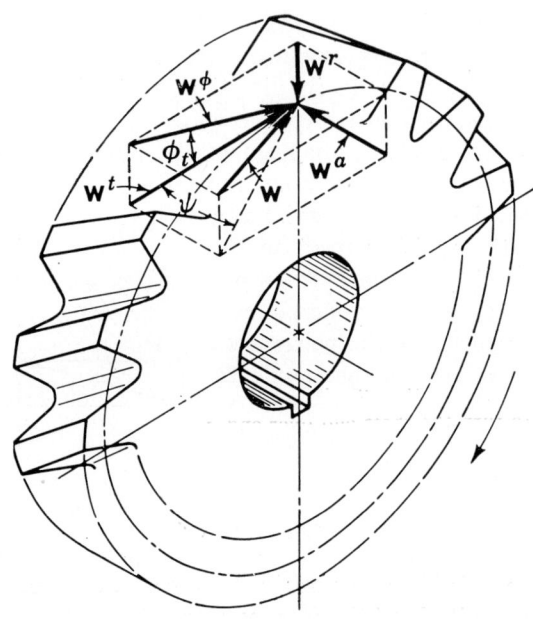

Fig. 15-10

tangential force is the transmitted force and is the one which is effective in transmitting torque. Designating the transverse pressure angle as ϕ_t and the helix angle as ψ, the following relations are evident from Fig. 15-10:

$$\mathbf{W} = \mathbf{W}^a + \mathbf{W}^r + \mathbf{W}^t \tag{15-9}$$
$$W^a = W^t \tan \psi \tag{15-10}$$
$$W^r = W^t \tan \phi_t \tag{15-11}$$

It is also expedient to make use of the resultant of $\mathbf{W}^r$ and $\mathbf{W}^t$. We shall designate this force as $\mathbf{W}^\phi$. It is defined by the equation

$$\mathbf{W}^\phi = \mathbf{W}^r + \mathbf{W}^t \tag{15-12}$$

EXAMPLE 15-1. A gear train is composed of three helical gears with the shaft centers in line. The driver is a right-hand helical gear having a pitch radius of 2 in., a transverse pressure angle of 20°, and a helix angle of 30°. An idler gear in the train has the teeth cut left hand and has a pitch radius of 3.25 in. The idler transmits no power to its shaft. The driven gear in the train has the teeth cut right hand and has a pitch radius of 2.50 in. If the transmitted force is 600 lb, find the shaft forces acting on each gear.

Solution. First, we consider only the axial components as previously suggested. For each mesh the axial component of the reaction is, from Eq. (15-10),

$$W^a = W^t \tan \psi = 600 \tan 30° = 347 \text{ lb}$$

Figure 15-11a is a top view of the three gears looking down on the plane formed by the three axes of rotation. For each gear, rotation is considered to be about the z axis for this problem. In Fig. 15-11b free-body diagrams of each of the three gears are drawn in perspective and the three coordinate axes are shown. As indicated, the idler exerts a force $\mathbf{W}_{32}{}^a$ on the driver. This is resisted by the axial shaft force $\mathbf{F}_{12}{}^a$. The forces $\mathbf{F}_{12}{}^a$ and $\mathbf{W}_{32}{}^a$ form a couple which is resisted by the moment $\mathbf{T}_2{}^a$. Note that this moment is negative about the y axis, and consequently, it is a moment which tends to rotate the driver shaft end over end. The magnitude of this moment is

$$T_2{}^a = W_{32}{}^a r_2 = (347)(2) = 694 \text{ lb-in.}$$

Turning next to the idler we see from Fig. 15-11a and b that the axial force of the shaft on the idler is zero. The axial component of the driver on the idler is $\mathbf{W}_{23}{}^a$, and that of the driven gear on the idler is $\mathbf{W}_{43}{}^a$. These two forces are equal and form a couple, tending to turn the shaft end over end, which is resisted by the moment $\mathbf{T}_3{}^a$ of magnitude

$$T_3{}^a = W_{23}{}^a(2r_3) = (347)(2)(3.25) = 2{,}260 \text{ lb-in.}$$

The driven gear has the axial force component $\mathbf{W}_{34}{}^a$ due to the idler acting at its pitch line which is resisted by the axial shaft reaction $\mathbf{F}_{14}{}^a$. As shown, these forces are equal and form a couple tending to turn the shaft end over end which is resisted by the moment $\mathbf{T}_4{}^a$. Since $W_{34}{}^a = 347$ lb, the magnitude of this moment, which is negative about the y axis, is

$$T_{14}{}^a = W_{34}{}^a r_4 = (347)(2.5) = 867 \text{ lb-in.}$$

It is emphasized again that the three resisting moments $\mathbf{T}_2{}^a$, $\mathbf{T}_3{}^a$, and $\mathbf{T}_4{}^a$ are due solely to the axial components of the reactions between the gear teeth. They produce static bearing reactions and have no effect on the amount of power transmitted.

Now that all the reactions due to the axial components have been found, we turn our attention to the remaining force components and examine their effect as if they were operating independently of the axial forces.

Free-body diagrams showing the forces in the plane of rotation for the driver, idler, and driven gears are shown, respectively, in Fig. 15-11c, d, and e. The forces can be obtained graphically as shown or by applying Eqs. (15-11) and (15-12). It is not necessary to combine the components to find the resultant forces because, in machine design, the component forces are exactly those which are desired.

15-8. Straight Bevel Gears. In determining the tooth forces on bevel gears it is customary to use the forces which would occur at the mid-point of the tooth on the pitch cone. The resultant tangential force probably occurs somewhere between the mid-point and the large end of the tooth,

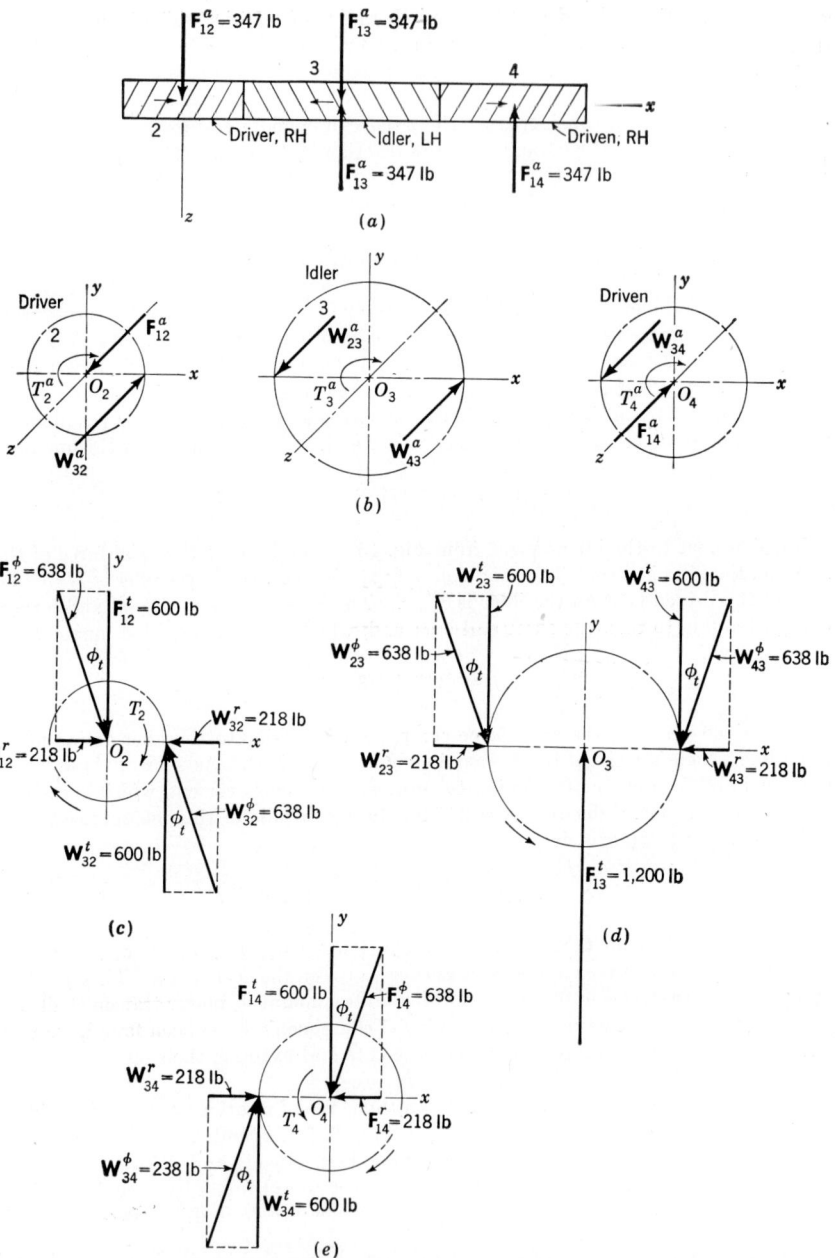

Fig. 15-11. (a, b) Axial forces. (c) Driver. (d) Idler. (e) Driven.

but there will be only a small error in making this assumption. The tangential or transmitted force is given by

$$W^t = \frac{T}{r_{av}} \qquad (15\text{-}13)$$

where r_{av} is the average radius of the pitch cone as shown in Fig. 15-12 and T is the torque.

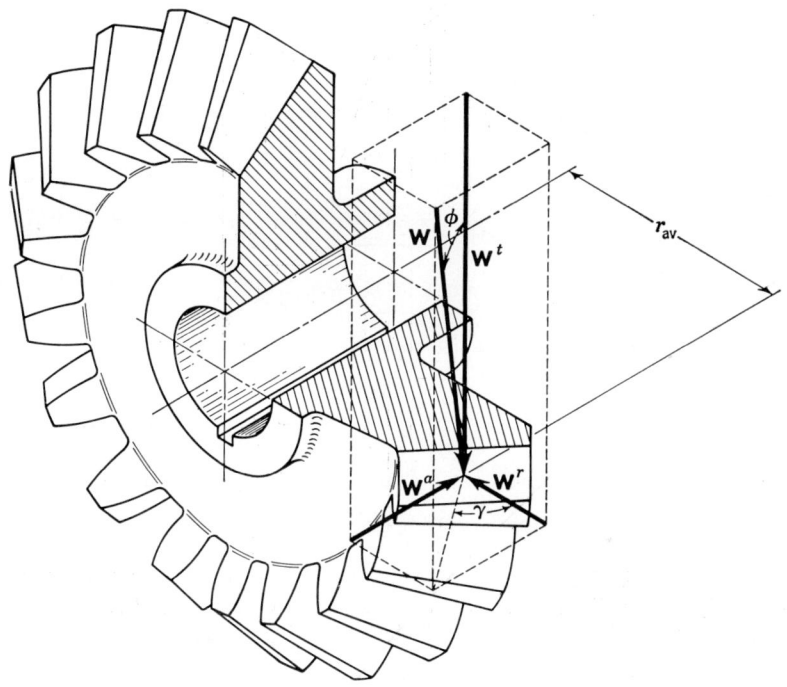

Fig. 15-12

Figure 15-12 also shows all the components of the resultant force acting at the mid-point of the tooth. The following relationships can be derived from an inspection of the figure:

$$\mathbf{W} = \mathbf{W}^a + \mathbf{W}^r + \mathbf{W}^t \qquad (15\text{-}14)$$
$$W^r = W^t \tan \phi \cos \gamma \qquad (15\text{-}15)$$
$$W^a = W^t \tan \phi \sin \gamma \qquad (15\text{-}16)$$

Note, as in the case of helical gears, that the axial force $\mathbf{W}^a$ results in a couple on the shaft which tends to turn it end over end.

EXAMPLE 15-2. The bevel pinion shown in Fig. 15-13 rotates at 600 rpm in the direction shown and transmits 5 hp to the gear. The mounting distances are shown together with the location of the bearings on each shaft. Bearings A and C are

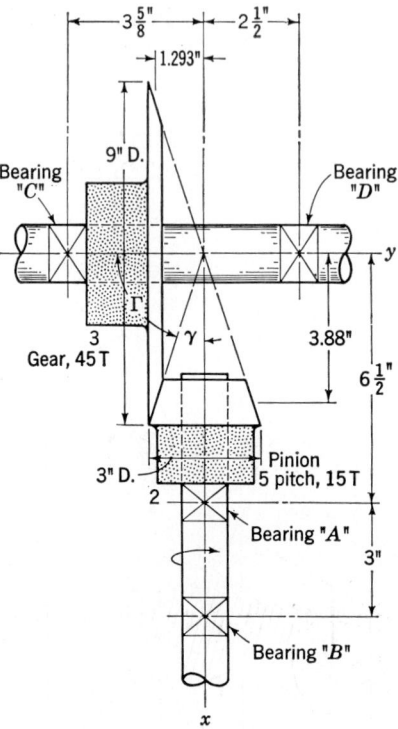

FIG. 15-13

capable of taking both radial and axial loads, while bearings B and D are built to receive only pure radial loads. The teeth of the gears have a 20° pressure angle. Find the components of the forces which the bearings exert on the shafts in the x, y, and z directions.

Solution. The pitch angles for the pinion and gear are

$$\gamma = \tan^{-1} 3/9 = 18.4°$$
$$\Gamma = \tan^{-1} 9/3 = 71.6°$$

The radii to the mid-point of the teeth are shown on the drawing and are $r_{2,\mathrm{av}} = 1.293$ in. and $r_{3,\mathrm{av}} = 3.88$ in. for the pinion and gear, respectively.

Let us determine the forces acting upon the pinion first. The tangential force is obtained from Eq. (15-7) using the average pinion radius.

$$W^t = \frac{(33,000)(12)\,\mathrm{hp}}{2\pi r_{2,\mathrm{av}} n_2} = \frac{(33,000)(12)(5)}{2\pi(1.293)(600)} = 406\text{ lb}$$

This force acts in the negative z direction. (In Fig. 15-13 the z axis is positive out of the paper for a right-handed system.) The radial and axial components are obtained from Eqs. (15-15) and (15-16).

$$W^r = W^t \tan\phi \cos\gamma = 406 \tan 20° \cos 18.4° = 140\text{ lb}$$
$$W^a = W^t \tan\phi \sin\gamma = 406 \tan 20° \sin 18.4° = 46.6\text{ lb}$$

determined; the dimensions, the torque T_2, and the force W are the given elements of the problem. In order to find F_B we shall take a summation of moments about point A. This requires the designation of the position of P relative to A and of B relative to A. These relative position vectors are

$$R_{PA} = -2.62I - 1.293J$$
$$R_{BA} = 3I$$

Then, the moment of W about A is

$$R_{BA} \times W = \begin{vmatrix} I & J & K \\ -2.62 & -1.293 & 0 \\ 46.6 & 140 & -406 \end{vmatrix} = 525I - 1{,}064J - 308K$$

And the moment of F_B about A is

$$R_{BA} \times F_B = \begin{vmatrix} I & J & K \\ 3 & 0 & 0 \\ 0 & F_B{}^y & F_B{}^z \end{vmatrix} = -3F_B{}^z J + 3F_B{}^y K$$

since $F_B{}^x$ is given as zero. Then, writing $\Sigma T_A = 0$ gives

$$T_2 + R_{BA} \times F_B + R_{PA} \times W = 0$$

or, substituting the appropriate values,

$$(-525I) + (-3F_B{}^z J + 3F_B{}^y K) + (525I - 1{,}064J - 308K) = 0$$

and the solution is

$$F_B = 102J - 355K$$
or $$F_B = 370 \text{ lb}$$

We can now find F_A from the equation $\Sigma F = 0$. Thus

$$W + F_A + F_B = 0$$

Substituting the known values,

$$(46.6I + 140J - 406K) + F_A + (102J - 355K) = 0$$
so that $$F_A = -46.6I - 242J + 761K$$
or $$F_A = 798 \text{ lb}$$

The calculations for the pinion shaft are now complete, and the results are shown in Fig. 15-14b. The procedure for obtaining the forces on the gear shaft is similar, although not shown here. The results are included in Fig. 15-14c for the benefit of readers who would like to check them.

15-9. Resisting Forces. The purpose of this section is to describe the effects of a rather complex variety of forces having in common the characteristic of always being opposed to the motion of a machine element. These forces might, rather loosely, be described as frictional forces, yet we shall see that each has a character of its own. It is also true that we cannot say that these forces are undesirable ones simply because they resist motion. In fact, in the study of automatically controlled machines

Here W^r acts in the positive y direction and W^a in the positive x direction. These three forces are components of the force **W**. Thus

$$\mathbf{W} = 46.6\mathbf{I} + 140\mathbf{J} - 406\mathbf{K}$$

The torque applied to the shaft must be

$$\mathbf{T}_2 = -(406)(1.293)\mathbf{I} = -525\mathbf{I} \quad \text{lb-in.}$$

A sketch of the free-body diagram of the combined pinion and pinion shaft is represented schematically in Fig. 15-14a. The bearing reactions $\mathbf{F}_A$ and $\mathbf{F}_B$ are to be

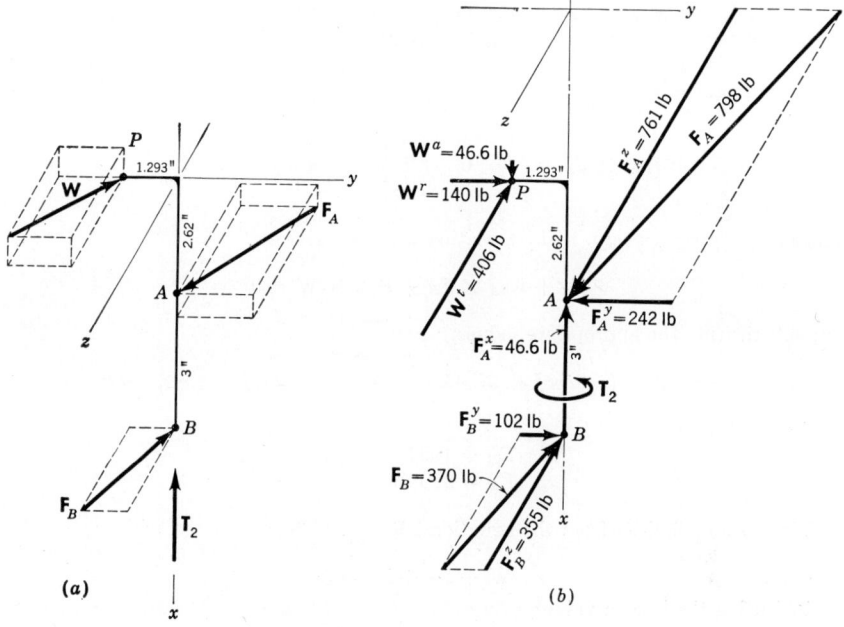

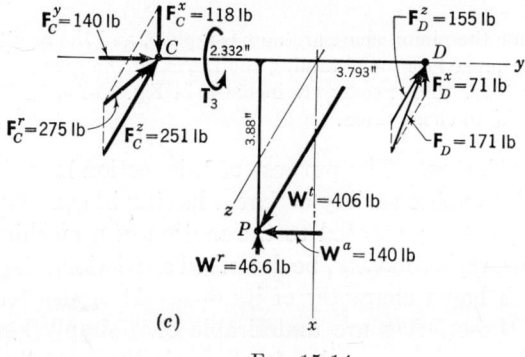

Fig. 15-14

in Chap. 22, it will be found that some of these resisting forces should have very significant values to assure stability of control. While it is true that the nature of these resisting forces is imperfectly understood, we can recognize them by their effects, and experimental investigations have taught us to account for them with acceptable engineering accuracy.

Static Frictional Forces. Consider two bodies forced into contact with each other and having no relative motion between them, as in Fig. 15-15. A force **P**, applied in an effort to slide one body relative to the other, is resisted by another force **F** acting in the plane of contact. The force **F** is called the *force of static friction* if there is no relative motion between the two bodies. A *normal force* **N** perpendicular to the plane of contact is related to the frictional force **F** by the relation

$$F = -\mu N \qquad (15\text{-}17)$$

where μ is defined here as the *coefficient of static friction*. The negative sign is used to indicate that the frictional force opposes the direction in which motion is attempted.[1]

Experiments with bodies having unlubricated contacting surfaces show that the frictional force with sliding impending is independent of the area of contact and proportional to the normal force existing between the two bodies. Equation (15-17) is the mathematical statement of this proportionality.

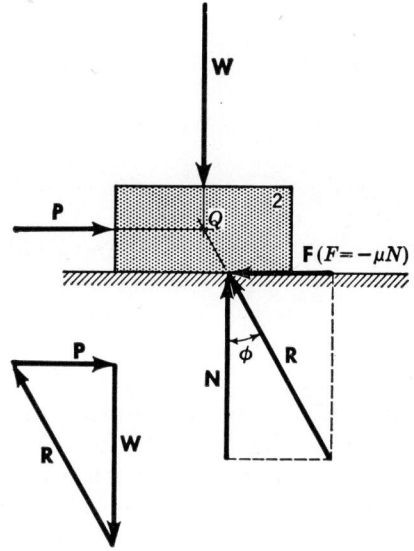

FIG. 15-15

The reason for the existence of static friction is still not clear, as evidenced by the fact that research to explain the mechanism of friction is being carried out today. Since it is impossible to manufacture a perfectly flat surface, we can probably conclude that, if the surfaces are examined under a powerful microscope, they will be seen as very rough surfaces consisting of many hills and valleys. Some of the hills are sure to be in contact with one another, and the normal force, acting on such minute areas of contact, will deform the hills and weld them together. While this is only a conjecture, it helps to explain the static frictional force as that force which is required to shear the welds apart. It is also prob-

---

[1] It is noted that μ of Eq. (15-17) has no real meaning unless sliding is impending. Thus, if sliding is not impending, then $F = -P$ and Eq. (15-17) is *not* valid.

ably true that the frictional force depends in part on the molecular structure of the two surfaces, their chemistry, and their cleanliness.

Sliding Frictional Forces. Experiments indicate that the frictional force **F** existing between the two bodies when the relative motion between the two is one of sliding is less than the frictional force when motion is impending. Equation (15-17) applies equally well for sliding motion, but now μ is the *coefficient of sliding friction*. Investigation shows that the coefficient of sliding friction is independent of the area of contact and also is independent of the relative velocity. Sliding friction is frequently described as *Coulomb friction* in order to differentiate it from other kinds of mechanical resistance.

Viscous Damping Forces. A viscous damping force is any resisting force which is proportional to velocity. It is defined by the relation

$$\mathbf{F} = -c\mathbf{V} \qquad (15\text{-}18)$$

where c is the *coefficient of viscous damping*, sometimes called the *damping factor*, or the *viscous damping constant*. Its units are pound-seconds per inch if the velocity **V** is expressed in inches per second. Equation (15-18) is a vector equation and expresses the fact that the viscous damping force is opposed to the velocity. As an example of viscous damping consider the forces which are necessary to oscillate a plunger up and down in a tank of water. For very slow velocities the plunger moves easily and only a small force is required, but if the speed of oscillation is increased, then the force required is increased correspondingly. However, water has a very low viscosity. Suppose the water is replaced with a thick sirup. This has a much higher viscosity, and consequently, the force necessary to oscillate the plunger is much greater.

Solid Frictional Forces. Still another form of friction is called *solid* or *internal friction*. This type is associated with the heating of materials which are subjected to rapidly alternating stresses such as springs or ball bearings. The force can be determined from the equation

$$\mathbf{F} = -\mu\mathbf{X} \qquad (15\text{-}19)$$

where **X** is the displacement and **F** and μ are, as before, the *frictional force* and, in this case, the *coefficient of solid friction*, respectively. Solid friction will not be discussed or used in this book as such, because its effects may be combined either with viscous damping or with Coulomb friction. It is sufficient to indicate that it is this kind of friction which causes a spring oscillating in a vacuum eventually to come to rest.

Higher-order Resisting Forces. It is probable that resisting forces exist which are proportional to acceleration and perhaps even to jerk. However, if these exist, they are of a very small order and need not be included here. It is sufficient to note that unexplained forces may be found when the oscilloscope records of operating machines are taken and examined. It is common to describe as "noise" anything which appears on an oscilloscope record unless it can be accounted for in some manner. Undoubtedly some of this "noise" is due to higher-order resisting forces.

Combined Resisting Forces. We have discussed several varieties of resisting forces in this section and found that these forces come into existence whenever a machine is operated. When a machine is carefully analyzed, it will usually be found that either the sliding friction or the viscous damping predominates. The solution of dynamics problems can be simplified considerably if the predominant resisting force is used to account for all the resisting forces whatever kind they may be. Thus it is quite customary in analyzing many machines to assume that sliding friction is present but that the higher-order forces and those due to viscous damping and internal friction are all zero. Such an assumption has the effect of "lumping" or combining all the resisting forces into a single "frictional force." In other cases only the viscous damping force is used in analysis and the others are assumed to be zero. Admittedly these assumptions are rather crude, and sometimes the calculated results differ substantially from the true ones. Nevertheless an engineer has a job to do, and if he can use such assumptions to design satisfactory machines, then the use of the assumptions is justified.

15-10. Force Analysis Using Coulomb Friction. In analyzing the frictional forces in machine parts it is customary to consider the friction as Coulomb as previously indicated. When the surfaces of a bearing are perfectly lubricated and the parts of the bearing moved relative to one another, the resisting forces which exist are nearly all due to viscous friction. But these conditions are obtained, usually, only when the other forces acting between the parts are so large that the frictional force is a relatively small part of the total. Furthermore, a sliding bearing is never perfectly lubricated when it is just starting to move. Consequently, under starting conditions, the friction is almost all sliding friction. In addition, great numbers of bearings exist in which imperfect lubrication may be present at all times, and for these it is better to assume sliding friction. Finally, if the relative velocity between the mating surfaces is constant, then both sliding and viscous friction can be treated in exactly the same manner.

In Fig. 15-15 let **W** be a known force pressing block 2 against the stationary surface 1. It is required to find the force **P** necessary to maintain uniform motion of the block. The forces **W** and **P** are resisted by the normal force **N** and the frictional force **F**, respectively. The resultant of **F** and **N** is **R**, and the three forces **R**, **W**, and **P** must be in equilibrium. Point *Q* is the point of concurrency and defines the direction of **R**. The force polygon can now be drawn, as shown, and the magnitude of the force **P** found. It is evident that the relationship between the frictional force and the normal force is

$$\tan \phi = \frac{F}{N} = \frac{\mu N}{N} = \mu \tag{15-20}$$

The angle ϕ between the resultant force and the normal force is called the *friction angle*.

The relationships shown in Fig. 15-15 also apply when motion is impending. Under these conditions μ is the static coefficient of friction and **P** is the force necessary to cause motion to begin.

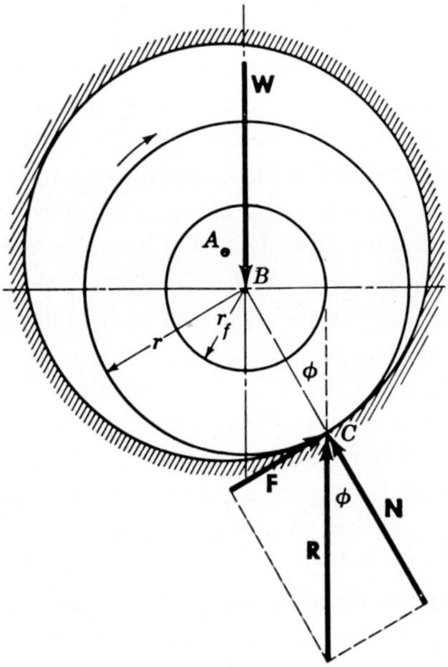

The Friction Circle. The friction circle is a convenient device to use in making a force analysis of a machine having turning joints. In Fig. 15-16 the journal with center at B rotates clockwise and, because of friction, rolls up the right side of the bearing. Sliding occurs when equilibrium is reached, but contact remains on the right side of the bearing.[1] This means that the normal force **N** will not have the same line of action as the load **W**. A frictional force $F = \mu N$ exists because of the normal force. The resultant of the frictional force **F** and the normal force **N** is **R**, equal and opposite to **W** and having its point of application at C. The circle constructed with center at B, tangent to the line of action of **R**, is called the *friction circle*. Its radius is

Fig. 15-16

$$r_f = r \sin \phi = r \tan \phi = \mu r \tag{15-21}$$

where for small angles $\sin \phi = \tan \phi$.

[1] It is interesting to know that the reverse situation exists in a perfectly lubricated bearing and that, after equilibrium is reached, the journal is nearest the left side of the bearing. During the starting of a perfectly lubricated journal and bearing, the journal climbs up the right side because it is in contact with the bearing at the beginning of its motion. However, the rotation of the journal pumps the lubricant around in a clockwise direction, and it is the oil pressure generated by the rotating journal which forces the journal to the left side of the bearing.

STATIC-FORCE ANALYSIS 369

Equation (15-21) shows that the radius of the friction circles does not depend upon the magnitude of the forces but only upon the radius of the journal and the coefficient of friction. Both of these are usually constant for a given problem regardless of the load, and consequently, the friction circle radius is constant.

The friction circle is used to locate the line of action of the force between the journal and the bearing. The circle is constructed first, and then the direction of the force must always be tangent to it. Note that these circles are necessarily quite small. For example, a 2-in.-diameter journal with a coefficient of friction of 0.08 would have a friction circle 0.16 in. in diameter, or less than $3/16$ in.

15-11. Forces in the Slider-crank Mechanism with Friction. We shall now make a static force analysis of the slider-crank or engine mechanism in order to illustrate the application of the principles investigated in Sec. 15-10. Figure 15-17a illustrates the phase of the mechanism chosen for analysis. The force **P** is taken as the driving force exerted on the piston, and this force causes crank 2 to rotate counterclockwise. A crankshaft torque T_2 resists the action of **P**.

The radii of the friction circles are first calculated for each turning joint, using Eq. (15-21). These circles can then be constructed on the drawing or sketch of the mechanism and are shown in the figure. In this example the size of the circles has been greatly exaggerated in order to illustrate the analysis better.

Friction exists between the piston and cylinder walls, that is, between the sliding block 4 and the stationary guide 1. Therefore, with the use of the coefficient of friction for these surfaces and Eq. (15-20) the friction angle ϕ can be calculated.

Figure 15-17b is a free-body diagram of the slider showing the driving force **P**, the frame reaction $\mathbf{F}_{14}$, and the connecting-rod reaction $\mathbf{F}_{34}$. The direction and magnitude of **P** are given. The magnitudes of $\mathbf{F}_{14}$ and $\mathbf{F}_{34}$ are both unknown. The motion of the slider is to the left, and the frictional component of $\mathbf{F}_{14}$ must oppose this. Therefore, the angle ϕ is measured to the left of the perpendicular, giving the direction of $\mathbf{F}_{14}$. The direction of $\mathbf{F}_{34}$ is determined from the free-body diagram of link 3 in Fig. 15-17c. The rotation of slider 4 relative to link 3 is clockwise, so $\mathbf{F}_{43}$ must be tangent to the bottom of the friction circle at B. At A the rotation of crank 2 relative to link 3 is counterclockwise, so the force $\mathbf{F}_{23}$ must be tangent to the bottom of the friction circle at A. This determines the line of action of $\mathbf{F}_{34}$, and the force polygon is now constructed as shown in Fig. 15-17b, and the magnitudes of $\mathbf{F}_{34}$ and $\mathbf{F}_{14}$ found.

Figure 15-17d is the free-body diagram of the crank 2. The force $\mathbf{F}_{12}$ must be tangent to the top of the friction circle at O_2 in order for the joint friction to oppose the counterclockwise rotation of link 2. The torque

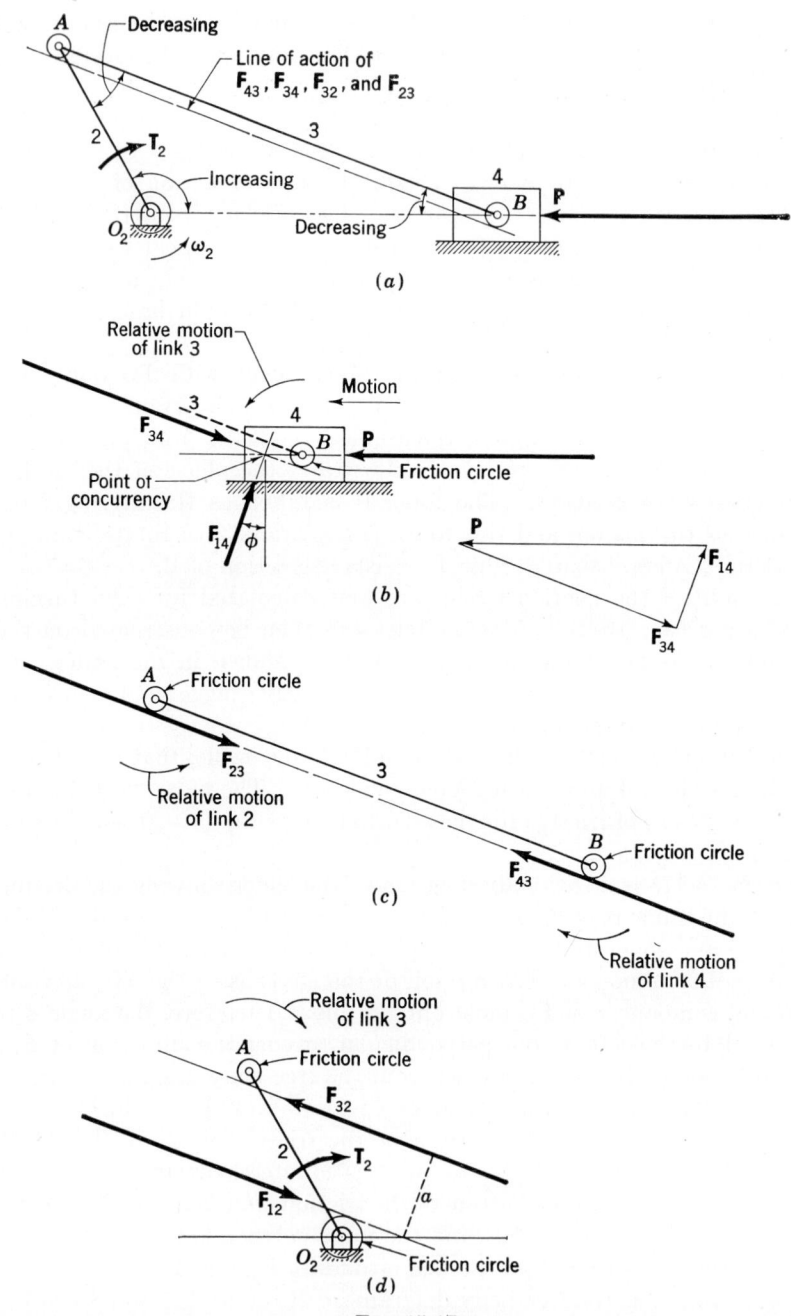

Fig. 15-17

STATIC-FORCE ANALYSIS 371

T_2 must, of course, equal the moment of the couple formed by the parallel forces F_{12} and F_{32}.

Figure 15-18 illustrates the four cases which arise in determining the location of the line of action for link 3.

15-12. Efficiency. *Efficiency* is defined as the ratio of the amount of work done by a machine to the amount of work supplied to the machine during any time interval. This definition of efficiency is one with which

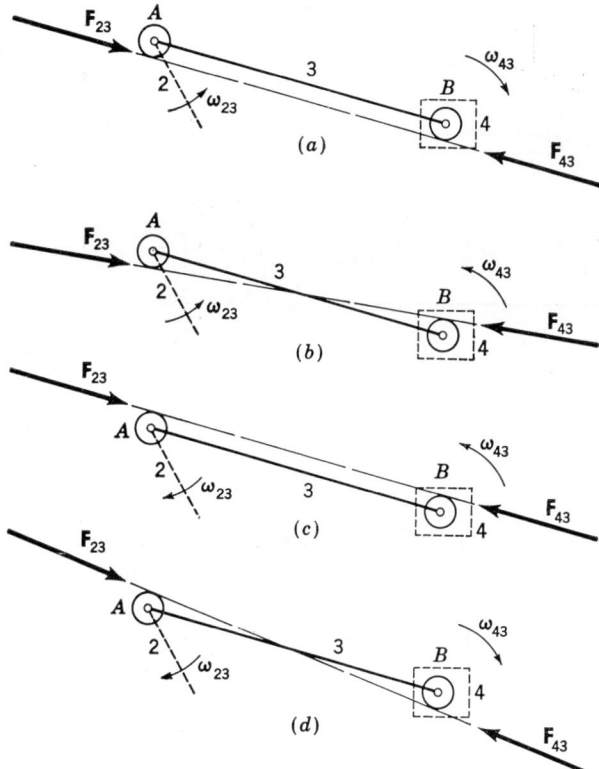

Fig. 15-18. The location of the line of action depends upon the direction of the relative angular velocities of the connecting links.

the reader is familiar, and we shall find occasion to use it in the following pages. The reader should particularly note that the idea of time is contained in the definition, and so this is necessarily *average* efficiency.

In the analysis of forces it is often desirable to define what is called *instantaneous efficiency*. Instantaneous efficiency does not include the idea of time and can be defined by the equation

$$e = \frac{F_0}{F} = \frac{T_0}{T} \qquad (15\text{-}22)$$

where F_0 and T_0 are the magnitudes of the input force and torque vectors, respectively, which would be required to operate the machine if no friction were present and F and T represent the actual magnitudes of the input force and torque vectors, that is, the values necessary to overcome friction and operate the machine.

15-13. Friction in Gear Teeth.[1] A precise analysis of the frictional forces between gear teeth is beyond the scope of this book, but the reader

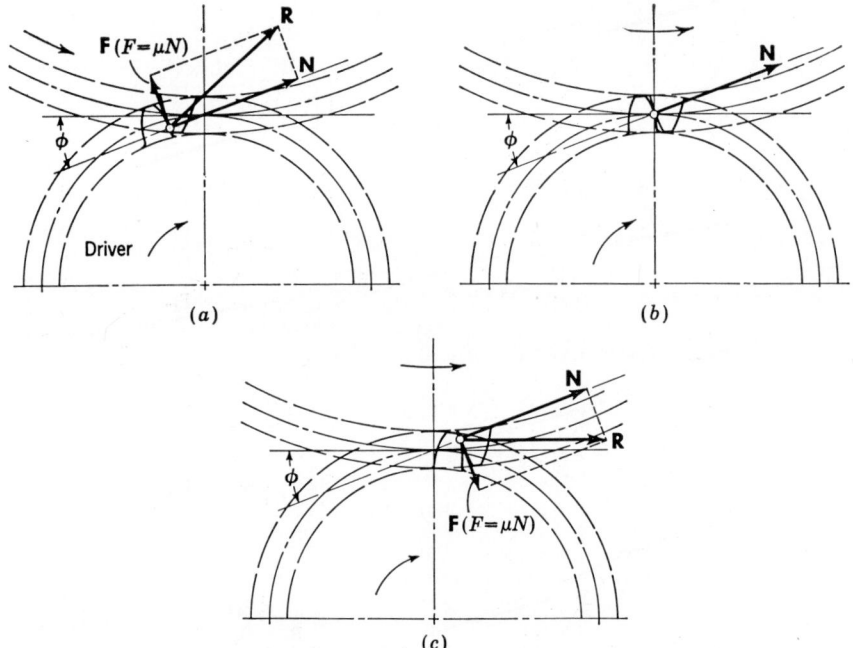

Fig. 15-19. Frictional forces in the meshing of spur-gear teeth. (a) Teeth coming into contact. (b) Teeth in contact at the pitch point. (c) Teeth going out of contact. Note that the angle ϕ is the pressure angle, not the friction angle.

can appreciate the magnitude of the problem from an examination of Fig. 15-19. Here a pinion rotates clockwise and drives a gear in a counterclockwise direction. The forces shown are those exerted by the pinion against the gear. In (a) the teeth have just begun contact and the flank of the pinion tooth is sliding down the face of the gear tooth, producing a frictional force in the direction shown. In (b) contact is at the pitch point; there is no sliding and, consequently, no sliding friction. In (c) the teeth are just about to leave contact. The face of the pinion

[1] See Sanae Wada, "Contributions to the Tooth Friction and Tip Modification of Tooth in Spur Gears," Waseda University, Tokyo, 1957.

tooth is sliding up the flank of the gear tooth. The resulting friction force is now approximately opposite in direction to that shown in (a).

We can now see that the friction force varies in magnitude and direction as the teeth go through the meshing cycle. If this were the only problem, it would not be too difficult to determine and plot the magnitude and direction of the forces through an entire meshing cycle, but most gearsets have more than one pair of teeth in contact during at least a portion of the cycle. An equal distribution of the forces between two pairs of meshing teeth is almost impossible to obtain because it will depend upon the accuracy of cutting the teeth and upon how much the teeth are instantaneously deflected or bent by the forces. In addition the sliding velocity is changing, and so is the effect of the lubricant. Consequently, we must expect a highly variable coefficient of friction.

All these things add up to the fact that we are up against a knowledge barrier and can do no more with this line of attack until someone hurdles it and makes a solution available. To an engineer this does not mean that a solution cannot be found but only that a different and more empirical approach must be employed in obtaining that solution.

An *engineering* approach to the problem is made by overlooking the frictional forces and making the force analysis using the *input* forces to the gearset rather than the *output* forces. In such an analysis errors will exist, and the designer accounts for these using *service factors* and *safety factors* which have, preferably, been determined experimentally. This is often very good engineering. It bypasses a very troublesome and lengthy problem and produces good, usable answers in a hurry.

In the case of straight spur gears and bevel gears the efficiency is usually about 99 per cent, and so the error is quite small. This is also quite frequently true for helical and spiral gears. The frictional forces are sometimes quite high for crossed-helical and worm gears. Consequently, we usually do not neglect the friction forces in analyzing these gears.

15-14. Worm Gears. Figure 15-20 is a simplified drawing of a worm showing the various components of the force exerted upon it by the worm gear. Using the subscript 2 to designate the worm (since it is always the driver) and 3 for the gear, the force magnitudes are

$$W_{32}{}^t = \frac{T_2}{r_2} \tag{15-23}$$

$$W_{32}{}^r = W_{32}{}^t \frac{\tan \phi_t}{\tan \lambda} \tag{15-24}$$

$$W_{32}{}^a = \frac{W_{32}{}^t}{\tan \lambda} \tag{15-25}$$

where
$$\mathbf{W}_{32} = \mathbf{W}_{32}{}^a + \mathbf{W}_{32}{}^r + \mathbf{W}_{32}{}^t \tag{15-26}$$

and where T_2 is the input torque to the worm. The reader should note that we are using the character ϕ to designate friction angles as well as pressure angles in gear teeth (because this is standard notation) and that ϕ_t in the above expressions is the transverse pressure angle. The angle λ is the lead angle of the worm, which is usually specified instead of the helix angle.

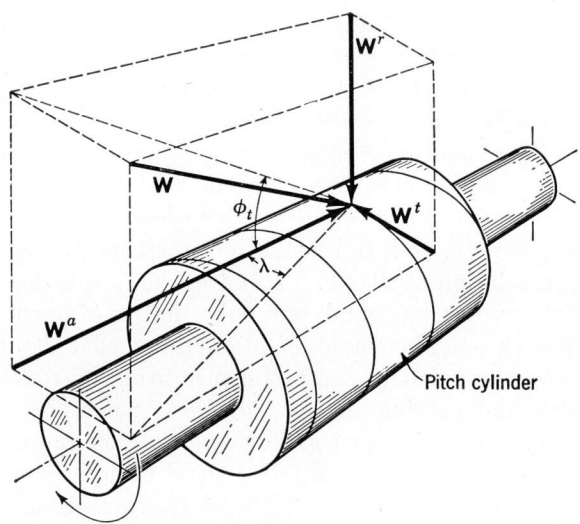

FIG. 15-20. Drawing of the pitch cylinder of a worm showing the forces exerted upon it by the worm gear.

Usually (but not necessarily) the gear axis is at right angles to the worm axis. Under these conditions the forces $\mathbf{W}^t$ and $\mathbf{W}^a$ exchange places and the force components acting on the gear are

$$W_{23}{}^r = W_{32}{}^r \qquad (15\text{-}27)$$
$$W_{23}{}^t = W_{32}{}^a \qquad (15\text{-}28)$$
$$W_{23}{}^a = W_{32}{}^t \qquad (15\text{-}29)$$

As indicated in the previous section, the forces determined here are in some error because of the difficulty of accounting for friction. Due allowance must be made when these forces are employed in design calculations.

PROBLEMS

15-1. The figure shows four mechanisms and the external forces or torques exerted on or by the mechanism. Sketch (freehand) the free-body diagram of each part of each mechanism, including the frame. Do not attempt to show the magnitudes of the forces, except roughly, but do sketch them in the proper directions. Assume no friction.

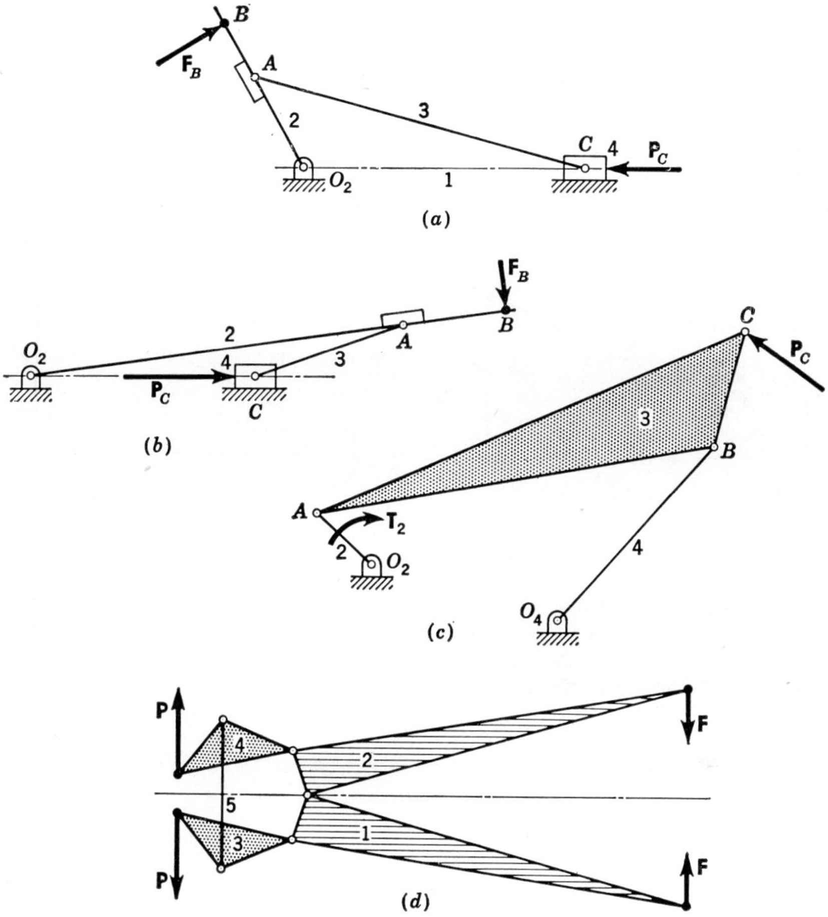

PROB. 15-1

15-2. What couple T_2 must be applied to link 2 in the figure in order to maintain equilibrium? Neglect friction forces.

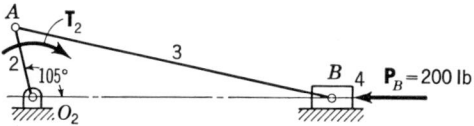

PROB. 15-2. $O_2A = 3$ in.; $AB = 14$ in.

15-3. If the friction forces for the linkage shown in the figure are assumed negligible, what force **P** is necessary to maintain equilibrium? *Ans.:* **P** = 329/−180° lb.

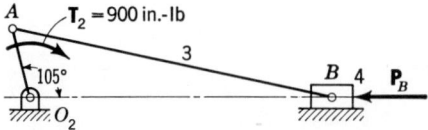

PROB. 15-3. O_2A = 3 in.; AB = 14 in.

15-4. Construct free-body diagrams for each element of the linkage shown in the figure, and determine the magnitude and direction of the forces acting on each part. The friction forces may be neglected. What input torque T_2 must be applied for equilibrium?

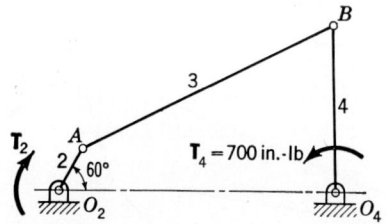
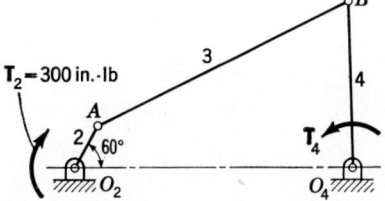

PROB. 15-4. O_2A = 2 in.; AB = O_2O_4 = 12 in.; O_4B = 7 in.

FIG. 15-5. O_2A = 2 in.; AB = O_2O_4 = 12 in.; O_4B = 7 in.

15-5. Determine the torque T_4 necessary to maintain equilibrium of the linkage shown in the figure. Construct free-body diagrams of each link, and show all the forces acting upon them in proper direction and magnitude. Assume the frictional forces to be zero. *Ans.:* T_4 = 1,700 lb-in.

15-6. Find the frame reactions and the torque T_2 necessary to maintain equilibrium of the linkage shown in the figure. Ignore friction.

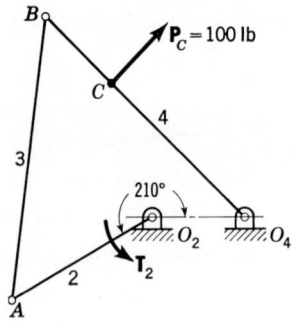
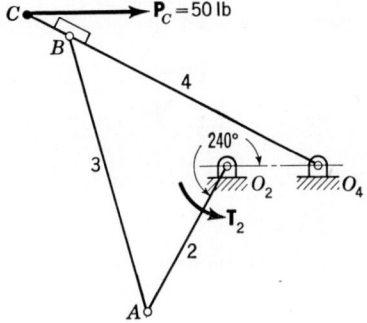

PROB. 15-6. O_2A = 3½ in.; AB = O_4B = 6 in.; O_4C = 4 in.; O_2O_4 = 2 in.

PROB. 15-7. O_2A = 3½ in.; AB = O_4B = 6 in.; O_4C = 7 in.; O_2O_4 = 2 in.

15-7. Determine the torque necessary to maintain equilibrium of the linkage shown in the figure if friction is neglected. Draw the free-body diagrams of links 1 and 4, and show all the forces acting in their proper direction and magnitude. *Ans.:* T_2 = 90.9 lb-in.

15-8. Find the force F_A necessary for equilibrium of the linkage shown in the figure. Construct free-body diagrams of links 2 and 4, and show all the forces acting upon them in the proper direction and magnitude. Neglect friction.

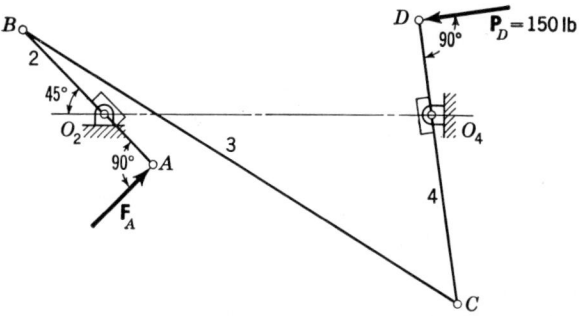

PROB. 15-8. $O_2A = 3$ in.; $AB = O_4C = 8$ in.; $BC = 22$ in.; $CD = 12$ in.; $O_2O_4 = 14$ in.

15-9. Neglect friction and find the force **P** necessary for equilibrium of the linkage shown. Construct free-body diagrams of each link and show all the forces. *Ans.:* $F_{16} = 37.4/\!-\!90°$ lb.

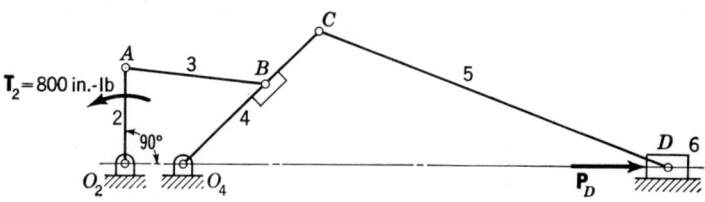

PROB. 15-9. $O_2A = 4$ in.; $AB = 6$ in.; $O_4B = 5$ in.; $O_4C = 8$ in.; $CD = 16$ in.; $O_2O_4 = 2\frac{1}{2}$ in.

15-10. Determine the torque T_2 required to drive the slider 6 in the figure against a load $P = 100$ lb for a crank angle of $\theta = 30°$. Neglect friction.

15-11. The same as Prob. 15-10 except that the crank angle is 60°. *Ans.:* $T_2 = 451$ lb-in.

15-12. Determine the torque T_2 required to drive the slider against a force $P = -20$ lb for a crank angle $\theta = 240°$. Neglect all friction forces.

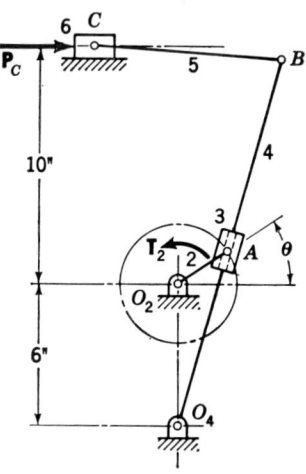

PROBS. 15-10 TO 15-12. $O_2A = 2\frac{1}{2}$ in.; $O_4B = 16$ in.; $BC = 8$ in.

15-13. Neglecting friction, determine the magnitude and direction of the couple which must be applied to link 2 to drive the linkage against the forces shown. Draw a free-body diagram of each link and show all the forces acting. *Ans.*: $T_2 = 282$ lb-in.

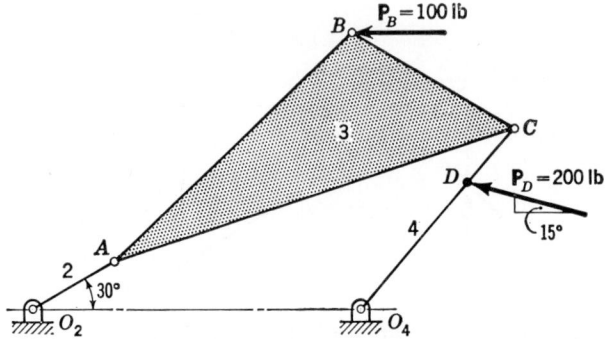

PROB. 15-13. $O_2A = 4$ in.; $AB = 14$ in.; $AC = 18$ in.; $BC = 8$ in.; $O_4D = 7$ in.; $O_4C = 10$ in.; $O_2O_4 = 14$ in.

15-14. The figure shows a four-bar linkage with external forces applied at points B and C. Neglecting friction, find the couple which must be applied to link 2 to maintain equilibrium. Draw a free-body diagram of each link, including the ground link, and show all the forces acting upon each.

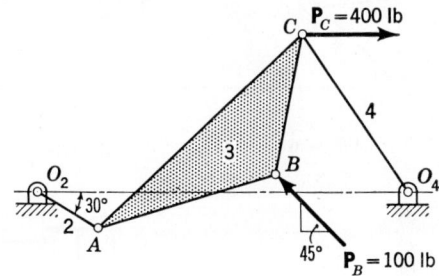

PROB. 15-14. $O_2A = 3$ in.; $AB = O_4C = 8$ in.; $AC = 12$ in.; $BC = 6$ in.; $O_2O_4 = 16$ in.

15-15. What torque T_2 must be applied to link 2 of the four-bar linkage in the figure for equilibrium? Find the forces acting upon links 1 and 3. Friction may be neglected. *Ans.*: $T_2 = 1{,}069$ lb-in.

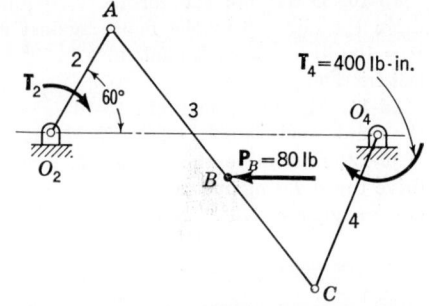

PROB. 15-15. $O_2A = 10$ in.; $AB = 16$ in.; $AC = O_2O_4 = 28$ in.; $O_4C = 14$ in.

15-16. Draw free-body diagrams of each link shown in the figure, and find the magnitude and direction of all the forces acting. Neglect friction. What are the magnitude and direction of the couple which must be applied to link 2 to retain equilibrium?

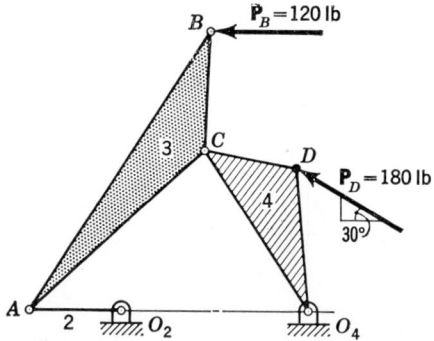

PROB. 15-16. $O_2A = 4$ in.; $AB = 14$ in.; $AC = 10$ in.; $BC = 5$ in.; $O_4C = O_2O_4 = 8$ in.; $CD = 4$ in.; $O_4D = 6$ in.

15-17. Determine the magnitude and direction of the forces which must be applied to link 2 to maintain equilibrium. Neglect friction. *Ans.:* $\mathbf{F}_{32} = 112/153°$ lb.

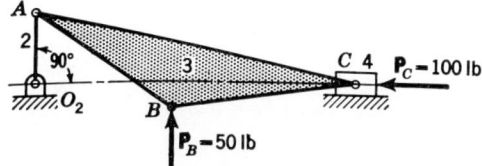

PROB. 15-17. $O_2A = 3$ in.; $AB = 7$ in.; $AC = 14$ in.; $BC = 8$ in.

15-18. A 5-diametral-pitch, 20°-pressure-angle, 15-tooth spur pinion rotates at 600 rpm and drives a 60-tooth gear. The horsepower transmitted is 25. Construct a free-body diagram of each gear showing upon it the tangential and radial components of the forces and their proper directions.

15-19. The gears shown in the figure are 3 diametral pitch, 20° pressure angle and are in the same plane. The pinion rotates at 500 rpm and transmits 35 hp through the idler to the 28-tooth gear. Find the magnitude and direction of the forces acting on each shaft. Calculate the torque available at shaft 4. *Ans.:* $\mathbf{F}_3 = 4{,}250/{-45°}$ lb.

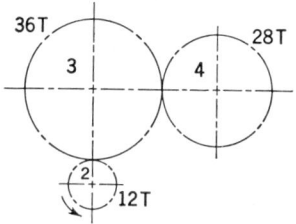

PROB. 15-19

15-20. The 16-tooth pinion on shaft 2 rotates at 1,720 rpm and transmits 5 hp to the double-reduction gear train. All gears have a 20° pressure angle. The distances between the centers of the bearings and gears for shaft 3 are shown in the figure. Find the magnitude and direction of the radial force which each bearing exerts against this shaft.

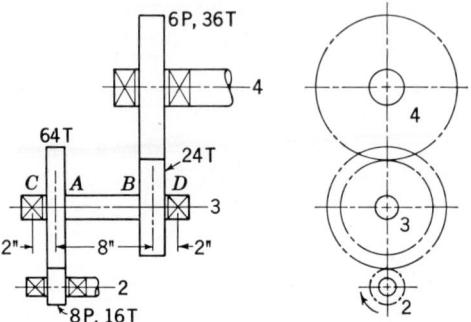

Prob. 15-20

15-21. The 24-tooth, 8-pitch, 20° pinion shown in the figure rotates clockwise at 900 rpm and transmits 3 hp to the planetary gear train. What torque can the arm 3 deliver to its output shaft? Draw a free-body diagram of the arm and of each gear, and show all forces which act upon them. *Ans.:* $T_3 = 928$ lb-in.

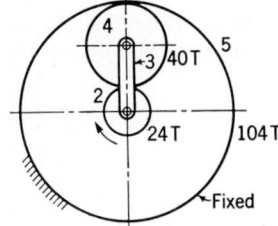

Prob. 15-21

15-22. The 16-tooth pinion with center at O_2 drives the double-reduction gear train shown in the figure. All gears are $14\frac{1}{2}°$ pressure angle. The pinion rotates

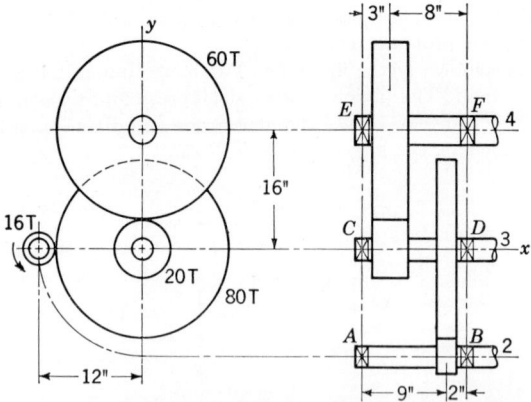

Prob. 15-22

counterclockwise at 1,200 rpm and transmits 50 hp to the gear train. Calculate the magnitude and direction of the radial force exerted by each bearing on its shaft.

15-23. The figure shows a double-reduction helical gearset. Pinion R is the driver, and it receives a torque $T_2 = 1,200$ lb-in. from its shaft in the direction shown. Pinion R has a normal diametral pitch of 8, 14 teeth, and a normal pressure angle of 20° and is cut right-handed with a helix angle of 30°. The mating gear S on shaft 3 has 36 teeth. Gear T, which drives the second pair of gears in the train, has a normal diametral pitch of 5, 15 teeth, and a normal pressure angle of 20° and is cut right-handed with a helix angle of 15°. Mating gear U has 45 teeth. Find the magnitude and direction of the force exerted by the bearings at C and D on shaft 3 if bearing C can take only radial load while bearing D is mounted to take both radial and thrust load. *Ans.*: $\mathbf{F}_C = 1,680\mathbf{I} + 365\mathbf{J}$ lb; $\mathbf{F}_D = 1,494\mathbf{I} - 76\mathbf{J} + 154\mathbf{K}$ lb.

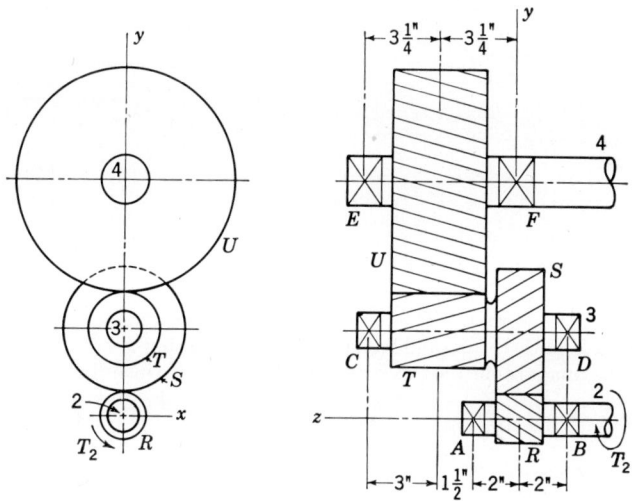

Prob. 15-23

15-24. The double-reduction helical gearset shown in the figure is driven through shaft 2 and receives $7\frac{1}{2}$ hp at a speed of 900 rpm. Gears R and S are 10 diametral pitch (transverse), 30° helix angle and have a 20° transverse pressure angle. Pinion R is cut with 14 teeth, left-handed; gear S has 54 teeth. The second pair of gears in the train, T and U, are 6 diametral pitch (transverse), 23° helix angle, and 20° transverse pressure angle. Gear T is left hand and has 16 teeth; gear U has 36 teeth. The gears are supported by bearings located as shown in the figure. In machine design it is usually convenient to locate the thrust bearing so that the shaft is loaded in compression. The thrust bearing then takes both radial and thrust loads, while the second bearing on the shaft is subject to pure radial load. In accordance with this convention determine the magnitude and direction of the radial and thrust loads which bearings C and D exert on shaft 3.

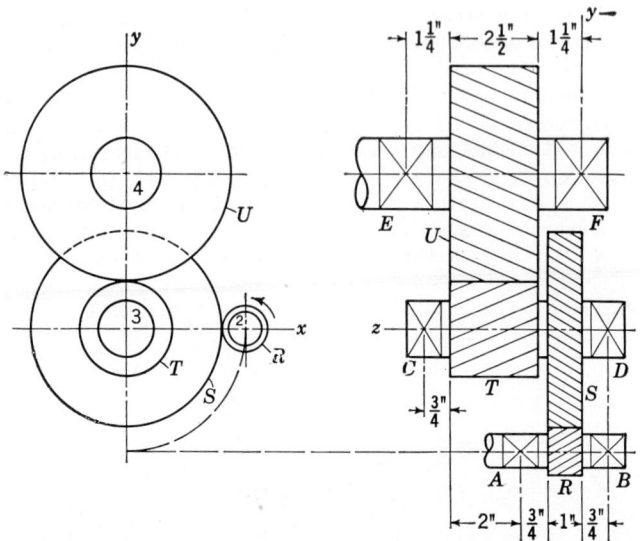

PROB. 15-24

15-25. The figure shows a gear train composed of a pair of helical gears and a pair of straight bevel gears. Shaft 4 is the output of the train, and it delivers 6 hp to the load at a speed of 370 rpm. The bevel gears have a pressure angle of 20°. If bearing E is to take both thrust and radial load while bearing F is to take only radial, determine the magnitude and direction of the forces which these bearings exert against shaft 4. *Ans.*: $\mathbf{F}_E = 163\mathbf{I} - 191.5\mathbf{J} + 355\mathbf{K}$ lb; $\mathbf{F}_F = 110\mathbf{J} + 145\mathbf{K}$ lb.

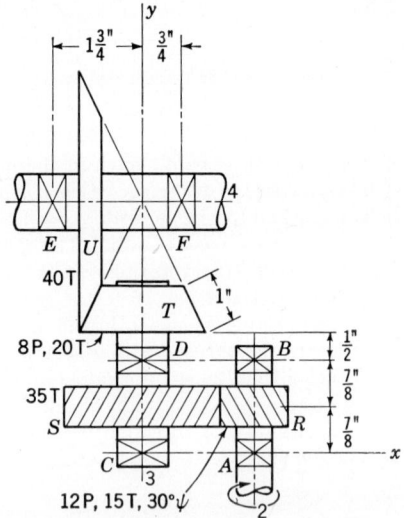

PROB. 15-25

15-26. Using the data of Prob. 15-25 find the forces exerted by bearings C and D on shaft 3. Which of these bearings should take the thrust load if the shaft is to be loaded in compression? The helical gears are 20° transverse pressure angle.

15-27. The gear train shown in the figure is composed of a right-hand helical pinion on shaft 2 which is cut with a $17\frac{1}{2}°$ normal pressure angle, the mating helical gear on shaft 3, a straight-tooth miter gear also on shaft 3, and its mating miter gear on shaft 4. The miter gears operate at a 20° pressure angle. The torque input to shaft 2 is 875 lb-in. in the direction shown. Find the forces exerted by bearings C and D on shaft 3. Specify which bearing is to take the thrust load. *Ans.:* $\mathbf{F}_D = 563\mathbf{I} - 239\mathbf{J} + 1{,}250\mathbf{K}$ lb; $\mathbf{F}_C = 239\mathbf{I} + 87\mathbf{K}$ lb.

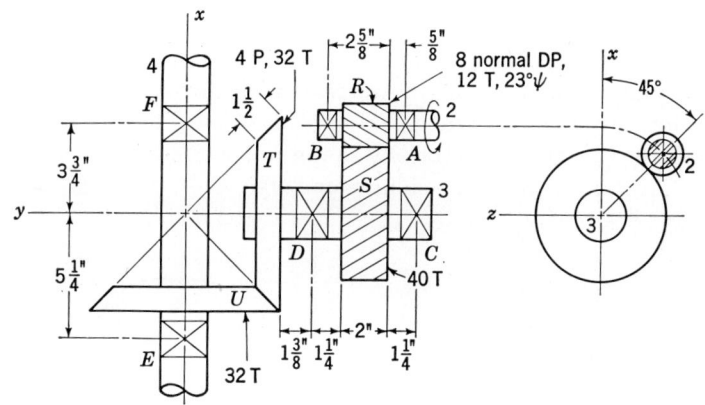

Prob. 15-27

15-28. Using the data of Prob. 15-27 find the forces exerted by bearings at E and F on shaft 4. Which of these two bearings should normally take the thrust component?

CHAPTER 16

DYNAMIC–FORCE ANALYSIS—
PLANE-MOTION MECHANISMS

In this chapter a study of the forces which are necessary to impart acceleration to the various members of a machine will be made. These forces will then be added vectorially to the static forces in order to obtain the resultant forces on the machine members.

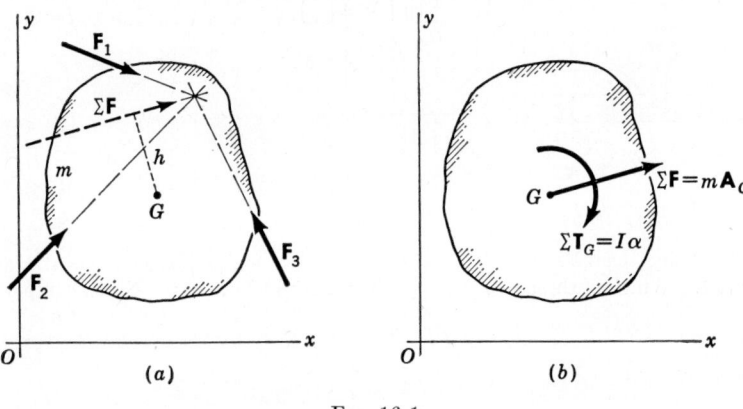

Fig. 16-1

The investigations to follow are limited to a study of plane-motion mechanisms. This means that the motion of the mechanism is restricted to two dimensions and that the mass of the parts is assumed to be concentrated in a single plane. This is seldom true; yet for a great many machines, the secondary effects are small and the simplification in analysis justifies the assumption.

16-1. D'Alembert's Principle. Consider a moving rigid body of mass m acted upon by any system of forces, say F_1, F_2, and F_3, as shown in Fig. 16-1a. Designate the center of mass of the body as point G, and find the resultant of the force system from the equation

$$\Sigma F = F_1 + F_2 + F_3 \qquad (a)$$

In the general case the line of action of this resultant will *not* be through

384

the mass center but will be displaced by some distance, such as the distance h, as shown in the figure. In the study of mechanics it is shown that the effect of this unbalanced force system is to produce linear and angular accelerations whose values are given by

$$\Sigma F = m\mathbf{A}_G \qquad (16\text{-}1)$$
$$\Sigma T_G = I\alpha \qquad (16\text{-}2)$$

in which $\mathbf{A}_G$ is the acceleration of the mass center and α is the angular acceleration of m (Fig. 16-1b). The quantity ΣF is the resultant of all the external forces acting upon the body, and ΣT_G is the sum of the external torques together with the moments of the external forces taken about G in the plane of motion. The mass moment of inertia is designated as I and is taken with reference to the mass center G also.

Equations (16-1) and (16-2) show that when an unbalanced system of forces acts upon a rigid body, the body experiences a linear acceleration $\mathbf{A}_G$ of its mass center in the same direction as the resultant force $\Sigma \mathbf{F}$ and that the body also experiences an angular acceleration α, due to the moments of the forces and torques about the center of mass, in the same direction as the resultant moment $\Sigma \mathbf{T}_G$. If the forces and moments are known, then Eqs. (16-1) and (16-2) can be employed to determine the resulting accelerations.

In machine design the motion of the machine members is usually specified in advance by other machine requirements. The problem then is: Given the motion of the machine elements, what forces are required to produce these motions? The problem therefore first requires a kinematic analysis in order to determine the linear and angular accelerations of the various members. Second, it requires a definition of the actual shape, dimensions, and material of the members; otherwise the masses and moments of inertia could not be determined. The motion analysis is made as shown in Part 1. In the examples to be demonstrated here only the results of the kinematic analysis will be presented. The selection of the materials, shape, and many of the dimensions of machine members is a subject of machine design and will not be discussed here either.[1]

Since, in the dynamic analysis of machines, the acceleration vectors are usually known, an alternative form of Eq. (16-1) and (16-2) is often convenient in determining the forces required to produce these known accelerations. Thus, we can write

$$\Sigma F - m\mathbf{A}_G = 0 \qquad (16\text{-}3)$$
$$\Sigma T_G - I\alpha = 0 \qquad (16\text{-}4)$$

[1] See Joseph E. Shigley, "Machine Design," McGraw-Hill Book Company, Inc., New York, 1956.

Both of these equations are vector equations applying to the plane motion of a rigid body. Equation (16-3) states that the vector sum of all the external forces acting upon the body plus the fictitious force $-m\mathbf{A}_G$ are zero. The fictitious force $-m\mathbf{A}_G$ is called an *inertia force*. It has the same line of action as $\mathbf{A}_G$ but is opposite in sense. Equation (16-4) states that the sum of the moments of all the external forces about an axis through G perpendicular to the plane of motion and the external torques acting upon the body plus a fictitious couple $-I\alpha$ is zero. The fictitious couple $-I\alpha$ is called an *inertial couple* or an *inertia torque*. The inertia torque is opposite in sense to the angular acceleration vector α. Equations (16-3) and (16-4) are extremely useful in studying the dynamics of machinery because they enable us to add inertia forces and torques to the external system of forces and to solve the resulting problem using the methods of statics.

The equations above are known as *D'Alembert's principle* because he was the first to call attention to the fact that addition of inertia forces to the real force system enabled a solution to be obtained from equations of equilibrium. We might note that the equations can also be written

$$\Sigma \mathbf{F} = 0 \qquad \Sigma \mathbf{T} = 0 \qquad (16\text{-}5)$$

where it is to be understood that both the external and the inertia forces and torques are to be included as terms in $\Sigma \mathbf{F}$ and $\Sigma \mathbf{T}$. Equation (16-5) is useful because it permits us to take a summation of the moments about any axis perpendicular to the plane of motion.

D'Alembert's principle is summarized as follows: *The vector sum of all the external forces and the inertia forces acting upon a rigid body is zero. The vector sum of all the external moments and the inertia torques acting upon a rigid body is also separately zero.*

16-2. Inertia Forces. In this section we shall show that the inertia force and inertia torque acting upon a rigid body can have their combined effect replaced by a single force. Such a substitution is unnecessary when the forces are to be determined analytically, but when a graphical solution is to be made, the substitution permits a solution to be obtained using only the force polygon.

In Fig. 16-2a link 3 is acted upon by the external forces $\mathbf{F}_{23}$ and $\mathbf{F}_{43}$ due to the action of other connecting links not shown on the drawing. When these two forces are summed, the resultant is

$$\Sigma \mathbf{F} = \mathbf{F}_{23} + \mathbf{F}_{43}$$

which is shown on the figure. The resultant force $\Sigma \mathbf{F}$ produces an acceleration $\mathbf{A}_G$ of the mass center G of the link and an angular acceleration α_3 because the line of action of $\Sigma \mathbf{F}$ is not through the mass center. Suppose a third force, having the same magnitude and the same line of action as ΣF but opposite in direction, is introduced. Such a third force would then place the link in static equilibrium, and the resultant linear and angular acceleration would both be zero. Let us designate another fictitious force by the character $\mathfrak{F}$. The force $\mathfrak{F}$ is to have the same magnitude as ΣF and the same line of action but is to be opposite in direction. Thus

$$\mathfrak{F} = -\Sigma F$$

Since ΣF produces the accelerations $\mathbf{A}_G$ and α_3, the force $\mathfrak{F}$, if it was not fictitious, would produce the accelerations $-\mathbf{A}_G$ and $-\alpha_3$. As shown in Fig. 16-2a the moment arm of ΣF about G is h, and so the force $\mathfrak{F}$ must have the same moment arm.

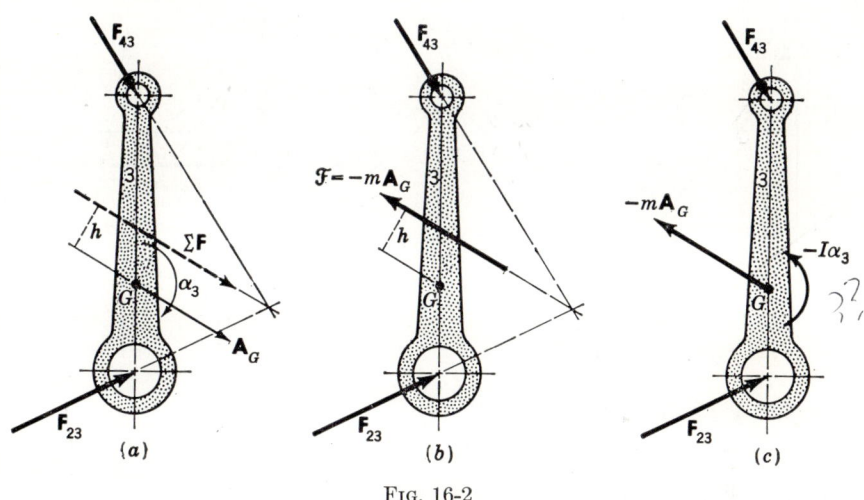

Fig. 16-2

The force $\mathfrak{F}$ can be obtained from Eq. (16-3) by substituting $-\mathfrak{F}$ for ΣF. This gives

$$\mathfrak{F} = -ma_G$$

where $\mathfrak{F}$ is the magnitude of the fictitious force and a_G is the magnitude of the acceleration. The force $\mathfrak{F}$ is to have a moment of $h\mathfrak{F}$ about the mass center. Substituting $h\mathfrak{F}$ for ΣT in Eq. (16-4) gives

$$h\mathfrak{F} - I\alpha_3 = 0$$

or

$$h = \frac{I\alpha_3}{\mathfrak{F}} = \frac{I\alpha_3}{ma_G} \qquad (16\text{-}6)$$

As shown in Fig. 16-2b the force ΣF is replaced by another force equal to $-m\mathbf{A}_G$ but displaced from the mass center a distance h. This fictitious force will then replace the combined effects of the inertia torque and the inertia force shown in Fig. 16-2c.

16-3. Analysis of a Floating Link.

As an example of the principles discussed in the preceding sections, let us select link 3 of the mechanism in Fig. 16-3a for investigation. This is a floating link, since it is not fixed to the frame at any point, and it moves with combined translation and rotation. Thus the solution will be quite general, and the procedures used can be applied to a large number of problems.

In order to present the solution as clearly as possible we shall pretend that sliding blocks 2 and 4 are weightless and therefore have no mass or

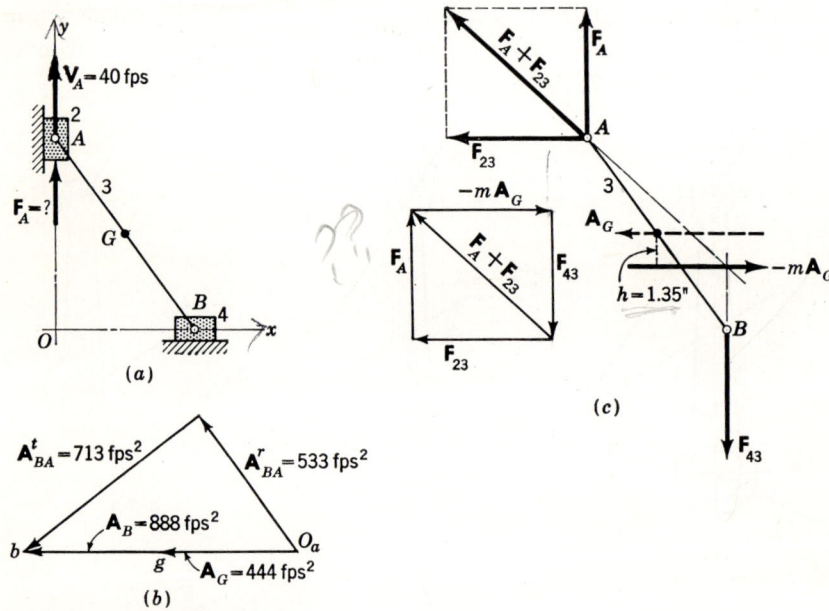

FIG. 16-3. $OB = 6$ in.; $OA = 8$ in.; $AG = 5$ in.

moment of inertia. It is also convenient to neglect friction and to assume that the static forces necessary to hold the link in its present position can be neglected.

With these facts cleared, let us give the velocity of A as $\mathbf{V}_A = 40$ fps in the positive y direction and state that the problem is to determine the force $\mathbf{F}_A$ which is required to produce this velocity for the position shown. It might be added that $\mathbf{V}_A$ is to be constant; that is, $\mathbf{A}_A = 0$. As other given elements of the problem we include the weight and moment of inertia of link 3 as $W_3 = 2.20$ lb and $I_3 = 0.0479$ lb-sec²-in.

Graphical Solution. The first step in the solution is the kinematic analysis. Since this has already been discussed in the earlier chapters of this book, only the results will be given here. Thus Fig. 16-3b shows the acceleration polygon from which the acceleration of the mass center G

is determined. The acceleration polygon also provides the information necessary to calculate α_3. Thus

$$\alpha_3 = \frac{a_{BA}{}^t}{BA} = \frac{713}{10\tfrac{1}{2}} = 855 \text{ rad/sec}^2 \qquad (a)$$

Now, using Eq. (16-6),

$$h = \frac{I\alpha_3}{ma_G} = \frac{(0.0479)(855)}{(2.20/32.2)(444)} = 1.35 \text{ in.}*$$

The free-body diagram and the resulting force polygon are shown in Fig. 16-3c. Notice that the inertia force $-m\mathbf{A}_G$ is displaced from G by the distance h so as to produce a moment about G of $-I\alpha_3$, also that $-m\mathbf{A}_G$ is in the opposite sense to $\mathbf{A}_G$. The reaction at B is $\mathbf{F}_{43}$ and is vertically downward because friction is neglected. The forces at A are the actuating force $\mathbf{F}_A$ and the block reaction $\mathbf{F}_{23}$, which is horizontal also because friction is neglected. The point of concurrency is the intersection of $-m\mathbf{A}_G$ and $\mathbf{F}_{43}$, the directions of both being known. The line of action of the total force at A, $\mathbf{F}_A + \mathbf{F}_{23}$, must pass through the point of concurrency. This fact permits construction of the force polygon. Then the unknown forces $\mathbf{F}_A$ and $\mathbf{F}_{23}$ are known in direction and are found as components of $\mathbf{F}_A + \mathbf{F}_{23}$, as shown in the figure. The actuating force is found by measurement to be

$$\mathbf{F}_A = 27 \text{ lb upward} \qquad Ans.$$

Analytical Solution. A kinematic analysis of the mechanism of Fig. 16-3 with $\mathbf{V}_A = 40\mathbf{J}$ as the given quantity gives

$$\mathbf{A}_G = -444\mathbf{I} \quad \text{fps}^2 \qquad \alpha_3 = -855\mathbf{K} \quad \text{rad/sec}^2$$

Thus, the inertia forces and torques are

$$-m\mathbf{A}_G = -\frac{2.20}{32.2}(-444\mathbf{I}) = 30.3\mathbf{I} \quad \text{lb}$$

$$-I_3\alpha_3 = -(0.0479)(-855\mathbf{K}) = 41.0\mathbf{K} \quad \text{lb-in.}$$

* Note that

$$I\alpha_3 = (\text{lb-sec}^2\text{-in.})\left(\frac{\text{rad}}{\text{sec}^2}\right) = \text{lb-in.}$$

and

$$ma_G = \left(\frac{\text{lb-sec}^2}{\text{ft}}\right)\left(\frac{\text{ft}}{\text{sec}^2}\right) = \text{lb}$$

This shows that the units of h are inches. In general, it is expedient to calculate I in inch units; then the torque will be in units of pound-inches. For this purpose use $g = 386$ in./sec². On the other hand the mass is best calculated in units of pounds, seconds, and feet, because the acceleration is usually expressed in terms of feet and seconds. So, for calculating mass, use $g = 32.2$ fps². Of course, both of these values for g include the assumption of sea level and no correction for latitude.

390   DYNAMIC ANALYSIS OF MACHINES

These forces, together with the unknowns, are shown on the free-body diagram of Fig. 16-4. As the first step we shall take moments about point A. This requires a solution to the equation

$$\Sigma T_A = R_{GA} \times (-mA_G) + R_{BA} \times F_{43} + (-I_3\alpha_3) = 0 \quad (b)$$

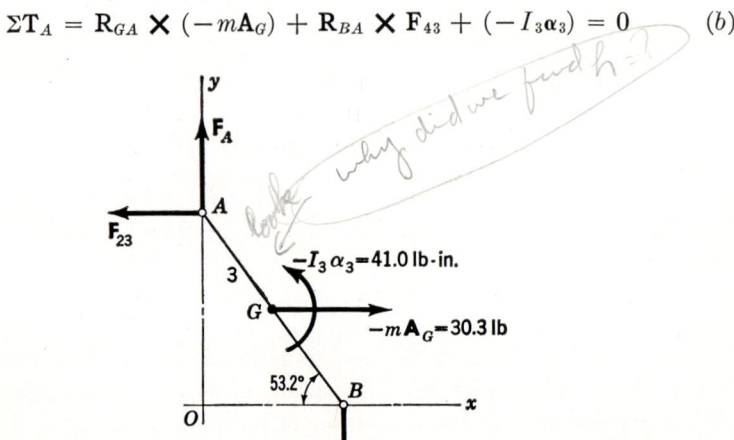

FIG. 16-4. $AG = 5$ in.; $AB = 10$ in.

The terms for this equation are as follows:

$$R_{GA} = 5 /\!-\!53.2° = 3\mathbf{I} - 4\mathbf{J}$$

$$R_{GA} \times (-mA_G) = \begin{vmatrix} \mathbf{I} & \mathbf{J} & \mathbf{K} \\ 3 & -4 & 0 \\ 30.3 & 0 & 0 \end{vmatrix} = 121.2\mathbf{K}$$

$$R_{BA} = 10 /\!-\!53.2° = 6\mathbf{I} - 8\mathbf{J}$$

$$F_{43} = F_{43} /\!-\!90° = -F_{43}\mathbf{J}$$

$$R_{BA} \times F_{43} = \begin{vmatrix} \mathbf{I} & \mathbf{J} & \mathbf{K} \\ 6 & -8 & 0 \\ 0 & -F_{43} & 0 \end{vmatrix} = -6F_{43}\mathbf{K}$$

Substituting these values in Eq. (b) gives

$$\Sigma T_A = 121.2\mathbf{K} - 6F_{43}\mathbf{K} + 41.0\mathbf{K} = 0$$
$$F_{43} = 27 \text{ lb} \quad \text{or} \quad F_{43} = -27\mathbf{J}$$

Now, writing the equation

$$\Sigma F = F_A + F_{23} + (-mA_G) + F_{43} = 0 \quad (c)$$

Since $\mathbf{F}_A = F_A\mathbf{J}$ and $\mathbf{F}_{23} = -F_{23}\mathbf{I}$, Eq. (c) becomes

$$\Sigma \mathbf{F} = F_A\mathbf{J} - F_{23}\mathbf{I} + 30.3\mathbf{I} - 27\mathbf{J} = 0$$

or $F_A = 27$ lb $\mathbf{F}_A = 27\mathbf{J}$ lb   *Ans.*
    $F_{23} = 30.3$ lb $\mathbf{F}_{23} = -30.3\mathbf{I}$ lb   *Ans.*

16-4. Rotation. The previous sections have dealt with the general case of dynamic forces for a rigid body having a combined motion of translation and rotation. It is important to emphasize that the equations

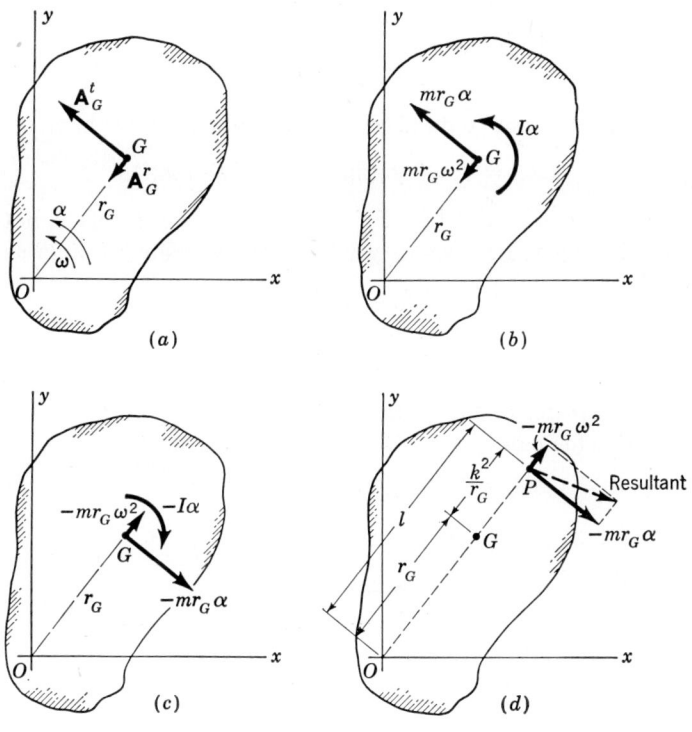

Fig. 16-5

and the methods of analysis investigated in these sections are general and apply to *all* problems of plane motion. It will be interesting now to study the application of these methods to a rigid body rotating about a *fixed* center.

Let us suppose a rigid body constrained to rotate about some fixed center O not coincident with the center of mass G (Fig. 16-5a). A system of forces (not shown) is to be applied to the body, causing it to have an angular acceleration α. We also include the fact that the body is rotating with an angular velocity ω. This motion of the body means that the mass center will have transverse and radial components of acceleration $\mathbf{A}_G{}^t$ and $\mathbf{A}_G{}^r$ whose magnitudes are

$$a_G{}^t = r_G\alpha \qquad a_G{}^r = r_G\omega^2 \qquad (a)$$

Thus, if we resolve the resultant external force into transverse and radial components, these components must have magnitudes

$$\Sigma F^t = mr_G\alpha \quad\text{and}\quad \Sigma F^r = mr_G\omega^2 \qquad (b)$$

in accordance with Eq. (16-1). In addition, Eq. (16-2) states that an external torque must exist to create the angular acceleration and that the magnitude of this torque is

$$T_G = I\alpha \qquad (c)$$

If we now sum the moments of these forces about O, we have

$$\Sigma T_O = I\alpha + r_G(mr_G\alpha) = (I + mr_G{}^2)\alpha \qquad (d)$$

But the moment of inertia of a rigid body about some axis O, not coincident with the centroidal axis, is given by the transfer formula as

$$I_O = I + mr^2 \qquad (e)$$

where I_O = moment of inertia about axis O
I = moment of inertia about centroidal axis G
r = transfer distance, that is, distance OG

Therefore Eq. (d) can be written in vector form as

$$\Sigma \mathbf{T}_O = I_O\boldsymbol{\alpha} \qquad (16\text{-}7)$$

Equations (16-3) and (16-4) then become

$$\Sigma \mathbf{F} - m\mathbf{A}_G = 0 \qquad (16\text{-}8)$$
$$\Sigma \mathbf{T}_O - I_O\boldsymbol{\alpha} = 0 \qquad (16\text{-}9)$$

by including the inertia force $-m\mathbf{A}_G$ and inertia torque $-I_O\boldsymbol{\alpha}$ (Fig. 16-5c). We observe particularly that the system of forces does *not* reduce to a single couple because of the existence of the inertia force component $-mr_G\omega^2$, which has no moment arm about O. Thus both Eqs. (16-8) and (16-9) are necessary.

A particular case arises when $\alpha = 0$. Then the external torque T_O is zero and the only inertia force is, from Fig. 16-5c, the centrifugal force $-mr_G\omega^2$.

A second special case exists under starting conditions when $\omega = 0$, but α is not zero. Under these conditions the only inertia force is $-mr_G\alpha$, and the system reduces to a single couple.

Since moment of inertia has the dimensions of mass multiplied by the square of a length, it is expedient to define another length called *radius of gyration.* When the mass of a body is known, the moment of inertia of the body about an axis is the product of the mass and the square of the radius of gyration with respect to the axis. Thus, the equation

$$I = mk^2 \tag{16-10}$$

defines the radius of gyration k.

When a rigid body has a motion of pure translation, the resultant inertia force and the resultant external force have the same line of action and this line of action passes through the mass center of the body. When a rigid body has rotation and angular acceleration, the resultant inertia force and the resultant external force have the same line of action but the line of action *does not* pass through the mass center. Let us now locate a point on the line of action of the resultant of the inertia forces of Fig. 16-5c.

The resultant of the inertia forces will pass through some point P on the line OG of Fig. 16-5c or an extension of it. This force can be resolved into two components, one of which will be the component $-mr_G\omega^2$ acting along the line OG. The other component will be $-mr_G\alpha$ acting perpendicular to OG but not through point G. The distance, designated as l, to the unknown point P can be found by equating the moment of the component $-mr_G\alpha$ through P to the sum of the inertia torque and the moment of the inertia forces which act through G. Thus, taking moments about O, we have

$$(-mr_G\alpha)l = -I\alpha + (-mr_G\alpha)r_G$$

or
$$l = \frac{I}{mr_G} + r_G \tag{f}$$

Substituting the value of I from Eq. (16-10) gives

$$l = \frac{k^2}{r_G} + r_G \tag{16-11}$$

The point P located by Eq. (16-11) and shown in Fig. 16-5d is called the *center of percussion.* As shown, the resultant inertia force passes through P, and consequently, the inertia force has zero moment about the center of percussion. If an external force is applied at P, an angular acceleration α will result but the bearing reaction at O will be zero except for the radial component due to the inertia force $-mr_G\omega^2$. It is the usual practice in shock-testing machines to apply the force at the center of percussion in order to eliminate the transverse bearing reaction due to the external force.

Equation (16-11) shows that the location of the center of percussion is independent of the values of ω and α.

If the axis of rotation is coincident with the mass center, then $r_G = 0$ and Eq. (16-11) shows that $l = \infty$. Under these conditions there is no resultant inertia force but, instead, a resultant inertia couple $-I\alpha$.

In closing this section it is observed that, by the methods of Chap. 12, the transverse and radial components of the acceleration of G can be written

$$\mathbf{A}_G{}^t = \boldsymbol{\alpha} \times \mathbf{R}_G \qquad (16\text{-}12)$$
$$\mathbf{A}_G{}^r = \boldsymbol{\Omega} \times (\boldsymbol{\Omega} \times \mathbf{R}_G) \qquad (16\text{-}13)$$

where $\boldsymbol{\Omega}$ is the vector angular velocity and $\mathbf{R}_G$ is the position vector of point G. Equations (b) can now be expressed in vector form:

$$\Sigma \mathbf{F}^t = m\boldsymbol{\alpha} \times \mathbf{R}_G \qquad (16\text{-}14)$$
$$\Sigma \mathbf{F}^r = m\boldsymbol{\Omega} \times (\boldsymbol{\Omega} \times \mathbf{R}_G) \qquad (16\text{-}15)$$

The resultant external force defined in terms of the transverse and radial components as given by these equations is often very useful in analysis.

16-5. An Example of Dynamic-force Analysis. We have now demonstrated all the principles necessary for making a complete dynamic-force analysis of a plane-motion mechanism. The steps in making such an analysis can be summarized as follows:

1. Make a kinematic analysis of the mechanism to find the angular acceleration of each link or element. Locate the center of mass of each link and determine the acceleration of these points.

2. Using the given value or values of the force or torque which the follower must deliver, make a complete static-force analysis of the mechanism. The results of this analysis will then include the magnitudes and directions of the forces and torques acting upon each member. Observe particularly that this is a static-force analysis and that it does not include inertia forces or torques.

3. Employing the given values of the masses and moments of inertia and the angular and linear accelerations found in step 1, calculate the inertia forces and inertia torques for each link or element of the mechanism. Taking these as applied forces make a free-body analysis of each member of the entire mechanism to find the total effect of all the inertia forces and torques.

4. Vectorially add the results of steps 2 and 3 to obtain the resultant forces and torques for every machine element.

EXAMPLE 16-1. Make a complete dynamic-force analysis of the four-bar linkage illustrated in Fig. 16-6. The given quantities are included in the figure caption.

Graphical Solution. The first step is to make the kinematic analysis of the mechanism. This step is not included here, but the acceleration polygon resulting from such an analysis is reproduced in Fig. 16-6. A number of the resulting quantities are given on the polygon to enable the reader to verify the results if he so desires. Upon completion of the polygon the angular acceleration of links 3 and 4 can be calculated.

PLANE-MOTION MECHANISMS 395

These are
$$\alpha_3 = 148 \text{ rad/sec}^2 \circlearrowleft \qquad \alpha_4 = 604 \text{ rad/sec}^2 \circlearrowright$$

A major portion of the analysis is concerned with links 3 and 4 because the center of mass of link 2 is located at O_2. Free-body diagrams of links 4 and 3 are shown separately in Figs. 16-7 and 16-8, respectively. Notice also that these diagrams are arranged in "equation form" for simpler reading. Thus, in each illustration, the forces in (a) plus those in (b) and in (c) produce the resultants shown in (d). The two sets of illustrations are also correlated; for example, F'_{34} in Fig. 16-7a is equal to minus

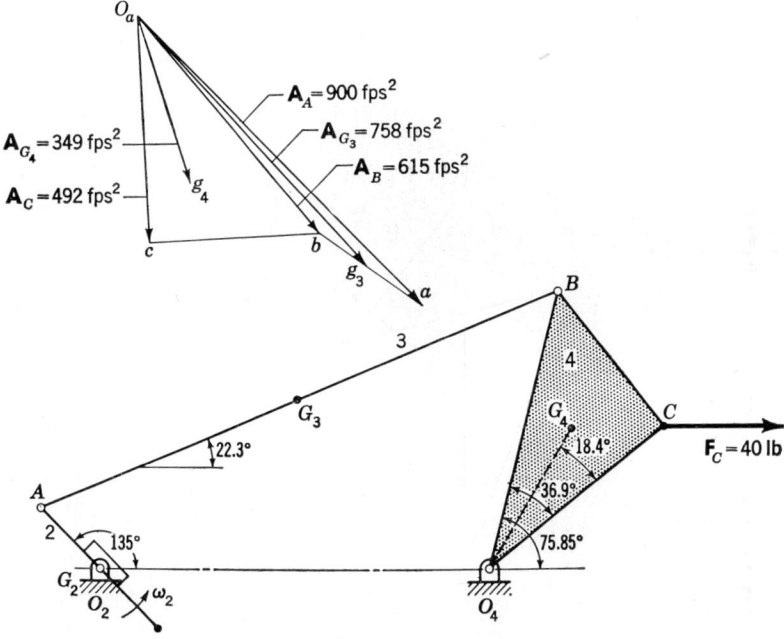

FIG. 16-6. $O_2A = 3$ in.; $AB = 20$ in.; $O_4B = 10$ in.; $O_2O_4 = 14$ in.; $O_4G_4 = 5^{11}/_{16}$ in.; $AG_3 = 10$ in.; $BC = 6$ in.; $O_4C = 8$ in.; $\omega_2 = 60$ rad/sec; $\alpha_2 = 0$ rad/sec^2; $W_3 = 7.13$ lb; $I_3 = 0.625$ lb-sec^2-in.; $W_4 = 3.42$ lb; $I_4 = 0.037$ lb-sec^2-in. The angular position of the various links have been calculated for the given position of link 2 and are given on the figure.

F'_{43} in Fig. 16-8a, etc. The following analysis is not difficult but it is complicated, and so the thoughtful reader will do well to read slowly and examine the illustrations carefully—detail by detail.

Let us start with link 4 in Fig. 16-7a. Proceeding in accordance with our earlier investigations we make the following calculations:

$$I_4\alpha_4 = (0.037)(604) = 22.3 \text{ lb-in.}$$
$$m_4 a_{G_4} = \frac{3.42}{32.2}(349) = 37.1 \text{ lb}$$
$$h_4 = \frac{I_4\alpha_4}{m_4 a_{G_4}} = \frac{22.3}{37.1} = 0.602 \text{ in.}$$

Now the force $-m_4 a_{G_4} = 37.1$ lb is placed on the free-body diagram opposite in direction to A_{G_4} and offset from G_4 by the distance h_4. The direction of the offset is that

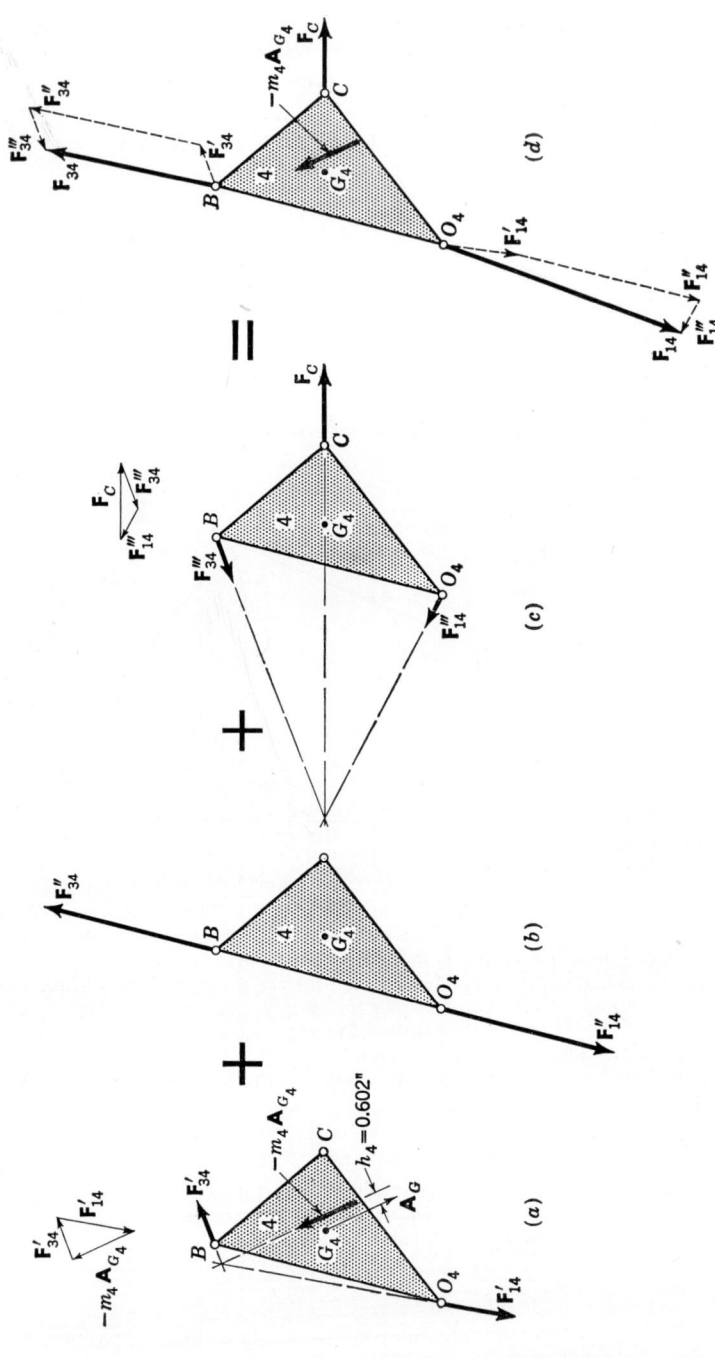

Fig. 16-7. Free-body diagrams of link 4. $-m_4 A_{G_4} = 37.1$ lb; $F'_{34} = 24.3$ lb; $F'_{14} = 44.3$ lb; $F''_{34} = -F''_{14} = -F''_{43} = 94.8$ lb; $F_C = 40$ lb; $F'''_{34} = 25$ lb; $F'''_{14} = 19.3$ lb; $F_{34} = 94.3$ lb; $F_{14} = 132$ lb.

required to produce a torque about G_4 opposed to $I_4\alpha_4$. The direction of $\mathbf{F}'_{34}$ is taken along link 3. The intersection of $\mathbf{F}'_{34}$ and $-m_4\mathbf{A}_{G_4}$ gives the point of concurrency and establishes the direction of $\mathbf{F}'_{14}$. The force polygon can now be constructed and the magnitudes of $\mathbf{F}'_{34}$ and $\mathbf{F}'_{14}$ found. These values are given in the figure caption.

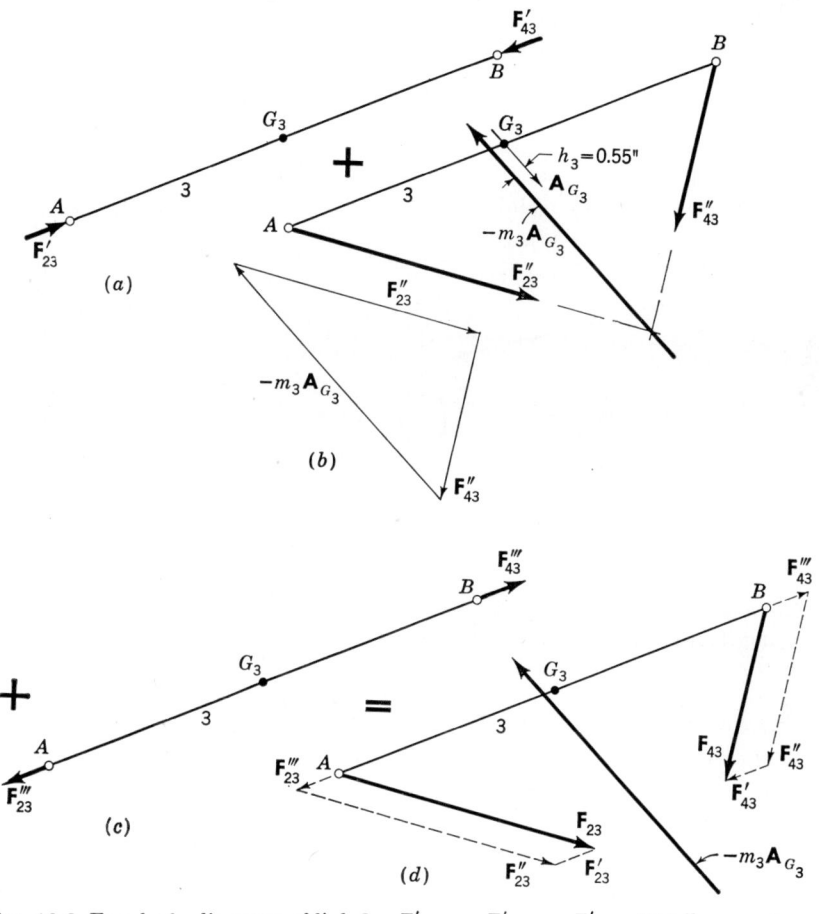

FIG. 16-8. Free-body diagrams of link 3. $\mathbf{F}'_{43} = -\mathbf{F}'_{23} = -\mathbf{F}'_{34} = 24.3$ lb; $-m_3\mathbf{A}_{G_3} = 168$ lb; $\mathbf{F}''_{43} = 94.8$ lb; $\mathbf{F}''_{23} = 145$ lb; $\mathbf{F}'''_{43} = -\mathbf{F}'''_{23} = -\mathbf{F}'''_{34} = 25$ lb; $\mathbf{F}_{43} = -\mathbf{F}_{34} = 94.3$ lb; $\mathbf{F}_{23} = 145$ lb.

Proceed next to link 3 in Fig. 16-8a. The forces $\mathbf{F}'_{43}$ and $\mathbf{F}'_{23}$ are now known from the preceding analysis.

Now go to Fig. 16-8b and link 3, and make the calculations

$$I_3\alpha_3 = (0.625)(148) = 92.5 \text{ lb-in.}$$
$$m_3 a_{G_3} = \frac{7.13}{32.2}(758) = 168 \text{ lb}$$
$$h_3 = \frac{I_3\alpha_3}{m_3 a_{G_3}} = \frac{92.5}{168} = 0.550 \text{ in.}$$

398   DYNAMIC ANALYSIS OF MACHINES

Locate the inertia force $-m_3\mathbf{A}_{G_3} = 168$ lb on the free-body diagram opposite in direction to $\mathbf{A}_{G_3}$ and offset a distance h_3 from G_3 so as to produce a torque about G_3 opposite in direction to $I_3\alpha_3$. The direction of $\mathbf{F}''_{43}$ is along the line BO_4. The forces $-m_3\mathbf{A}_{G_3}$ and $\mathbf{F}''_{43}$ intersect and determine the point of concurrency. Thus the direction of $\mathbf{F}''_{23}$ is known and the force polygon can be constructed. The resulting values of $\mathbf{F}''_{43}$ and $\mathbf{F}''_{23}$ are included in the caption.

In Fig. 16-7b the forces $\mathbf{F}''_{34}$ and $\mathbf{F}''_{14}$, acting on link 4, are now known from the preceding analysis.

Figures 16-7c and 16-8c show the results of the static-force analysis with $\mathbf{F}_C = 40$ lb as the given quantity. The force polygon in Fig. 16-7c determines the values of the forces acting on link 4, and from these the direction and magnitude of the forces operating on link 3 are found.

The next step is a vector addition of the results already obtained as shown in (d) of each figure.

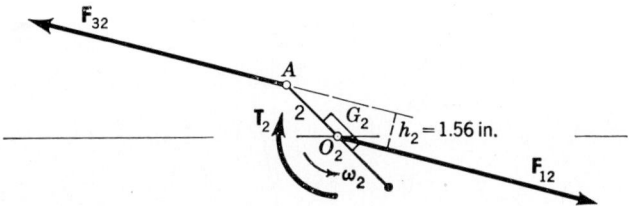

Fig. 16-9. Free-body diagram of link 2. $\mathbf{F}_{32} = -\mathbf{F}_{23} = -\mathbf{F}_{12} = 145$ lb; $T_2 = 226$ lb-in.

The analysis is completed by taking the resultant force $\mathbf{F}_{23}$ from Fig. 16-8d and applying its negative, $\mathbf{F}_{32}$, to link 2. This is done in Fig. 16-9. The distance h_2 is found by measurement. The external torque to be applied to link 2 is

$$T_2 = h_2 F_{32} = (1.56)(145) = 226 \text{ lb-in.}$$

It is interesting to observe that the applied torque is opposed to the direction of rotation of link 2. The reader should be able to explain what elements are necessary in a rotating machine in order that it can apply a torque opposite to the direction of rotation.

Analytical Solution. The results of an analytical acceleration analysis of the linkage are

$$\alpha_3 = 148\mathbf{K} \qquad \alpha_4 = -604\mathbf{K} \qquad \mathbf{A}_{G_3} = 519\mathbf{I} - 551\mathbf{J}$$
$$\mathbf{A}_{G_4} = 134\mathbf{I} - 323\mathbf{J}$$

Taking link 4 first, we shall write the vector equation

$$\Sigma \mathbf{T}_{O_4} = \mathbf{R}_{G_4} \times (-m_4\mathbf{A}_{G_4}) + (-I_4\alpha_4) + \mathbf{R}_B \times \mathbf{F}'_{34} = 0 \tag{1}$$

where $\mathbf{F}'_{34}$ is the force identified by Fig. 16-7a. The elements of Eq. (1) are found from an examination of Fig. 16-6 and are

$$\mathbf{R}_{G_4} = 5.68 \underline{/57.5°} = 3.05\mathbf{I} + 4.80\mathbf{J}$$
$$-m_4\mathbf{A}_{G_4} = -\frac{3.42}{32.2}(134\mathbf{I} - 323\mathbf{J}) = -14.25\mathbf{I} + 34.3\mathbf{J}$$

so
$$\mathbf{R}_{G_4} \times (-m_4 \mathbf{A}_{G_4}) = \begin{vmatrix} \mathbf{I} & \mathbf{J} & \mathbf{K} \\ 3.05 & 4.80 & 0 \\ -14.25 & 34.3 & 0 \end{vmatrix} = 173.2\mathbf{K}$$

$$-I_4 \alpha_4 = -(0.037)(-604\mathbf{K}) = 22.3\mathbf{K}$$
$$\mathbf{R}_B = 10/\underline{75.85°} = 2.44\mathbf{I} + 9.70\mathbf{J}$$
$$\mathbf{F}'_{34} = F'_{34}/\underline{22.3°}$$

Then
$$\mathbf{R}_B \times \mathbf{F}'_{34} = \begin{vmatrix} \mathbf{I} & \mathbf{J} & \mathbf{K} \\ 2.44 & 9.70 & 0 \\ F'^x_{34} & F'^y_{34} & 0 \end{vmatrix} = (2.44 F'^y_{34} - 9.70 F'^x_{34})\mathbf{K}$$

Substituting these values in Eq. (1) gives

$$\Sigma \mathbf{T}_{O_4} = 173.2\mathbf{K} + 22.3\mathbf{K} + (2.44 F'^y_{34} - 9.70 F'^x_{34})\mathbf{K} = 0 \quad (2)$$

But
$$\frac{F'^y_{34}}{F'^x_{34}} = \tan 22.3° = 0.41 \quad (3)$$

Solving Eqs. (2) and (3) simultaneously yields

$$\mathbf{F}'_{34} = 22.5\mathbf{I} + 9.2\mathbf{J}$$

Then, for link 4, we write the equation

$$\Sigma \mathbf{F} = \mathbf{F}'_{34} + (-m_4 \mathbf{A}_{G_4}) + \mathbf{F}'_{14} = 0$$

Substituting the known values and solving for $\mathbf{F}'_{14}$ give

$$\Sigma \mathbf{F} = (22.5\mathbf{I} + 9.2\mathbf{J}) + (-14.25\mathbf{I} + 34.3\mathbf{J}) + \mathbf{F}'_{14} = 0$$
$$\mathbf{F}'_{14} = -8.25\mathbf{I} - 43.5\mathbf{J}$$

A similar procedure is followed for link 3. The equation for a summation of moments about point A is

$$\Sigma \mathbf{T}_A = \mathbf{R}_{GA} \times (-m_3 \mathbf{A}_{G_3}) + (-I_3 \alpha_3) + \mathbf{R}_{BA} \times \mathbf{F}''_{43} = 0 \quad (4)$$

where the force $\mathbf{F}''_{43}$ is identified in Fig. 16-8b. The parts of Eq. (4) are as follows:

$$\mathbf{R}_{GA} = 10/\underline{22.3°} = 9.25\mathbf{I} + 3.80\mathbf{J}$$
$$-m_3 \mathbf{A}_{G_3} = -\frac{7.13}{32.2}(519\mathbf{I} - 551\mathbf{J}) = -119\mathbf{I} + 126\mathbf{J}$$

so
$$\mathbf{R}_{GA} \times (-m_3 \mathbf{A}_{G_3}) = \begin{vmatrix} \mathbf{I} & \mathbf{J} & \mathbf{K} \\ 9.25 & 3.80 & 0 \\ -119 & 126 & 0 \end{vmatrix} = 1{,}618\mathbf{K}$$

$$-I_3 \alpha_3 = -0.625(148\mathbf{K}) = -92.5\mathbf{K}$$
$$\mathbf{F}''_{43} = F''_{43}/\underline{75.85°}$$
$$\mathbf{R}_{BA} = 20/\underline{22.3°} = 18.5\mathbf{I} + 7.60\mathbf{J}$$

$$\mathbf{R}_{BA} \times \mathbf{F}''_{43} = \begin{vmatrix} \mathbf{I} & \mathbf{J} & \mathbf{K} \\ 18.5 & 7.60 & 0 \\ F''^x_{43} & F''^y_{43} & 0 \end{vmatrix} = (18.5 F''^y_{43} - 7.60 F''^x_{43})\mathbf{K}$$

Substituting these terms in Eq. (4),

$$\Sigma \mathbf{T}_A = 1{,}618\mathbf{K} - 92.5\mathbf{K} + (18.5 F''^y_{43} - 7.60 F''^x_{43})\mathbf{K} = 0 \quad (5)$$

But the components of $\mathbf{F}''_{43}$ are related by the equation

$$\frac{F''_{43}{}^y}{F''_{43}{}^x} = \tan 75.85° = 3.97 \tag{6}$$

Solving Eqs. (5) and (6) simultaneously gives

$$\mathbf{F}''_{43} = -23.2\mathbf{I} - 92\mathbf{J}$$

Next, for link 3, we write

$$\Sigma \mathbf{F} = -m_3 \mathbf{A}_{G_3} + \mathbf{F}''_{43} + \mathbf{F}''_{23} = 0$$

Substituting and solving for $\mathbf{F}''_{23}$,

$$\Sigma \mathbf{F} = (-119\mathbf{I} + 126\mathbf{J}) + (-23.2\mathbf{I} - 92\mathbf{J}) + \mathbf{F}''_{23} = 0$$

Hence
$$\mathbf{F}''_{23} = 142\mathbf{I} - 34\mathbf{J}$$

The static-force analysis of link 4 is made next. We desire to solve the equation

$$\Sigma \mathbf{T}_{O_4} = \mathbf{R}_C \times \mathbf{F}_C + \mathbf{R}_B \times \mathbf{F}'''_{34} = 0 \tag{7}$$

where the direction of $\mathbf{F}'''_{34}$ is along link 3. The terms for Eq. (7) are

$$\mathbf{R}_C = 8\underline{/38.95°} = 6.21\mathbf{I} + 5.04\mathbf{J}$$
$$\mathbf{F}_C = 40\underline{/0°} = 40\mathbf{I}$$

so that
$$\mathbf{R}_C \times \mathbf{F}_C = \begin{vmatrix} \mathbf{I} & \mathbf{J} & \mathbf{K} \\ 6.21 & 5.04 & 0 \\ 40 & 0 & 0 \end{vmatrix} = -202\mathbf{K}$$

$$\mathbf{R}_B \times \mathbf{F}'''_{34} = \begin{vmatrix} \mathbf{I} & \mathbf{J} & \mathbf{K} \\ 2.44 & 9.70 & 0 \\ F'''_{34}{}^x & F'''_{34}{}^y & 0 \end{vmatrix}$$
$$= (2.44 F'''_{34}{}^y - 9.70 F'''_{34}{}^x)\mathbf{K}$$

Equation (7) becomes

$$\Sigma \mathbf{T}_{O_4} = -202\mathbf{K} + (2.44 F'''_{34}{}^y - 9.70 F'''_{34}{}^x)\mathbf{K} = 0 \tag{8}$$

But
$$\frac{F'''_{34}{}^y}{F'''_{34}{}^x} = \tan 22.3° = 0.41 \tag{9}$$

Solving Eqs. (8) and (9) simultaneously yields

$$\mathbf{F}'''_{34} = -23.2\mathbf{I} - 9.5\mathbf{J}$$

Then, to find the frame reaction on link 4, we write

$$\Sigma \mathbf{F} = \mathbf{F}'''_{34} + \mathbf{F}_C + \mathbf{F}'''_{14} = 0$$
or
$$\Sigma \mathbf{F} = (-23.2\mathbf{I} - 9.5\mathbf{J}) + (40\mathbf{I}) + \mathbf{F}'''_{14} = 0$$
and
$$\mathbf{F}'''_{14} = -16.8\mathbf{I} + 9.5\mathbf{J}$$

The forces are now completely determined, and it is only a matter of summing the proper components to get the total forces acting on each link. All the results are summarized below for convenience.

For link 4:

$$\begin{aligned}\mathbf{F}_{14} &= \mathbf{F}'_{14} + \mathbf{F}''_{14} + \mathbf{F}'''_{14} \\ &= (-8.2\mathbf{I} - 43.5\mathbf{J}) + (-23.2\mathbf{I} - 92\mathbf{J}) + (-16.8\mathbf{I} + 9.5\mathbf{J}) \\ &= -48.2\mathbf{I} - 126\mathbf{J} = 132\underline{/-110.9°} \text{ lb}\end{aligned}$$

$$\begin{aligned}\mathbf{F}_{34} &= \mathbf{F}'_{34} + \mathbf{F}''_{34} + \mathbf{F}'''_{34} \\ &= (22.5\mathbf{I} + 9.2\mathbf{J}) + (23.2\mathbf{I} + 92\mathbf{J}) + (-23.2\mathbf{I} - 9.5\mathbf{J}) \\ &= -22.5\mathbf{I} + 91.7\mathbf{J} = 94.3\underline{/76.2°} \text{ lb}\end{aligned}$$

For link 3:

$$\mathbf{F}_{43} = -\mathbf{F}_{34} = -22.5\mathbf{I} - 91.7\mathbf{J} = 94.3\underline{/-103.8°} \text{ lb}$$

$$\begin{aligned}\mathbf{F}_{23} &= \mathbf{F}'_{23} + \mathbf{F}''_{23} + \mathbf{F}'''_{23} \\ &= (22.5\mathbf{I} + 9.2\mathbf{J}) + (142\mathbf{I} - 34\mathbf{J}) + (-23.2\mathbf{I} - 9.5\mathbf{J}) \\ &= 141\mathbf{I} - 34.3\mathbf{J} = 145\underline{/-13.7°} \text{ lb}\end{aligned}$$

For link 2:

$$\mathbf{F}_{32} = -\mathbf{F}_{12} = -\mathbf{F}_{23} = -141\mathbf{I} + 34.3\mathbf{J} = 145\underline{/166.3°} \text{ lb}$$

The torque to be applied to link 2 is found from the equation

$$\Sigma \mathbf{T}_{O_2} = \mathbf{R}_A \times \mathbf{F}_{32} + \mathbf{T}_2 = 0$$

Since $\mathbf{R}_A = 3\underline{/135°} = -2.12\mathbf{I} + 2.12\mathbf{J}$, we have

$$\mathbf{R}_A \times \mathbf{F}_{32} = \begin{vmatrix} \mathbf{I} & \mathbf{J} & \mathbf{K} \\ -2.12 & 2.12 & 0 \\ -141 & 34.3 & 0 \end{vmatrix} = 226\mathbf{K}$$

so

$$\mathbf{T}_2 = -226\mathbf{K} \quad \text{lb-in.}$$

16-6. Shaking Forces. Of especial interest to the designer are the forces transmitted to the frame or foundation of the machine owing to the inertia of moving links and other machine members. When these forces vary in magnitude or direction, they tend to shake or vibrate the machine, and consequently they are called *shaking forces*. Thus the shaking forces are the forces which act upon the frame of a machine owing only to the inertia forces of the moving parts.

If we consider a four-bar linkage, as an example, with links 2, 3, and 4 as the moving members and link 1 as the frame, then the inertia forces associated with the moving members are $-m_2\mathbf{A}_{G_2}$, $-m_3\mathbf{A}_{G_3}$, and $-m_4\mathbf{A}_{G_4}$. Therefore, taking the moving members as a free body, one can immediately write

$$\Sigma \mathbf{F} = \mathbf{F}_{12} + \mathbf{F}_{14} + (-m_2\mathbf{A}_{G_2}) + (-m_3\mathbf{A}_{G_3}) + (-m_4\mathbf{A}_{G_4}) = 0 \quad (a)$$

Using $\mathbf{F}_S$ as the resultant shaking force,

$$\mathbf{F}_S = \mathbf{F}_{21} + \mathbf{F}_{41} \quad (b)$$

Therefore
$$\mathbf{F}_S = -(m_2\mathbf{A}_{G_2} + m_3\mathbf{A}_{G_3} + m_4\mathbf{A}_{G_4}) \quad (c)$$

Thus, a general equation for the shaking forces in any machine is

$$\mathbf{F}_S = -\sum_{2}^{n} m_n \mathbf{A}_{G_n} \quad (16\text{-}16)$$

where it is understood that link 1 is always the frame and where n is the number of members making up the machine.

16-7. The Method of Virtual Work. It is well known that energy methods of analysis can frequently be used to short-cut a great deal of the algebraic or arithmetical detail of conventional analytical approaches. In this section we shall find that this is also true in the analysis of forces in mechanisms; in fact, the greatest timesaving is obtained with the most complicated mechanisms. The solution to be presented here is called the *method of virtual work*. Its main advantage is in the fact that it eliminates the usual link-to-link treatment and, instead, permits an examination of the whole mechanism at one time.

Let us first define the term *work* as it is to be used here. In Fig. 16-10a a force **F** operates on a particle while it moves from A to A'. The work done is the *scalar product of the displacement and the component of the*

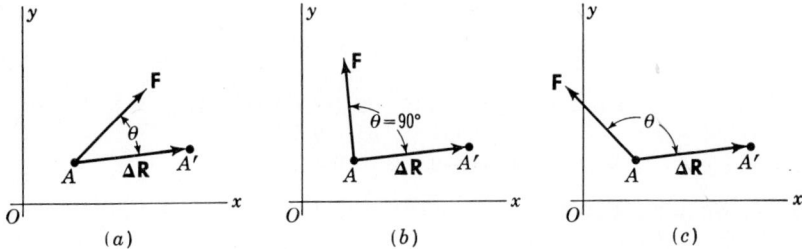

FIG. 16-10. (a) The work of a force. (b) Zero work. (c) Negative work.

force in the direction of the displacement. Designating the displacement vector by **ΔR** and its magnitude by Δr, the work is

$$\Delta u = (F)(\Delta r) \cos \theta \tag{a}$$

Equation (a) shows that the work Δu is positive for $-90° < \theta < 90°$, zero for $\theta = \pm 90°$, and negative for $90° < \theta < 270°$, as shown in Fig. 16-10b and c.

Comparison of Eq. (a) with Eq. (12-5) shows that we can write the work as the dot or scalar product of the force and displacement vectors. Thus

$$\Delta u = \mathbf{F} \cdot \mathbf{\Delta R} \tag{16-17}$$

Therefore the work is a scalar quantity which may be either positive or negative.

There are quite a number of forces existing between the members of various mechanisms which do no work because the force is perpendicular to the displacement, because there is no displacement, or because two forces act and produce equal amounts of positive and negative work.

Consider, for example, two rigid links connected by a frictionless pin joint. The pin joint undergoes a displacement which, when multiplied by one of the pair of forces, produces positive work. However, if the displacement is multiplied by the reacting force of the pair, an equal amount of negative work is obtained and the net work done by this pair of forces is zero. Or, again, the normal reacting forces to a block sliding along a frictionless surface produce zero work, for there is no displacement in the direction of the force. If a wheel rolls along a track without sliding, the frictional force of the track against the wheel at the point of contact does no work because the point of contact is instantaneously at rest.

Consider the linkage of Fig. 16-11. No matter what displacement is given to the linkage, forces $\mathbf{F}_{12}$ and $\mathbf{F}_{14}$ do no work because points O_2 and O_4 have no displacement (friction is neglected). Two forces $\mathbf{F}_{23}$ and $\mathbf{F}_{32}$ (not shown) exist at point A, for example, but no work is done at this

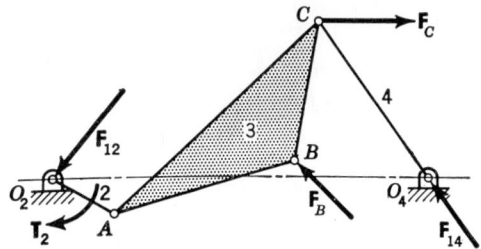

Fig. 16-11

constraint because the two forces produce equal amounts of positive and negative work. On the other hand if the linkage is given a small displacement, the forces $\mathbf{F}_B$ and $\mathbf{F}_C$ and the moment $\mathbf{T}_2$ do produce work.

The method of virtual work utilizes an imaginary small displacement of the mechanism called a *virtual displacement*. Such movements must be consistent with the constraints of the mechanism. That is, if link 2 of Fig. 16-11 is given a small motion $\Delta\theta_2$, then the virtual displacements of B and C must be those which would actually obtain. The work done by these virtual displacements is called the *virtual work*. We observe again that the displacements, and consequently the work done by them, are purely hypothetical.

If we suppose that the linkage of Fig. 16-11 is in static equilibrium under the action of the applied forces and torques, then the work done with a virtual displacement must be zero. This is the basis of the method of virtual work. Give the system a virtual displacement, and if the system is to be in equilibrium under the action of the applied forces, the total virtual work must be zero. If inertia forces and inertia torques are treated as applied forces, then the method of virtual work applies equally well for mechanisms in motion.

The virtual displacement of points in a moving mechanism will all be proportional to the velocities of those points. This is so because the virtual displacements can take place in the same interval dt of time.

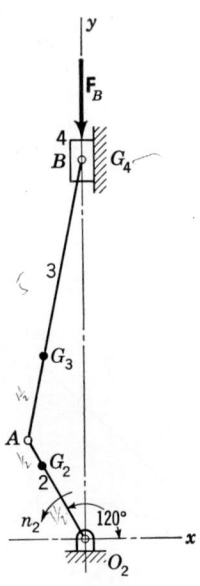

FIG. 16-12. $O_2A = 2$ in.; $O_2G_2 = 1\frac{1}{2}$ in.; $AB = 5$ in.; $AG_3 = 1\frac{1}{2}$ in.; $W_4 = 1.20$ lb; $W_3 = 1.80$ lb; $I_2 = 0.0010$ lb-sec^2-in.; $I_3 = 0.0121$ lb-sec^2-in.; $F_B = 150$ lb; $n_2 = 2,400$ rpm; $W_2 = 0.60$ lb.

Thus, for any mechanism composed of n members, the method of virtual work can be written

$$\Sigma T_n \cdot \Omega_n + \Sigma F_n \cdot V_n + \Sigma(-m_n A_{G_n} \cdot V_{G_n}) + \Sigma(-I_n \alpha_n \cdot \Omega_n) = 0 \quad (16\text{-}18)$$

For the usual mechanism Eq. (16-18) can normally be solved for one quantity. Since the terms are vector quantities, the solution will include both the magnitude and the direction of the unknown. Especial care must be used in defining the units of the terms; in particular we note that if F_n and V_n are in the customary units of pounds and feet per second, then the units of torque (T_n and $-I_n \alpha_n$) must be in foot-pounds.

Since Eq. (16-18) contains only the applied forces and torques, it cannot be used in its present form to solve for internal forces or the reactions between members of the mechanism.

The method of analysis developed in this section can be applied either graphically or analytically. In solving graphically, care must be used to be sure that the forces and velocities are both in the same direction. With the analytical approach, however, the scalar product takes care of this automatically. Since the method of virtual work appears at its greatest advantage in analytical solutions, we shall not utilize it for graphical solutions in this book.[1]

EXAMPLE 16-2. In Fig. 16-12 one phase of the slider-crank or engine mechanism is shown together with other data necessary for a kinematic and dynamic analysis of the machine. Using the method of virtual work, find the external torque necessary to drive the machine at the required speed and to overcome the external force F_B for the position given in the illustration.

Solution. A kinematic analysis yields the following data in vector form:

$\Omega_2 = 251\mathbf{K}$ rad/sec $\qquad \Omega_3 = -88.8\mathbf{K}$ rad/sec
$\alpha_2 = 0$ rad/sec^2 $\qquad \alpha_3 = 11{,}240\mathbf{K}$ rad/sec^2
$V_{G_3} = -25.4\mathbf{I} - 23.1\mathbf{J}$ fps $\qquad V_{G_2} = -21.8\mathbf{I} - 12.6\mathbf{J}$ fps
$A_{G_2} = 3{,}940\mathbf{I} - 6{,}830\mathbf{J}$ fps^2 $\qquad V_B = -28.3\mathbf{J}$ fps
$A_B = -11{,}380\mathbf{J}$ fps^2 $\qquad A_{G_3} = 3{,}650\mathbf{I} - 9{,}770\mathbf{J}$ fps^2

Equation (16-18) for this mechanism becomes

$$T_2 \cdot \Omega_2 + F_B \cdot V_B + (-m_2 A_{G_2}) \cdot V_{G_2} + (-m_3 A_{G_3}) \cdot V_{G_3} + (-m_4 A_B) \cdot V_B + (-I_3 \alpha_3) \cdot \Omega_3 = 0 \quad (1)$$

The terms for Eq. (1) are

$T_2 \cdot \Omega_2 = (T_2 \mathbf{K}) \cdot (251\mathbf{K}) = 251 T_2$
$F_B \cdot V_B = (-150\mathbf{J}) \cdot (-28.3\mathbf{J}) = 4{,}250$ ft-lb
$(-m_2 A_{G_2}) \cdot V_{G_2} = -\dfrac{0.60}{32.2}(3{,}940\mathbf{I} - 6{,}830\mathbf{J}) \cdot (-21.8\mathbf{I} - 12.6\mathbf{J})$
$\qquad = 0$ ft-lb
$(-m_3 A_{G_3}) \cdot V_{G_3} = -\dfrac{1.80}{32.2}(3{,}650\mathbf{I} - 9{,}770\mathbf{J}) \cdot (-25.4\mathbf{I} - 23.1\mathbf{J})$
$\qquad = -7{,}320$ ft-lb
$(-m_4 A_B) \cdot V_B = -\dfrac{1.20}{32.2}(-11{,}380\mathbf{J}) \cdot (-28.3\mathbf{J}) = -12{,}000$ ft-lb
$(-I_3 \alpha_3) \cdot \Omega_3 = -\dfrac{0.0121}{12}(11{,}240\mathbf{K}) \cdot (-88.8\mathbf{K}) = 1{,}006$ ft-lb

Substituting these values in Eq. (1) and solving for the torque give

$$T_2 = \frac{-4{,}250 + 0 + 7{,}320 + 12{,}000 - 1{,}006}{251} = 56.0 \text{ lb-ft}$$

or $\qquad T_2 = (12)(56.0\mathbf{K}) = 672\mathbf{K} \qquad$ lb-in. $\qquad Ans.$

[1] A good discussion of the method of virtual work appears in Ferdinand P. Beer and E. Russell Johnston, Jr., "Mechanics for Engineers," chap. 10, McGraw-Hill Book Company, Inc., New York, 1957. An application of the method to the solution of problems in dynamics of machinery can be found in W. G. Green, "Theory of Machines," pp. 292–297, Blackie & Son, Ltd., Glasgow, 1955. In the latter book a four-bar linkage is solved by a graphical method.

PROBLEMS

16-1. Find the external torque which must be applied to link 2 of the mechanism illustrated in the figure to drive it at the given velocity. *Ans.:* $T_2 = -190K$ lb-in.

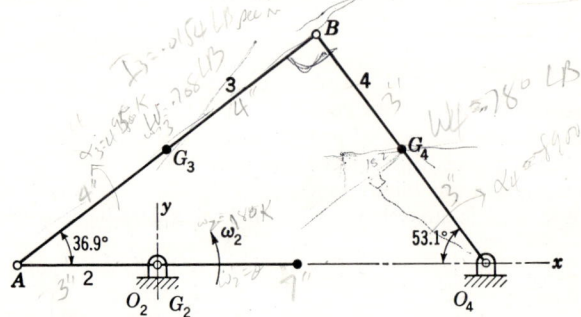

Prob. 16-1. $O_2A = 3$ in.; $AG_3 = 4$ in.; $AB = 8$ in.; $O_4G_4 = 3$ in.; $O_4B = 6$ in.; $O_2O_4 = 7$ in.; $\Omega_2 = 180K$ rad/sec; $W_3 = 0.708$ lb; $W_4 = 0.780$ lb; $I_3 = 0.0154$ lb-sec^2-in.; $I_4 = 0.0112$ lb-sec^2-in.; $\alpha_2 = 0$ rad/sec^2; $\alpha_3 = 4{,}950K$ rad/sec^2; $\alpha_4 = -8{,}900K$ rad/sec^2; $A_{G_3} = 6{,}320I + 750J$ fps^2; $A_{G_4} = 2{,}280I + 750J$ fps^2.

16-2. Crank 2 of the four-bar linkage in the illustration is balanced. For the given angular velocity of link 2, find the forces acting at each pin joint and the external torque which must be applied to link 2.

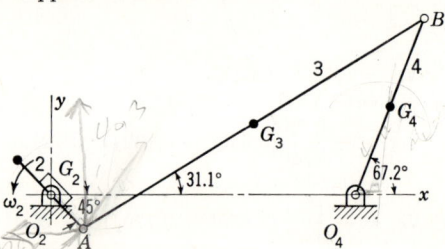

Prob. 16-2. $O_2A = 2$ in.; $AG_3 = 8\frac{1}{2}$ in.; $AB = 17$ in.; $O_4G_4 = 4$ in.; $O_4B = 8$ in.; $O_2O_4 = 13$ in.; $\Omega_2 = 200K$ rad/sec; $W_3 = 2.65$ lb; $W_4 = 6.72$ lb; $I_3 = 0.0606$ lb-sec^2-in.; $I_4 = 0.531$ lb-sec^2-in.; $\alpha_2 = 0$ rad/sec^2; $\alpha_3 = -6{,}530K$ rad/sec^2; $\alpha_4 = -240K$ rad/sec^2; $A_{G_3} = -3{,}160I + 262J$ fps^2; $A_{G_4} = -800I - 2{,}110J$ fps^2.

16-3. For the given angular velocity of crank 2 in the figure, find the reactions at each pin joint and the external torque to be applied to the crank. *Ans.:* $F_{14} = 300/-90°$ lb; $F_{34} = 755/156.6°$ lb; $F_{32} = 1{,}535/7.95°$ lb; $T_2 = 2{,}780K$ lb-in.

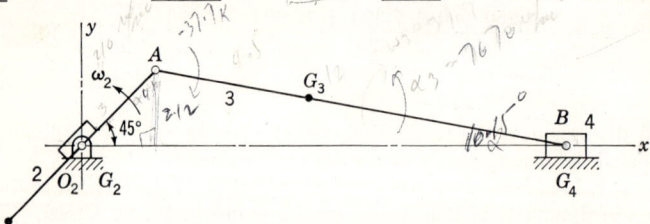

Prob. 16-3. $O_2A = 3$ in.; $AG_3 = 4\frac{1}{2}$ in.; $AB = 12$ in.; $\Omega_2 = 210K$ rad/sec; $\Omega_3 = -37.7K$ rad/sec; $\alpha_2 = 0$ rad/sec^2; $\alpha_3 = 7{,}670K$ rad/sec^2; $A_{G_3} = -7{,}820I - 4{,}876J$ fps^2; $A_B = -7{,}850I$ fps^2; $W_3 = 3.40$ lb; $W_4 = 2.85$ lb; $I_3 = 0.1085$ lb-sec^2-in.

16-4. The same as Prob. 16-3 except that $\theta_2 = 180°$. Find the kinematic quantities first and then the required forces, using only analytical methods.

16-5. The figure shows an engine mechanism with an external force F_B applied to the piston. For the given crank velocity find all the pin reactions and the crank torque. *Ans.*: $F_{14} = 25.6 / 90°$ lb; $F_{34} = 313 / -4.7°$ lb; $F_{23} = 378 / 144.4°$ lb; $T_2 = 112K$ rad/sec.

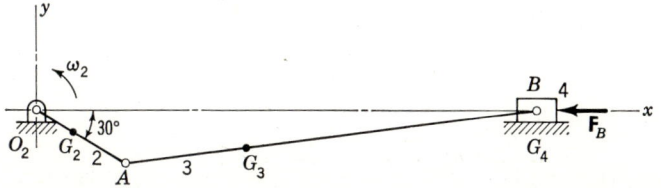

PROB. 16-5. $O_2G_2 = 1\frac{1}{4}$ in.; $O_2A = 3$ in.; $AG_3 = 3\frac{1}{2}$ in.; $AB = 12$ in.; $\Omega_2 = 160K$ rad/sec; $\Omega_3 = -35K$ rad/sec; $\alpha_2 = 0$ rad/sec²; $\alpha_3 = -3,090K$ rad/sec²; $A_{G_2} = 2,640 / 150°$ fps²; $A_{G_3} = 6,130 / 158.3°$ fps²; $A_B = 6,280 / 180°$ fps²; $W_2 = 0.95$ lb; $W_3 = 3.50$ lb; $W_4 = 2.50$ lb; $I_2 = 0.00369$ lb-sec²-in.; $I_3 = 0.110$ lb-sec²-in.; $F_B = 800 / 180°$ lb.

16-6. The figure shows a slider-crank mechanism and a given angular velocity for the crank. Find the crank torque necessary to drive the linkage in the direction shown. The crank is balanced.

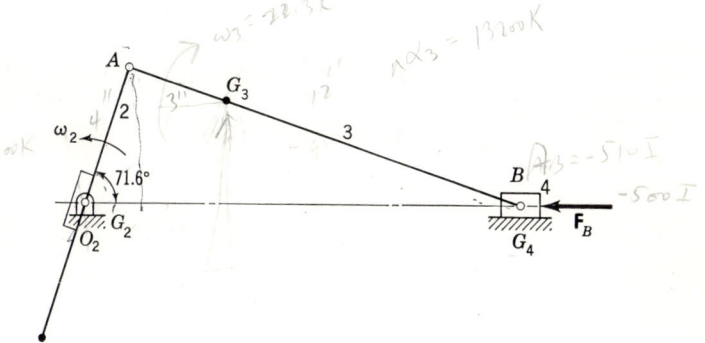

PROB. 16-6. $O_2A = 4$ in.; $AG_3 = 3$ in.; $AB = 12$ in.; $\Omega_2 = 200K$ rad/sec; $\Omega_3 = -22.2K$ rad/sec; $\alpha_2 = 0$ rad/sec²; $\alpha_3 = 13,200K$ rad/sec²; $W_3 = 1.50$ lb; $W_4 = 2.00$ lb; $I_3 = 0.0476$ lb-sec²-in.; $A_{G_3} = -3,285I - 9,480J$ fps²; $A_B = -510I$ fps²; $F_B = -500I$ lb.

16-7. Solve Prob. 16-6 using the method of virtual work.

16-8. The mechanism of Prob. 16-6 is to be used at altitudes where the acceleration due to gravity is only 140 in./sec². Solve for the crank torque under these conditions.

16-9.[1] Cranks 2 and 4 of the crossed linkage shown in the figure are balanced. If

---

[1] The data for the position, velocity, and acceleration for the remaining problems in this chapter were obtained by computation on the University of Michigan's IBM 704 digital computer.

408 DYNAMIC ANALYSIS OF MACHINES

crank 2 is the driver, find the input torque necessary to drive the mechanism at the given velocity. Find, also, all the pin reactions.

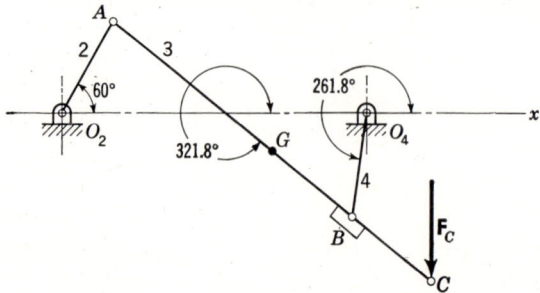

PROB. 16-9. $O_2A = 6$ in.; $AG = 12$ in.; $AB = 18$ in.; $AC = 24$ in.; $O_2O_4 = 18$ in.; $O_4B = 6$ in.; $\Omega_2 = 10\mathbf{K}$ rad/sec; $\Omega_3 = -1.43\mathbf{K}$ rad/sec; $\Omega_4 = -11.43\mathbf{K}$ rad/sec; $\alpha_2 = 0$ rad/sec^2; $\alpha_3 = 84.74\mathbf{K}$ rad/sec^2; $\alpha_4 = 84.74\mathbf{K}$ rad/sec^2; $W_3 = 4.0$ lb; $I_3 = 0.497$ lb-sec^2-in.; $I_4 = 0.063$ lb-sec^2-in.; $\mathbf{F}_C = -30\mathbf{J}$ lb.

16-10. Find the driving torque required for the mechanism of Prob. 16-9 if crank 4 is the driver. What are the pin reactions?

16-11. If crank 2 of the linkage of Prob. 16-9 is turned until θ_2 is 210°, the following quantities are obtained: $\theta_3 = 14.7°$; $\theta_4 = 164.7°$; $\Omega_3 = 4.73\mathbf{K}$ rad/sec; $\Omega_4 = -5.27\mathbf{K}$ rad/sec; $\alpha_3 = -10.39\mathbf{K}$ rad/sec^2; $\alpha_4 = -10.39\mathbf{K}$ rad/sec^2. No change occurs in the external force. Solve Prob. 16-9 for this phase of the linkage.

16-12 to 16-23. The figure shows a linkage with an extended coupler having an external force $\mathbf{F}_C$ which acts at point C on the coupler during a portion of the stroke. In the table below, the angular positions of the links and the kinematic quantities are tabulated for phases of the mechanism at intervals of 30°. For each problem calculate all the pin reactions and the input torque to crank 2. Assume that crank 2 is balanced and rotates at a constant angular velocity.

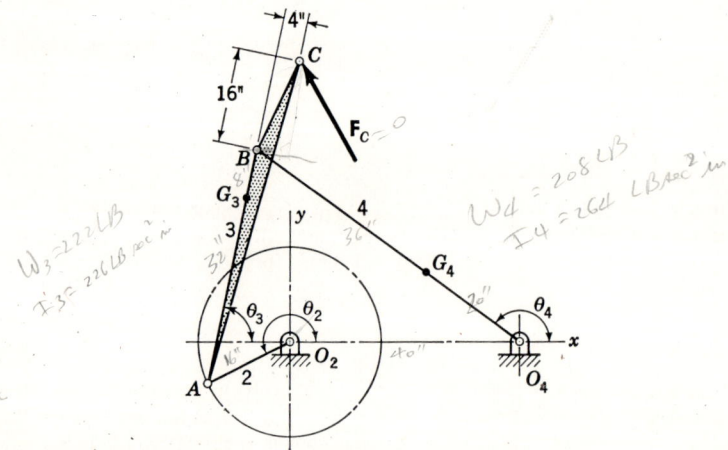

PROBS. 16-12 TO 16-23. $O_2A = 16$ in.; $AG_3 = 32$ in.; $AB = O_2O_4 = 40$ in.; $O_4G_4 = 20$ in.; $O_4B = 56$ in.; $\Omega_2 = 10\mathbf{K}$ rad/sec; $\mathbf{F}_C = -500\mathbf{I} + 866\mathbf{J}$ lb for $90° \leq \theta_2 \leq 300°$; otherwise $\mathbf{F}_C = 0$; $W_3 = 222$ lb; $W_4 = 208$ lb; $I_3 = 226$ lb-sec^2-in.; $I_4 = 264$ lb-sec^2-in.

Problem No.	θ_2, deg	θ_3, deg	θ_4, deg	ω_3, rad/sec	ω_4, rad/sec	α_3, rad/sec^2	α_4, rad/sec^2
16-12	0	120.0	141.8	-6.67	-6.67	-141.1	-64.1
16-13	30	94.1	121.2	-8.78	-5.64	51.0	86.1
16-14	60	73.2	111.4	-5.05	-1.05	70.3	73.0
16-15	90	62.9	112.9	-2.03	1.70	46.3	34.6
16-16	120	60.0	120.0	0	2.86	33.0	11.8
16-17	150	62.4	129.1	1.55	3.11	27.0	-1.0
16-18	180	69.0	138.1	2.86	2.86	22.8	-7.8
16-19	210	79.3	146.0	3.91	2.36	17.2	-10.8
16-20	240	92.2	152.2	4.61	1.76	9.1	-12.1
16-21	270	106.5	156.5	4.79	1.06	-3.6	-15.3
16-22	300	120.0	158.2	4.00	0	-30.5	-27.7
16-23	330	128.1	155.2	0.80	-2.33	-102.9	-67.9

16-24. The figure shows a motor geared to a shaft on which is mounted a flywheel. The moments of inertia of the parts are as follows: flywheel, $I = 2.73$ lb-sec^2-in.; flywheel shaft, $I = 0.0155$ lb-sec^2-in.; gear, $I = 0.172$ lb-sec^2-in.; pinion, $I = 0.00349$ lb-sec^2-in.; motor, $I = 0.0864$ lb-sec^2-in. If the motor has a starting torque of 75 lb-in., what is the angular acceleration of the flywheel shaft at the instant the motor switch is turned on?

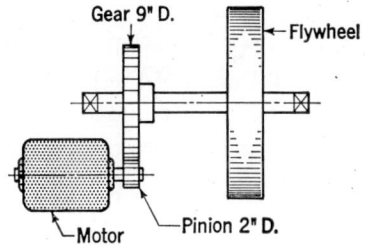

Prob. 16-24

CHAPTER 17

DYNAMIC-FORCE ANALYSIS—SPACE MECHANISMS

The methods of Chap. 16 can also be applied to the analysis of space mechanisms, but if an alternative method of analysis is known, it results in a better comprehension of the physical principles involved and, in addition, may also greatly simplify the solution. For these reasons it is desirable to investigate three-dimensional systems employing the concept of momentum.

17-1. Linear Impulse and Momentum. Let us consider a rigid body made up of a group of particles of masses m_A, m_B, m_C, etc., with the particles located by the position vectors $\mathbf{R}_A$, $\mathbf{R}_B$, $\mathbf{R}_C$, etc., relative to an origin fixed in an inertial system. Then the mass-center position of the rigid body is defined by the equation

$$\mathbf{R} = \frac{\Sigma m_N \mathbf{R}_N}{m} \tag{a}$$

where the summation is to be taken over all the N particles making up the body and where m is the total mass of the body.

Now if each particle of the body is acted upon by an external force $\mathbf{F}_A$, $\mathbf{F}_B$, $\mathbf{F}_C$, etc., then Newton's law permits us to write

$$\Sigma m_N \ddot{\mathbf{R}}_N = \Sigma \mathbf{F}_N \tag{17-1}$$

where, again, the summation is to be carried out for all the particles. Note that the internal forces due to interactions between particles are pairs of equal and opposite collinear forces and that the summation of these for a rigid body is always zero. If Eq. (a) is substituted into (17-1) and the resultant external force designated by $\mathbf{F}$, then

$$\mathbf{F} = m\ddot{\mathbf{R}} \tag{b}$$

which shows that the center of mass has the same motion that it would have if all the mass were concentrated there and the resultant force applied there.

If we now multiply both sides of Eq. (b) by dt and integrate between 0 and t, there results

$$\int_0^t \mathbf{F}\,dt = \int_0^t m\ddot{\mathbf{R}}\,dt = m\dot{\mathbf{R}} - m\dot{\mathbf{R}}_0 \tag{17-2}$$

where $\dot{\mathbf{R}}_0$ or $\mathbf{V}_0$ is the velocity of the mass center at $t = 0$ and $\dot{\mathbf{R}}(\mathbf{V})$ the velocity after an interval of time t. The vector

$$\mathbf{L} = m\dot{\mathbf{R}} = m\mathbf{V} \qquad (17\text{-}3)$$

defines the *linear momentum* of the body, and the vector $\int_0^t \mathbf{F}\,dt$ the *linear impulse*. Equation (17-2) states that the change in linear momentum is due to the impulse of the resultant external force. If we make the time interval an infinitesimal, then Eq. (17-2) can also be written

$$\mathbf{F}\,dt = (\mathbf{L} + d\mathbf{L}) - \mathbf{L}$$
or
$$\mathbf{F} = \dot{\mathbf{L}} \qquad (17\text{-}4)$$

Therefore the resultant external force acting upon a rigid body equals the time rate of change of linear momentum. If there is no external force acting upon a body, then there can be no change in the linear momentum. Thus

$$\dot{\mathbf{L}} = 0 \quad \text{or} \quad \mathbf{L} = \text{const} \qquad (c)$$

is a statement of the law of conservation of linear momentum.

17-2. Moment of Momentum. In a rather broad sense the quantities which describe the motion of a rotating body are analogous to those which describe the linear motion of a body. When a body moves linearly, we are interested in the forces that are acting upon the body; when it rotates, we are interested in the torques, or the moments of the forces acting upon the body. Similarly when a body has linear motion, its motion is described in terms of linear velocity and linear acceleration, but the motion of a rotating body is stated in terms of angular velocity and angular acceleration. When a body has linear motion, it is the mass, or simple inertia, of the body which measures its sluggishness or resistance to the application of forces. If the body rotates, it is the importance, or the moment, of the mass which is a measure of its resistance to the application of torques. Therefore, in dealing with the rotational motion of a body, we need to consider such terms as *angular impulse*, or *moment of linear impulse*, and *angular momentum*, or *moment of linear momentum*.

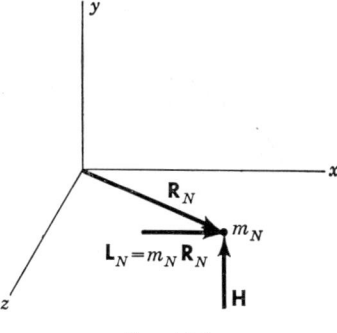

Fig. 17-1

The term *moment of momentum* for a particle of mass is designated as $\mathbf{H}$ and defined in Fig. 17-1. Here

$$\mathbf{H} = \mathbf{R}_N \times \mathbf{L}_N = \mathbf{R}_N \times m_N \dot{\mathbf{R}}_N \qquad (17\text{-}5)$$

where the position vector $\mathbf{R}_N$ locates a particle of mass m_N having a linear momentum designated by $\mathbf{L}_N$. A rigid body containing many particles of mass has a moment of momentum which is the vector sum of the moments of momentum of all the particles. This sum is

$$\mathbf{H} = \Sigma \mathbf{R} \times m\dot{\mathbf{R}} \qquad (a)$$

Taking the cross product of both sides of Eq. (17-2) by the position vector $\mathbf{R}$ gives

$$\int_0^t (\mathbf{R} \times \mathbf{F}) \, dt = \int_0^t (\mathbf{R} \times m\ddot{\mathbf{R}}) \, dt = \mathbf{R} \times m\dot{\mathbf{R}} - \mathbf{R}_0 \times m\dot{\mathbf{R}}_0$$

or
$$\int_0^t \mathbf{T} \, dt = \mathbf{H} - \mathbf{H}_0 \qquad (17\text{-}6)$$

This equation establishes the relationship between angular impulse and angular momentum. The term $\int_0^t \mathbf{T} \, dt$ is the moment of the impulse which occurs in the time interval between 0 and t, and it is called the *angular impulse*. The term on the right is the change in moment of momentum (angular momentum) which occurs as a result of the angular impulse. If we make the time interval very small in Eq. (17-6), then it can be written

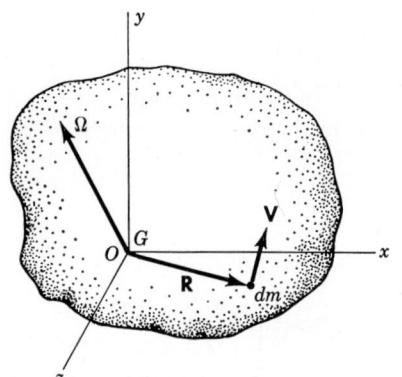

Fig. 17-2

$$\mathbf{T} \, dt = (\mathbf{H} + d\mathbf{H}) - \mathbf{H}$$
or
$$\mathbf{T} = \dot{\mathbf{H}} \qquad (17\text{-}7)$$

which states that the change in the moment of momentum is equal to the external torque acting upon the system. If there are no external torques, that is, if $\mathbf{T} = 0$, then this requires that

$$\dot{\mathbf{H}} = 0 \quad \text{or} \quad \mathbf{H} = \text{const} \qquad (17\text{-}8)$$

Equation (17-8) is an expression of the law of conservation of moment of momentum.

17-3. The Components of Moment of Momentum. Let us determine the moment of momentum for a rigid body in motion when the coordinate system does not rotate but has its origin coincident with the mass center of the body. Thus the reference system moves in translation with the center of mass but does not rotate. Such an arrangement is illustrated in Fig. 17-2. Here the mass center G and the origin O of the xyz system are coincident. We give the body an angular velocity $\boldsymbol{\Omega}$, causing any particle of mass, such as dm, to have a velocity relative to the translating origin of **V**. The moment of momentum is

$$\mathbf{H} = \int_m (\mathbf{R} \times \mathbf{V}) \, dm = \int_m \mathbf{R} \times (\mathbf{\Omega} \times \mathbf{R}) \, dm \qquad (a)$$

where the integration is to be carried out over all particles of mass of the body. Since

$$\mathbf{R} = x\mathbf{I} + y\mathbf{J} + z\mathbf{K}$$
$$\mathbf{\Omega} = \omega^x \mathbf{I} + \omega^y \mathbf{J} + \omega^z \mathbf{K}$$

and

then

$$\mathbf{\Omega} \times \mathbf{R} = \begin{vmatrix} \mathbf{I} & \mathbf{J} & \mathbf{K} \\ \omega^x & \omega^y & \omega^z \\ x & y & z \end{vmatrix}$$
$$= (z\omega^y - y\omega^z)\mathbf{I} + (x\omega^z - z\omega^x)\mathbf{J} + (y\omega^x - x\omega^y)\mathbf{K} \qquad (b)$$

and

$$\mathbf{R} \times (\mathbf{\Omega} \times \mathbf{R}) = \begin{vmatrix} \mathbf{I} & \mathbf{J} & \mathbf{K} \\ x & y & z \\ z\omega^y - y\omega^z & x\omega^z - z\omega^x & y\omega^x - x\omega^y \end{vmatrix}$$
$$= [\omega^x(y^2 + z^2) - \omega^y xy - \omega^z xz]\mathbf{I}$$
$$+ [-\omega^x yx + \omega^y(z^2 + x^2) - \omega^z yz]\mathbf{J}$$
$$+ [-\omega^x zx - \omega^y zy + \omega^z(x^2 + y^2)]\mathbf{K}$$

Substituting these values in Eq. (a) and separating the x, y, and z components produce

$$H^x = \omega^x \int (y^2 + z^2) \, dm - \omega^y \int xy \, dm - \omega^z \int xz \, dm$$
$$H^y = -\omega^x \int xy \, dm + \omega^y \int (z^2 + x^2) \, dm - \omega^z \int yz \, dm \qquad (17\text{-}9)$$
$$H^z = -\omega^x \int zx \, dm - \omega^y \int zy \, dm + \omega^z \int (x^2 + y^2) \, dm$$

The reader will readily identify the integrals in Eqs. (17-9) as *moments* and *products of inertia*. Thus

$$\int (y^2 + z^2) \, dm = I_x \qquad \int xy \, dm = I_{xy} \qquad \text{etc.}$$

where I_x, I_y, and I_z are the *moments of inertia* and I_{xy}, I_{xz}, etc., are the *products of inertia*. It is also observed that $I_{xy} = I_{yx}$, $I_{xz} = I_{zx}$, and $I_{yz} = I_{zy}$. Therefore

$$H^x = I_x \omega^x - I_{xy} \omega^y - I_{xz} \omega^z$$
$$H^y = -I_{xy} \omega^x + I_y \omega^y - I_{yz} \omega^z \qquad (17\text{-}10)$$
$$H^z = -I_{zx} \omega^x - I_{zy} \omega^y + I_z \omega^z$$

where

$$\mathbf{H} = H^x \mathbf{I} + H^y \mathbf{J} + H^z \mathbf{K} \qquad (17\text{-}11)$$

The reader should now read the beginning of this section again and observe how the reference axes were selected. Since they were chosen as a nonrotating system, this means that the values of the integrals in Eqs. (17-9) are functions of time. In other words the moments and products of inertia in Eqs. (17-10) are variables whose values depend upon the instantaneous location of the rigid body relative to the reference system. The calculation of these integrals is often very tedious, as we shall shortly see, and it is for this reason that we shall usually choose our reference system fixed to the moving body so that the axes rotate and translate with it. When this is done, it is also very convenient to fix the origin at the mass center as we have done in this section.

17-4. Motion of a Rigid Body. We have seen that the motion of the mass center of a rigid body is the same as that of a particle having the same mass and acted upon by the same forces. This means that translation of the mass center can be determined by the same procedure used in investigating the motion of a particle.

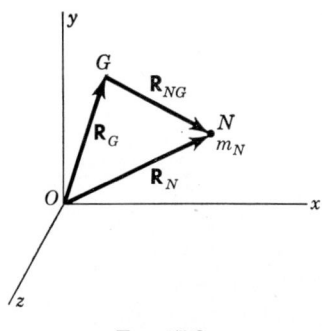

Fig. 17-3

In studying rotation of a rigid body we may desire to refer the motion to either of two coordinate systems:

1. A coordinate system fixed in the body with the origin attached to the mass center
2. A coordinate system with axes fixed in an inertial system

In Fig. 17-3 the xyz axes with origin at O are fixed in an inertial system. A rigid body with mass center at G has a particle of mass m_N located at N. The relative position equation is written

$$\mathbf{R}_N = \mathbf{R}_G + \mathbf{R}_{NG} \tag{a}$$

so

$$\dot{\mathbf{R}}_N = \dot{\mathbf{R}}_G + \dot{\mathbf{R}}_{NG} \tag{b}$$

The moment of momentum of the body about the fixed origin O is

$$\begin{aligned}\mathbf{H}_O &= \Sigma m_N(\mathbf{R}_G + \mathbf{R}_{NG}) \times \dot{\mathbf{R}}_N \\ &= m\mathbf{R}_G \times \dot{\mathbf{R}}_G + \mathbf{R}_G \times (\Sigma m_N \dot{\mathbf{R}}_{NG}) + (\Sigma m_N \mathbf{R}_{NG}) \times \dot{\mathbf{R}}_G \\ &\qquad + \Sigma m_N \mathbf{R}_{NG} \times \dot{\mathbf{R}}_{NG} \end{aligned} \tag{c}$$

where the summation is to be carried out for all the particles composing the body and where $m = \Sigma m_N$ is the mass of the whole body. Now, since $\mathbf{R}_{NG}$ is the vector from the mass center,

$$\Sigma m_N \mathbf{R}_{NG} = 0 \tag{d}$$

and

$$\Sigma m_N \dot{\mathbf{R}}_{NG} = \frac{d}{dt}(\Sigma m \mathbf{R}_{NG}) = 0 \tag{e}$$

But the momentum of the moving body about its mass center G is

$$\mathbf{H} = \Sigma m_N \mathbf{R}_{NG} \times \dot{\mathbf{R}}_{NG} \tag{f}$$

Substituting Eqs. (d), (e), and (f) in (c) shows that

$$\mathbf{H}_O = m\mathbf{R}_G \times \dot{\mathbf{R}}_G + \mathbf{H} \tag{17-12}$$

Next, we calculate the torque with respect to the fixed xyz system. Thus

$$\mathbf{T}_O = \Sigma(\mathbf{R}_G + \mathbf{R}_{NG}) \times \mathbf{F}_N$$
$$= \mathbf{R}_G \times \Sigma\mathbf{F}_N + \Sigma\mathbf{R}_{NG} \times \mathbf{F}_N \tag{g}$$

Here $\Sigma\mathbf{R}_{NG} \times \mathbf{F}_N = \mathbf{T}$, which is the torque about the mass center G. Therefore

$$\mathbf{T}_O = \mathbf{R}_G \times \Sigma\mathbf{F}_N + \mathbf{T} \tag{17-13}$$

Substituting the value of $\mathbf{H}_O$ from Eqs. (17-9) and the value of $\mathbf{T}_O$ from Eq. (17-13) into (17-7) gives

$$\mathbf{R}_G \times \Sigma\mathbf{F}_N + \mathbf{T} = \frac{d}{dt}(m\mathbf{R}_G \times \dot{\mathbf{R}}_G) + \dot{\mathbf{H}} \tag{h}$$

Now note that $\Sigma\mathbf{F}_N = m\ddot{\mathbf{R}}_G$ because the acceleration of the mass center is the same as if the entire mass were concentrated and acted upon by the same external force system. We also observe that

$$\mathbf{R}_G \times \Sigma\mathbf{F}_n = m\mathbf{R}_G \times \ddot{\mathbf{R}}_G \tag{i}$$

and
$$\frac{d}{dt}(m\mathbf{R}_G \times \dot{\mathbf{R}}_G) = m\mathbf{R}_G \times \ddot{\mathbf{R}}_G + m\dot{\mathbf{R}}_G \times \dot{\mathbf{R}}_G$$
$$= m\mathbf{R}_G \times \ddot{\mathbf{R}}_G$$

because the cross product $\dot{\mathbf{R}}_G \times \dot{\mathbf{R}}_G$ vanishes. Therefore the first term on the left side of Eq. (h) is equal to the first term on the right. Consequently,

$$\mathbf{T} = \dot{\mathbf{H}} \tag{17-14}$$

The significance of Eq. (17-14) is that the relation between torque and moment of momentum is valid no matter whether these quantities are referred to a moving system whose origin is coincident with the center of mass or are referred to a fixed system.

17-5. Moments and Products of Inertia. It is not our purpose here to demonstrate the integration of the expressions for the moments and products of inertia but rather to define them for other sets of axes when they have already been determined in a particular reference system. It is assumed that the reader has already practiced integrating these expressions in his previous studies, so that additional practice would be wasteful of his time.

In Fig. 17-4 a particle of mass dm of a rigid body is located in the xyz system by the position vector $\mathbf{R}$, where

$$\mathbf{R} = x\mathbf{I} + y\mathbf{J} + z\mathbf{K}$$

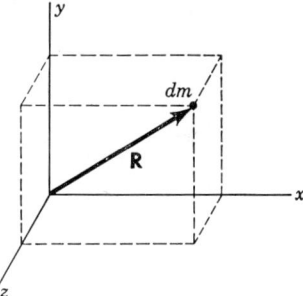

Fig. 17-4

The cross product $\mathbf{I} \times \mathbf{R}$ is

$$\mathbf{I} \times \mathbf{R} = \begin{vmatrix} \mathbf{I} & \mathbf{J} & \mathbf{K} \\ 1 & 0 & 0 \\ x & y & z \end{vmatrix} = y\mathbf{K} - z\mathbf{J} \qquad (a)$$

The scalar product of $\mathbf{I} \times \mathbf{R}$ by itself is

$$(\mathbf{I} \times \mathbf{R}) \cdot (\mathbf{I} \times \mathbf{R}) = (y\mathbf{K} - z\mathbf{J}) \cdot (y\mathbf{K} - z\mathbf{J}) = y^2 + z^2 \qquad (b)$$

Equation (b) suggests the notion of writing the inertia integrals in the form

$$\begin{aligned} I_x &= \int (y^2 + z^2)\, dm = \int (\mathbf{I} \times \mathbf{R})^2\, dm \\ I_y &= \int (x^2 + z^2)\, dm = \int (\mathbf{J} \times \mathbf{R})^2\, dm \\ I_z &= \int (x^2 + y^2)\, dm = \int (\mathbf{K} \times \mathbf{R})^2\, dm \end{aligned} \qquad (17\text{-}15)$$

where $(\mathbf{I} \times \mathbf{R})^2 = (\mathbf{I} \times \mathbf{R}) \cdot (\mathbf{I} \times \mathbf{R})$, etc. Using a similar procedure it is found that the products of inertia can be written as

$$\begin{aligned} I_{xy} &= \int xy\, dm = -\int (\mathbf{I} \times \mathbf{R}) \cdot (\mathbf{J} \times \mathbf{R})\, dm \\ I_{yz} &= \int yz\, dm = -\int (\mathbf{J} \times \mathbf{R}) \cdot (\mathbf{K} \times \mathbf{R})\, dm \\ I_{zx} &= \int zx\, dm = -\int (\mathbf{K} \times \mathbf{R}) \cdot (\mathbf{I} \times \mathbf{R})\, dm \end{aligned} \qquad (17\text{-}16)$$

Sometimes the moment of inertia is written in the form $I_x = mr_x^2$, $I_y = mr_y^2$, etc., where r_x, r_y, and r_z are called the *radii of gyration* and where m is the total mass of the body. Then

$$r_x = \sqrt{\frac{I_x}{m}} \qquad r_y = \sqrt{\frac{I_y}{m}} \qquad r_z = \sqrt{\frac{I_z}{m}} \qquad (17\text{-}17)$$

where only the positive sign has meaning. The signs of the products of inertia, however, may be either positive or negative.

The *principal axes* are defined as the particular set of coordinate axes for which the products of inertia in Eqs. (17-16) are all zero. If, for example, I_{xy} and I_{yz} are both zero, then the y axis is a principal axis. The moments of inertia given in Appendix III are all taken about the principal axes, and thus the products of inertia are zero.

It is often necessary to determine moments and products of inertia of bodies which are composed of several of the shapes shown in Appendix III. The easiest method of doing this is to compute the moments about the principal axes of each shape and then to shift the origin to the mass center of the composite body. This requires that we develop methods of defining moments and products of inertia (1) when the axes are translated to a new position such that the new axes are parallel to the old ones and (2) when the axes are rotated about the same origin to a new position.

17-6. Translation of Axes. In this section we shall assume that we know the moments and products of inertia with respect to an xyz system

SPACE MECHANISMS

and that we wish to determine these quantities with respect to a new system (designated as the x', y', z' system) located so that its axes are parallel to those of the old. It is expedient, too, to locate the origin of the new system at the mass center of the body.

Let $G(x_G, y_G, z_G)$ be the coordinates of the origin of the new system with respect to the old, and let a particle of mass dm have the coordinates x', y', z' in the new system and x, y, z in the old one. Then, since $x = x_G + x'$, $y = y_G + y'$, etc.,

$$I_x = \int [(y_G + y')^2 + (z_G + z')^2] \, dm$$
$$= \int (y'^2 + z'^2) \, dm + (y_G^2 + z_G^2) \int dm + 2y_G \int y' \, dm + 2z_G \int z' \, dm \quad (a)$$

The first term in this equation is the moment of inertia in the $x'y'z'$ system; the second term is the product of the mass of the body and the square of the perpendicular distance from the x axis to the origin G of the new system; the last two terms are both zero because the origin of the $x'y'z'$ system is at the mass center. Now we can write

$$\begin{aligned} I_x &= I_{x'} + (y_G^2 + z_G^2)m \\ I_y &= I_{y'} + (x_G^2 + z_G^2)m \\ I_z &= I_{z'} + (x_G^2 + y_G^2)m \end{aligned} \quad (17\text{-}18)$$

The reader will recognize Eqs. (17-18) as the transfer equations for transforming moment of inertia from the centroidal axes to another set of parallel axes or vice versa.

The same procedure is used for the products of inertia. Thus, for the same conditions

$$I_{xy} = \int (x_G + x')(y_G + y') \, dm$$
$$= \int x'y' \, dm + x_G y_G \int dm + x_G \int y' \, dm + y_G \int x' \, dm \quad (b)$$

so that
$$\begin{aligned} I_{xy} &= I_{x'y'} + x_G y_G m \\ I_{yz} &= I_{y'z'} + y_G z_G m \\ I_{zx} &= I_{z'x'} + z_G x_G m \end{aligned} \quad (17\text{-}19)$$

17-7. Rotation of Axes. Here we assume that we know the moments and products of inertia in the xyz system and that we wish to find them for a new system, designated as $x'y'z'$ as before, where both systems have the same origin. Let us designate the direction of one of the new axes by the unit vector

$$\mathbf{E} = u\mathbf{I} + v\mathbf{J} + w\mathbf{K} \quad (a)$$

Then the moment of inertia about this axis can be designated as

$$I = \int (\mathbf{E} \times \mathbf{R})^2 \, dm \quad (b)$$

If we now substitute the value of $\mathbf{E}$ and carry out the operation inside the integral sign, we obtain

$$(\mathbf{E} \times \mathbf{R})^2 = u^2(\mathbf{I} \times \mathbf{R})^2 + v^2(\mathbf{J} \times \mathbf{R})^2 + w^2(\mathbf{K} \times \mathbf{R})^2$$
$$+ 2uv(\mathbf{I} \times \mathbf{R}) \cdot (\mathbf{J} \times \mathbf{R}) + 2vw(\mathbf{J} \times \mathbf{R}) \cdot (\mathbf{K} \times \mathbf{R})$$
$$+ 2wu(\mathbf{K} \times \mathbf{R}) \cdot (\mathbf{I} \times \mathbf{R}) \quad (c)$$

Thus from Eqs. (17-15) and (17-16) we have

$$I = u^2 I_x + v^2 I_y + w^2 I_z - 2uv I_{xy} - 2vw I_{yz} - 2wu I_{zx} \quad (17\text{-}20)$$

The transformation is obtained by applying Eq. (17-20) three times, once each for the x', y', and z' axes.

The products of inertia can be transformed in the same manner, but in this case the unit vectors corresponding to all three of the new axes appear in each equation. Thus, let the unit vectors corresponding to the directions of the three new axes be

$$\mathbf{X} = u_x \mathbf{I} + v_x \mathbf{J} + w_x \mathbf{K}$$
$$\mathbf{Y} = u_y \mathbf{I} + v_y \mathbf{J} + w_y \mathbf{K} \quad (d)$$
$$\mathbf{Z} = u_z \mathbf{I} + v_z \mathbf{J} + w_z \mathbf{K}$$

Then, as an example

$$I_{x'y'} = -\int (\mathbf{X} \times \mathbf{R}) \cdot (\mathbf{Y} \times \mathbf{R}) \, dm \quad (e)$$

Expanding the term behind the integral gives

$$(\mathbf{X} \times \mathbf{R}) \cdot (\mathbf{Y} \times \mathbf{R}) = u_x u_y (\mathbf{I} \times \mathbf{R})^2 + v_x v_y (\mathbf{J} \times \mathbf{R})^2 + w_x w_y (\mathbf{K} \times \mathbf{R})^2$$
$$+ (u_y v_x + u_x v_y)(\mathbf{I} \times \mathbf{R}) \cdot (\mathbf{J} \times \mathbf{R})$$
$$+ (v_x w_y + v_y w_x)(\mathbf{J} \times \mathbf{R}) \cdot (\mathbf{K} \times \mathbf{R})$$
$$+ (w_x u_y + w_y u_x)(\mathbf{K} \times \mathbf{R}) \cdot (\mathbf{I} \times \mathbf{R})$$

Therefore, from Eqs. (17-15) and (17-16)

$$I_{x'y'} = -u_x u_y I_x - v_x v_y I_y - w_x w_y I_z + (u_y v_x + u_x v_y) I_{xy}$$
$$+ (v_x w_y + v_y w_x) I_{yz} + (w_x u_y + w_y u_x) I_{zx} \quad (17\text{-}21)$$

17-8. Measuring Moment of Inertia. Sometimes the shapes of machine parts are so complicated that it is extremely tedious and time-consuming to calculate the moments of inertia. There are several experimental methods which can be used to measure the moment of inertia if the machine part is available. These are great timesavers and offer a means of checking the analytical results too.

The Pendulum Method. This method is useful when the part is symmetrical about a single plane, that is, when the mass can be assumed to be concentrated in a single plane. The part must be suspended from a point fairly near, but not coincident with, the center of gravity and caused to oscillate. Then, when either the frequency or the period of oscillation is measured, the moment of inertia can be calculated. The procedure requires that the specimen be weighed and also that its mass center be located.

Figure 17-5 shows a body suspended about any point O not coincident with the mass center so that oscillation is possible. A link or connecting rod can usually be pivoted on a knife-edge through the hole nearest the mass center. A spoked wheel or gear can often be pivoted at a point underneath the rim. Designating the weight of the body by W the equation of motion is written

$$I\ddot{\theta} + T = 0$$

Since $T = Wr_G \sin \theta$,

$$\ddot{\theta} - \frac{Wr_G}{I}\theta = 0 \qquad (a)$$

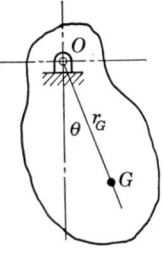

Fig. 17-5

where $\sin \theta$ has been replaced by θ because only small angles are involved. Equation (a) is a differential equation, and the solution is well known. It is

$$\theta = C_1 \sin \sqrt{\frac{Wr_G}{I}} t + C_2 \cos \sqrt{\frac{Wr_G}{I}} t \qquad (b)$$

where C_1 and C_2 are the constants of integration and depend for their value upon how the oscillation began. If the reader will differentiate Eq. (b) twice to find $\ddot{\theta}$ and then substitute both θ and $\ddot{\theta}$ in Eq. (a), he can easily satisfy himself as to the validity of the solution. The oscillation is usually started by pulling the body through a small angle, say θ_a, and releasing it. Therefore the initial or starting conditions are

$$\theta = \theta_a \quad \text{when } t = 0$$
$$\dot{\theta} = 0 \quad \text{when } t = 0$$

Substituting these conditions in Eq. (b) and solving for the constants produce $C_1 = 0$ and $C_2 = \theta_a$. Therefore, the solution is

$$\theta = \theta_a \cos \sqrt{\frac{Wr_G}{I}} t \qquad (17\text{-}22)$$

Equation (17-22) gives the angular position of the mass center at any time t after the beginning of the oscillation. Friction is considered to be negligible, and hence the equation states that the oscillation will endure forever. Any friction present will be that at the knife-edge plus air friction. Both of these are so small compared with observational errors of measuring the other quantities that they can be neglected.

Since a cosine wave repeats itself every 360°, the period of the oscillation is

$$\tau = 2\pi \sqrt{\frac{I}{Wr_G}}$$

where τ is the period in seconds. Therefore

$$I = Wr_G \left(\frac{\tau}{2\pi}\right)^2 \qquad (17\text{-}23)$$

If the pendulum oscillates rapidly, it is more convenient to measure the frequency. The period is then obtained from the equation

$$\tau = \frac{60}{n} \qquad (17\text{-}24)$$

where n is the frequency in oscillations per minute.

Equation (17-23) gives the mass moment of inertia about point O. The transfer equations [Eqs. (17-18)] must be used to find the moment of inertia at the mass center. Thus

$$I = I_G + \frac{W}{g} r_G^2$$

or

$$I_G = I - \frac{W}{g} r_G^2$$

Substituting the value of I from Eq. (17-23) gives

$$I_G = Wr_G \left(\frac{\tau^2}{4\pi^2} - \frac{r_G}{g}\right) \qquad (17\text{-}25)$$

For large values of r_G the two terms in parentheses in Eq. (17-25) are nearly equal, and so both τ and r_G must be measured accurately.

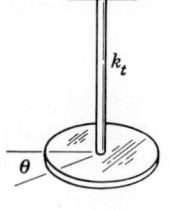

FIG. 17-6

The Torsional Pendulum Method. In Fig. 17-6 a disk of moment of inertia I is connected to a slender rod or wire. We define the torsional stiffness k_t of the rod as the torque required to twist it through an angle of 1 rad. Thus

$$k_t = \frac{T}{\theta} \qquad (17\text{-}26)$$

Now, if the disk of Fig. 17-6 is turned through any angle θ and then released, the following relation exists:

$$I\ddot{\theta} + k_t \theta = 0$$

Since this is similar to Eq. (a),

$$\theta = C_1 \sin \sqrt{\frac{k_t}{I}} t + C_2 \cos \sqrt{\frac{k_t}{I}} t$$

If the initial conditions are

$$\theta = \theta_a \quad \text{when } t = 0$$
$$\dot{\theta} = 0 \quad \text{when } t = 0$$

then $C_1 = 0$ and $C_2 = \theta_a$ and the solution is

$$\theta = \theta_a \cos \sqrt{\frac{k_t}{I}} \, t \qquad (17\text{-}27)$$

The period of the oscillation is then

$$\tau = 2\pi \sqrt{\frac{I}{k_t}}$$

or

$$I = k_t \left(\frac{\tau}{2\pi}\right)^2 \qquad (17\text{-}28)$$

If k_t is known for the torsion rod, then the moment of inertia of the specimen can be calculated after measuring the period of oscillation of the specimen using Eq. (17-28). If the torsional stiffness of the rod is unknown, then a mass of known moment of inertia can be mounted on the rod and its period observed; Eq. (17-28) is then used to determine k_t of the rod.

Sometimes there is no method of mounting the specimen on a torsion rod. In these cases it is best to construct a platform which is arranged to oscillate as a torsional pendulum. In Fig. 17-7 the torsion rod is rigidly connected at its lower end to a platform on which the specimen is placed. The periods of oscillation are observed first for the platform alone and then for the platform with the specimen loaded on it. The moment of inertia of the platform alone is, from Eq. (17-28),

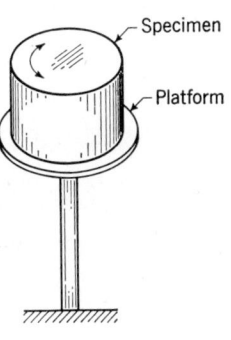

Fig. 17-7

$$I_p = \frac{k_t \tau_p{}^2}{4\pi^2} \qquad (c)$$

where I_p is the moment of inertia of the platform and supports and τ_p is the period of oscillation of the platform alone. The moment of inertia of the platform and specimen is

$$I + I_p = \frac{k_t \tau^2}{4\pi^2} \qquad (d)$$

Dividing both sides of Eq. (d) by I_p gives

$$\frac{I}{I_p} = \frac{k_t \tau^2}{4\pi^2} \frac{1}{I_p} - 1 \qquad (e)$$

$$= \frac{k_t \tau^2}{4\pi^2} \frac{4\pi^2}{k_t \tau_p^2} - 1$$

so
$$I = I_p \left(\frac{\tau^2}{\tau_p^2} - 1\right) \qquad (17\text{-}29)$$

The moment of inertia of the platform and supports is obtained first using the methods already described. Then the periods of oscillation of the platform alone and of the platform and specimen are observed, and Eq. (17-29) employed to find the moment of inertia of the specimen.[1]

17-9. Euler's Equations of Motion. We have seen that the general equation for the rotation of a rigid body is given by Eq. (17-7) as

$$\mathbf{T} = \dot{\mathbf{H}} = \frac{d}{dt}(H^x \mathbf{I} + H^y \mathbf{J} + H^z \mathbf{K}) \qquad (a)$$

where, from Eqs. (17-10),

$$\begin{aligned} H^x &= I_x \omega^x - I_{xy} \omega^y - I_{xz} \omega^z \\ H^y &= -I_{xy} \omega^x + I_y \omega^y - I_{yz} \omega^z \\ H^z &= -I_{zx} \omega^x - I_{zy} \omega^y + I_z \omega^z \end{aligned} \qquad (b)$$

We shall employ these relationships in a reference system having its origin attached to the mass center of the body and with the axes fixed in the body and moving with it. Thus the moments and products of inertia in Eqs. (b) are all constants. Note, however, that because the axes are fixed in the body, the unit vectors $\mathbf{I}, \mathbf{J}, \mathbf{K}$ in Eq. (a) are not fixed in direction and, for this reason, do have derivatives. For example,

$$\dot{\mathbf{I}} = \mathbf{\Omega} \times \mathbf{I} = \begin{vmatrix} \mathbf{I} & \mathbf{J} & \mathbf{K} \\ \omega^x & \omega^y & \omega^z \\ 1 & 0 & 0 \end{vmatrix} = \omega^z \mathbf{J} - \omega^y \mathbf{K}$$

Similarly
$$\dot{\mathbf{J}} = \omega^x \mathbf{K} - \omega^z \mathbf{I} \qquad \dot{\mathbf{K}} = \omega^y \mathbf{I} - \omega^x \mathbf{J}$$

Now, since
$$\dot{\mathbf{H}} = \dot{H}^x \mathbf{I} + H^x \dot{\mathbf{I}} + \dot{H}^y \mathbf{J} + H^y \dot{\mathbf{J}} + \dot{H}^z \mathbf{K} + H^z \dot{\mathbf{K}}$$

---

[1] For other means of measuring moment of inertia see Ferdinand P. Beer and E. Russell Johnston, Jr., "Mechanics for Engineers," pp. 653, 654, McGraw-Hill Book Company, Inc., New York, 1957. Also, Hamilton H. Mabie and Fred W. Ocvirk, "Mechanisms and Dynamics of Machinery," pp. 318–322, John Wiley & Sons, Inc., New York, 1957. Also see C. E. Crede, Determining Moment of Inertia, *Machine Design*, vol. 20, p. 138. The article, Bernard Brenner, Moment of Inertia, *Prod. Eng.*, vol. 31, no. 3, pp. 61–63, illustrates eight different methods of measuring moment of inertia.

then

$$\mathbf{T} = (\dot{H}^x - \omega^z H^y + \omega^y H^z)\mathbf{I} + (\dot{H}^y - \omega^x H^z + \omega^z H^x)\mathbf{J}$$
$$+ (\dot{H}^z - \omega^y H^x + \omega^x H^y)\mathbf{K} \quad (17\text{-}30)$$

where the values of H in Eqs. (b) are to be substituted into Eq. (17-30). However, the equations for H can be greatly simplified by selecting coordinate axes coincident with the principal axes of inertia of the body. This makes the products of inertia zero, and Eqs. (b) become

$$H^x = I_x \omega^x \qquad H^y = I_y \omega^y \qquad H^z = I_z \omega^z \quad (c)$$

Substituting these values into Eq. (17-30),

$$\mathbf{T} = (I_x \alpha^x - I_y \omega^y \omega^z + I_z \omega^y \omega^z)\mathbf{I} + (I_y \alpha^y - I_z \omega^x \omega^z + I_x \omega^x \omega^z)\mathbf{J}$$
$$+ (I_z \alpha^z - I_x \omega^y \omega^x + I_y \omega^y \omega^x)\mathbf{K} \quad (d)$$

and separating the components gives

$$\begin{aligned} T^x &= I_x \alpha^x + (I_z - I_y)\omega^y \omega^z \\ T^y &= I_y \alpha^y + (I_x - I_z)\omega^x \omega^z \\ T^z &= I_z \alpha^z + (I_y - I_x)\omega^x \omega^y \end{aligned} \quad (17\text{-}31)$$

These are Euler's equations of motion. It should be emphasized that they refer to any motion of a rigid body in which the origin of the reference system is fixed to the mass center and having coordinate axes coincident with the principal axes of inertia.

17-10. Rotation about a Fixed Axis. Mathematics can be an extremely useful tool to the engineer in that it produces correct results automatically— and, perhaps, unintelligently! The abstract and tedious mathematical analysis of the preceding sections of this chapter yields usable equations, but unless the engineer understands the physical significance of the solutions, he is in no position to modify his designs to produce better machines.

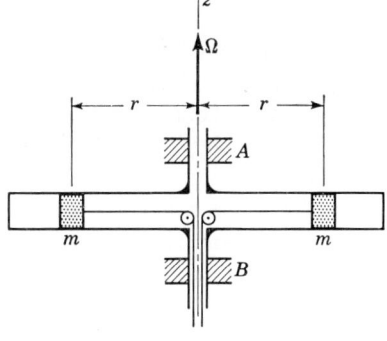

Fig. 17-8

Therefore, in this section we shall study the forces existing when masses rotate about fixed axes, but our emphasis shall be primarily upon obtaining a physical understanding of the problem.

We begin with the device of Fig. 17-8, which consists of a hollow shaft rotating in bearings A and B with a constant angular velocity Ω about the fixed z axis. Mounted upon the shaft are two tubular arms at right angles to the shaft and in line with each other which we assume to be without mass. Placed within the arms are two masses m which we pretend can slide without friction. Each mass is connected to a weightless wire which runs up the hollow shaft and radially outward to the mass. When each wire is positioned correctly, the two masses can rotate at equal distances r from the axis of rotation. Experience teaches us that tensile forces must exist in each wire. These are

$$\mathbf{F} = m\mathbf{\Omega} \times (\mathbf{\Omega} \times \mathbf{R}) \qquad \text{where } F = mr\omega^2 \qquad (a)$$

It is also within our experience that the magnitude of this force depends upon the mass, its distance from the axis of rotation, and the speed with which it is rotating, as indicated by Eq. (a).

Now suppose that Ω is zero and that we move the two masses in toward the axis of rotation so that r is very small. Under these conditions only a small amount of torque is required to give the system rotation, or, to put it another way, a small amount of torque will produce a substantial amount of angular velocity if the masses are located close to the axis of rotation. Nor is it difficult to see that a large amount of torque is needed to begin rotation if the masses are arranged so that r is quite large. This is simply another way of saying that, in the first case, the inertia is not very important because rotation takes very little effort to begin and, in the second case, inertia *is* very important because much more effort is needed to effect rotation. There are those who say that moment of inertia is an integral which appears in mathematical expressions often and so it is convenient to give it a name but that it really does not have any physical meaning! However, we have seen that when the masses are arranged near the axis of rotation, then the inertia is not very significant or influential, but as the radius increases, so does the importance (moment) of inertia. Thus the word "moment" has exactly the same meaning when applied to inertia or to force as it has when applied to, say, an occasion. So we speak of an occasion as being momentous, meaning that it is important or significant, and of the moment of a force or the moment of inertia, meaning the significance or importance of the force or the inertia.

Next let us arrange the two masses at an equal and fairly large distance r from the axis of rotation and apply torque to the shaft until Ω reaches an appreciable value. The product of the torque and the angular distance through which it is applied is the work done on the system, and this work must appear as kinetic energy of the rotating masses. If no friction forces are present, such as air friction or the sliding friction of the shaft in its bearings, the system will continue to rotate at its present angular velocity until another external torque is applied. The product of the original torque and the time interval during which it was applied is the angular impulse (moment of linear impulse) applied to the masses. Thus when the masses reach an angular velocity Ω, their angular momentum (moment of momentum) is

$$\mathbf{H} = I\mathbf{\Omega} \qquad (b)$$

where I is the moment of inertia of both masses. The law of conservation of moment of momentum states that the angular momentum remains constant unless an external torque acts upon the system. This, also, is within our frame of experience, for we can easily imagine the system of Fig. 17-8 continuing to coast at a constant angular velocity Ω if all friction is eliminated. Now, while the system continues to rotate at constant Ω, we must continue to exert a pull on each wire of magnitude $F = mr\omega^2$. Suppose the force F is increased sufficiently to cause both masses to slide closer to the axis of rotation so that, now, each mass is located at the new distance r' from the axis of rotation. Since the force F has no moment arm, we have exerted no external torque, and, according to the law of conservation of momentum, the moment of momentum cannot change. Accordingly, in Eq. (b), if I is decreased, then Ω must increase; otherwise the moment of momentum would not remain constant. Figure skaters take advantage of this by beginning a slow rotation and then by moving their arms and legs to the axis of their rotation (that is, closer to their bodies), causing them to spin or rotate very rapidly. Thus the effect of moving the two masses closer to the axis of rotation is to increase the angular velocity but without changing the angular momentum. Since the moment of inertia of each mass is mr^2 in the first instance, the angular momentum is $mr^2\omega = mrv$. Therefore, if r decreases, v must increase. When the force of the wire on the mass increases above $mr\omega^2$, the mass no longer moves in a circular path but now in a spiral path, and the increase in v is due to a component of the force tangent to the spiral. Movement of the masses through the distance $r - r'$ required the application of a force, and hence work has been done on the system. The effect is an increase in the kinetic energy of the rotating masses, although the angular momentum remains constant.

For the next investigation consider the rotating system of Fig. 17-9a. Here a shaft rotating with constant angular velocity Ω has fastened to it two arms to which are attached the masses m at distance r from the axis of rotation. The shaft is supported by bearings at A and B. As in the preceding examples, tensile forces exist in each arm of magnitude $F = mr\omega^2$ (Fig. 17-9b). Since these forces are equal, opposite, and parallel, they constitute a couple acting upon the shaft and must be

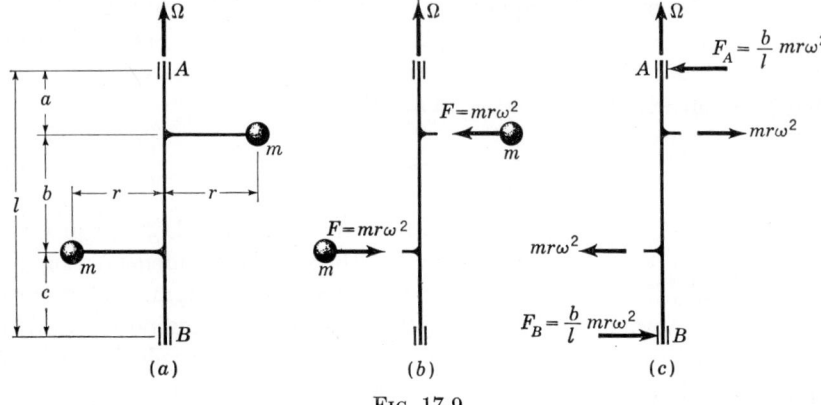

Fig. 17-9

resisted by radial forces at bearings A and B, as shown in Fig. 17-9c. The magnitude of the bearing forces is obtained by taking moments about A and then about B and is

$$F_A = F_B = \frac{b}{l} mr\omega^2 \qquad (c)$$

These radial bearing forces rotate with the shaft and cause the bearings to wear evenly around their circumference. Therefore, after a small interval of time, the masses will have turned through 180°, and at this instant the bearing forces will be opposite in direction to those shown in Fig. 17-9c.

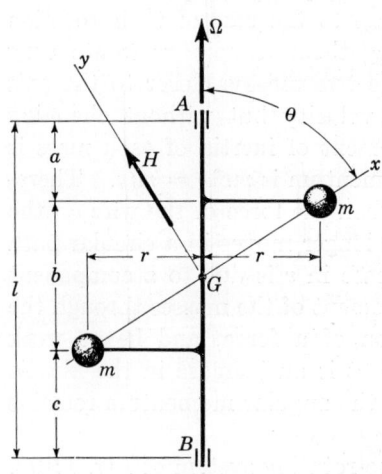

Fig. 17-10

The equations of Sec. 17-9 make this problem easy to analyze. We first choose the coordinate system with its origin at the center of gravity of the system and rotating with it, as illustrated in Fig. 17-10. The two masses, in this example, and the axis of rotation are contained by the xy plane, while the z axis is perpendicular to the axis of rotation. The angular velocity Ω therefore has two components ω^x and ω^y. The principal axes of the mass system are coincident with the coordinate axes; so the moments of inertia are

$$I_y = 2\left(r^2 + \frac{b^2}{4}\right)m \qquad I_z = 2\left(r^2 + \frac{b^2}{4}\right)m \qquad (d)$$

Applying Euler's equations [Eqs. (17-31)],

$$T^z = I_y \omega^x \omega^y \quad (e)$$

since α is zero. Now, because

$$\theta = \sin^{-1} \frac{r}{\sqrt{r^2 + b^2/4}} \quad \text{and} \quad \theta = \cos^{-1} \frac{b/2}{\sqrt{r^2 + b^2/4}}$$

then
$$\omega^x = \omega \cos \theta \qquad \omega^y = \omega \sin \theta$$

and Eq. (e) becomes

$$T^z = 2\left(r^2 + \frac{b^2}{4}\right) m\omega^2 \frac{rb/2}{r^2 + b^2/4} = bmr\omega^2 \quad (f)$$

Taking moments about A and then about B gives

$$F_A = F_B = \frac{b}{l} mr\omega^2$$

which is the same result obtained previously.

We note, too, from Eqs. (17-10) that

$$H^x = 0 \qquad H^y = I_y \omega^y \qquad H^z = 0 \quad (g)$$

so the moment of momentum is a vector $\mathbf{H}$ in the positive y direction. Although the magnitude of this vector is constant, the vector is fixed to the moving coordinate system, and consequently, it is rotating with a constant angular velocity $\mathbf{\Omega}$. Since $\mathbf{H}$ and $\mathbf{\Omega}$ are not in the same direction, this means that $\mathbf{H}$ is constantly undergoing a change of direction. Therefore the torque $\mathbf{T}$ exerted by the bearings on the shaft is a direct consequence of the change of direction of the moment of momentum vector.

EXAMPLE 17-1. A rectangular steel block is fastened at its center at an angle of 30° to a shaft which is rotating 1,500 rpm in the direction shown in Fig. 17-11. At

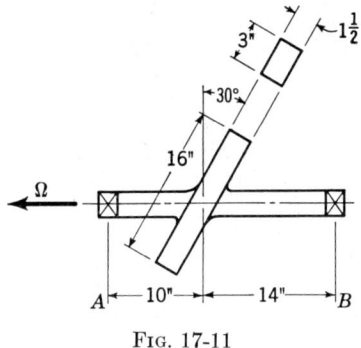

FIG. 17-11

428   DYNAMIC ANALYSIS OF MACHINES

the instant shown the shaft speed is being reduced at the rate of 100 rpm/sec. The shaft is supported by radial bearings at A and B. Using 0.282 lb/in.^{3} as the unit weight of steel and neglecting the weight and moment of inertia of the shaft, calculate the bearing reactions.

Solution. A schematic view of the system including the choice of axes is shown in Fig. 17-12. The shaft is in the xy plane with the y axis along the 16-in. dimension

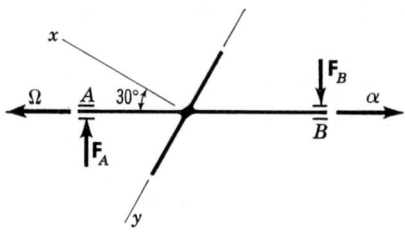

Fig. 17-12

of the bar. The angular velocity is

$$\Omega = \frac{2\pi(1,500)}{60} \underline{/30°} = 136\mathbf{I} + 78.5\mathbf{J} \qquad \text{rad/sec}$$

The angular acceleration is

$$\alpha = \frac{2\pi(100)}{60} \underline{/-120°} = -5.24\mathbf{I} - 9.07\mathbf{J} \qquad \text{rad/sec}^2$$

The moments of inertia are calculated from the formulas of Appendix III. The mass of the bar is found to be 0.0526 lb-sec^2-in., and the moments of inertia are

$$I_x = 1.162 \qquad I_y = 0.0494 \qquad I_z = 1.135 \qquad \text{lb-sec}^2\text{-in.}$$

Substituting these values in Eqs. (17-31) gives

$$\mathbf{T} = [I_x\alpha^x + (I_z - I_y)\omega^y\omega^z]\mathbf{I} + [I_y\alpha^y + (I_x - I_z)\omega^x\omega^z]\mathbf{J} + [I_z\alpha^z + (I_y - I_x)\omega^x\omega^y]\mathbf{K}$$
$$= [(1.162)(-5.24)]\mathbf{I} + [(0.0494)(-9.07)]\mathbf{J} + [(0.0494 - 1.162)(136)(78.5)]\mathbf{K}$$
$$= -6.09\mathbf{I} - 0.448\mathbf{J} - 11,900\mathbf{K} \qquad \text{lb-in.}$$

The **I** and **J** terms are due to the deceleration of the shaft; they represent components of torque which are externally applied to the shaft to cause it to slow down and are due to a decrease in the magnitude of the moment of momentum vector. The third term results from the fact that the moment of momentum vector is changing its direction. The bearing reactions are

$$F_A = F_B = \frac{11,900}{24} = 495 \text{ lb} \qquad Ans.$$

and they act in the direction shown in Fig. 17-12. Note particularly that these reactions do not include the static forces which resist the weight of the shaft and bar. Both $\mathbf{F}_A$ and $\mathbf{F}_B$ are contained in the xy plane and move with it as it rotates with the shaft.

17-11. Gyroscopes. The gyroscope of Fig. 17-13 is an instrument which has fascinated students of mechanics and applied mathematics for many years. In fact, once the rotor is set spinning, it appears to act as a device possessing intelligence. If we attempt to move some of its parts, it seems not only to resist this motion but even to evade it. We shall even see that it apparently fails to conform to the laws of statical equilibrium and of gravitation.

The uses of the gyroscope as turn-and-bank indicators, artificial horizons, and automatic pilots in aircraft and missiles are well known, as is its use in the gyrocompass. For many years it has been employed as a stabilizer in ships and torpedoes. One also becomes concerned with

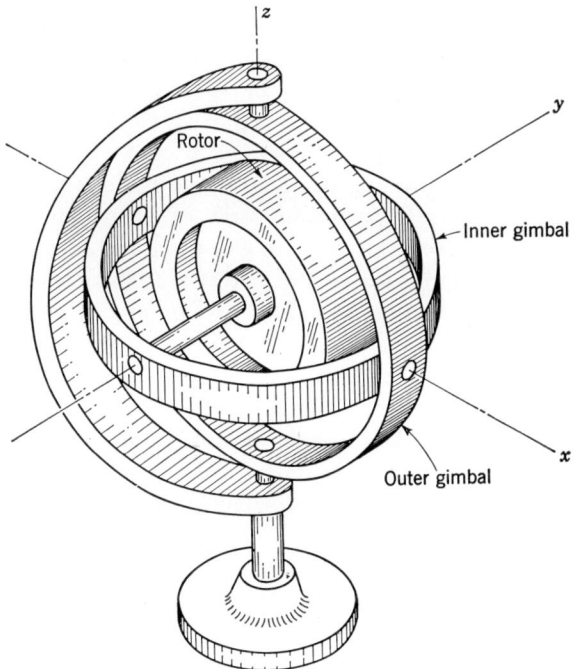

FIG. 17-13. A laboratory gyroscope.

gyroscopic effects in the design of machines—although not always intentionally. Such effects are present when a motorcycle or bicycle is being ridden; they are also present, owing to the rotating masses, when an airplane or automobile is making a turn. Sometimes these effects are desirable, but more often they are undesirable, and the designer must account for them in his selection of bearings and rotating parts. It is certainly true that as machine speeds increase to higher and higher values and as factors of safety decrease, we must stop neglecting gyroscopic forces in our machine designs because their values will be more significant. The general equations for the motion of a gyroscope are, indeed, not simple. Fortunately, in designing machines, only a few simple and approximate solutions are necessary. The reader engaged in the design of gyroscopes for guidance and control should acquire a much more thorough background than can be given here.[1]

The rotor of the gyroscope of Fig. 17-13 has a heavy rim and is fastened to a shaft which rotates in bearings in the inner gimbal. The inner gimbal is mounted on pivots so that it is free to rotate about an axis which is perpendicular to the rotational axis of the rotor. It is pivoted to an outer gimbal which can turn about a vertical axis through the frame perpendicular to the plane of the rotor and inner-gimbal axes for the position shown in the figure. Thus the rotor can rotate only about the y axis or, together with the inner gimbal, about the x axis or, with both gimbals, about the z axis. In fact the rotor can simultaneously have these three kinds of rotation. It will be convenient to designate the rotor axis, or y axis, as the *axis of spin*.

To provide a vehicle for the explanation of the simpler motions of a gyroscope it is expedient to perform a series of experiments with the one of Fig. 17-13. This may even serve to call to mind the results of similar experiments once made by the reader. In the following we assume that the rotor is spinning and that the pivot friction can be neglected.

1. If the z axis is kept in the vertical position, the gyroscope can be moved anywhere about a table or a room without altering the direction of the axis of spin. This is a consequence of the law of conservation of moment of momentum. If the axis of spin is to change its direction, then the moment of momentum vector must also change its direction, but this requires an external torque which, in this experiment, we have not supplied. We might, while the rotor is still spinning, lift the inner gimbal out of its bearings and move it about. We then find that it can be translated anywhere but that we meet with definite resistance when we attempt to rotate the axis of spin.

2. With the inner gimbal back in the bearings suppose that pressure is applied, say by a pencil, to the inner gimbal so as to cause it to turn about the x axis. Not only do we meet with resistance to the pressure of the pencil but it is also found that the outer gimbal begins to rotate slowly about the vertical z axis and that it continues this rotation until

---

[1] Harold Crabtree, "Spinning Tops and Gyroscopic Motion," Longmans, Green, & Co., Ltd., London, 1909.

Richard F. Deimel, "Mechanics of the Gyroscope," Dover Publications, Inc., New York, 1950.

Leigh Page, "Theoretical Physics," pp. 120–131, D. Van Nostrand Company, Inc., Princeton, N.J., 1928.

A. M. Worthington, "Dynamics of Rotation," Longmans, Green, & Co., Ltd., London, 1906.

James B. Scarborough, "The Gyroscope," Interscience Publishers, Inc., New York, 1958.

the pressure is released. The pressure of the pencil constitutes a torque on the inner gimbal with the parallel and opposite force of the couple coming from the pivots in the gimbal.

In order to study these effects carefully we might cause the rotor to spin in the positive direction, that is, with the angular-velocity vector pointing in the positive y direction. Then if we apply a positive torque to the inner gimbal (torque vector pointing in the positive x direction), the rotation of the outer gimbal is found to be in the negative z direction. The reader should note that these effects occur in a right-handed coordinate system. Either a negative spin velocity or a negative torque will cause the gimbal to rotate in the positive z direction for the set of axes shown. The rotation of the spin axis about an axis perpendicular to

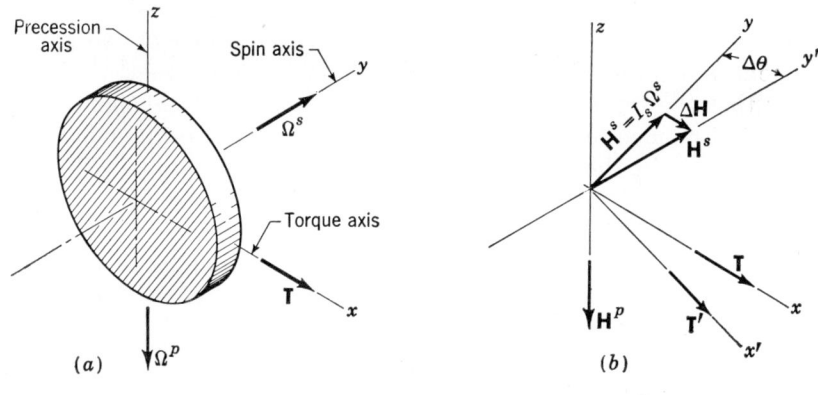

FIG. 17-14

that of a torque applied to it is called *precession*, and so the application of a torque to the spinning rotor causes it to *precess*. In this example the z axis is called the *axis of precession*.

3. As a third experiment we might apply a torque to the outer gimbal in an attempt to cause it to rotate about the z axis. Such an attempt meets with resistance and causes the inner gimbal with the spin axis to rotate. When the spin axis is in the vertical position, the gyroscope is in stable equilibrium though, and the outer gimbal can then be turned quite freely. Note in this as well as in the previous example that the moment of momentum vector is changing its direction because of an application of external torque.

In Fig. 17-14 suppose the rotor is spinning about its spin axis with an angular velocity Ω^s while at the same time the spin axis precesses with an angular velocity Ω^p. We can designate the moment of inertia of the rotor about the spin axis as I_s and about the x and z axes as I, since they are equal. Because the axes of the rotor are the principal axes of inertia, the component of the moment of momentum vector along the spin axis is $\mathbf{H}^s = I_s \mathbf{\Omega}^s$ and along the precession axis is $\mathbf{H}^p = I\mathbf{\Omega}^p$. After a small period of time Δt the spin axis has rotated through the angle $\Delta \theta$ to a new position indicated as y' in Fig. 17-14b. Thus the component of the moment of momentum along the spin axis is continually changing its direction during precession. In Chap. 12 we learned that any vector, such as $\mathbf{H}^s$, rotating with a constant angular velocity $\mathbf{\Omega}^p$ has a rate of change

$$\dot{\mathbf{H}}^s = \mathbf{\Omega}^p \times \mathbf{H}^s$$

Applying the equation of motion and substituting the value of $\mathbf{H}^s$ give

$$\mathbf{T} = I_s \mathbf{\Omega}^p \times \mathbf{\Omega}^s \tag{17-32}$$

The direction of the torque required to *maintain* the precession is shown in Fig. 17-14a. Figure 17-14b shows that the direction of the applied torque must continue to change in order to maintain precession. It also shows that the torque *does not* vary the precessional component of the moment of momentum. It *does* show that the change in the moment of momentum is in the *same* direction as the applied torque. We note further that Eq. (17-32) applies *only* to the maintenance of an existing motion and not to the beginning or ending of a precession. It might be noted, though not demonstrated here, that the beginning or ending of precession is accompanied by vibrations which, usually, are damped out quite rapidly by friction.

EXAMPLE 17-2. Figure 17-15 illustrates a hypothetical problem typical of the situations occurring in the design or analysis of machines in which gyroscopic forces

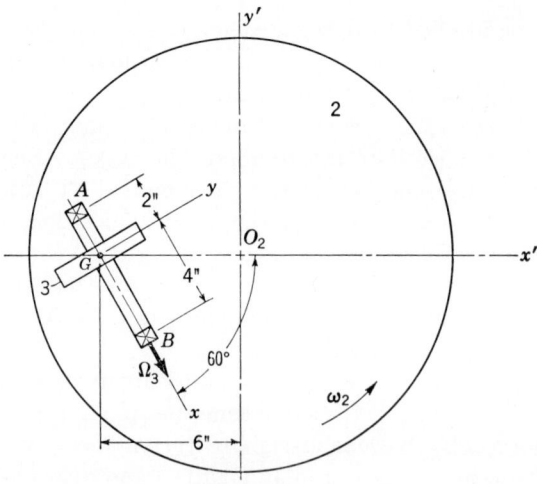

FIG. 17-15

must be considered. A round plate designated as 2 rotates about the z' axis with angular velocity ω_2. Mounted on this revolving plate are two bearings A and B which retain a shaft and mass 3 rotating at the vector angular velocity Ω_3. An xyz system is selected fixed to the shaft and mass and therefore rotating with it. The mass center G defines the origin of this system, and the x axis is coincident with the axis of the shaft rotation. The angular velocity Ω_3 is that which an observer stationed on the rotating plate would report the shaft as having. Let the weight of the mass be $W = 10$ lb, its radius of gyration be $r = 2$ in., and its angular velocity $\Omega_3 = 350\mathbf{I}$ rad/sec. Using $\omega_2 = 5$ rad/sec, in the direction shown, find the bearing reactions. Assume that the weight of the shaft is negligible and that bearing B is to take only radial load.

Solution. Since we are dealing with vectors, superposition can be used. Therefore the bearing reactions at A and B will be calculated first considering that Ω_3 is zero. Then to these components we shall add those due to gyroscope action.

When Ω_3 is zero, the methods of Chap. 16 apply. The results are

At A $\qquad\qquad \mathbf{F}_{23} = 1.94\mathbf{I} + 2.24\mathbf{J} + 6.67\mathbf{K} \qquad$ lb $\qquad\qquad(1)$

At B $\qquad\qquad \mathbf{F}_{23} = 1.12\mathbf{J} + 3.33\mathbf{K} \qquad$ lb $\qquad\qquad(2)$

where the vectors are referred to the xyz system.

The forces due to gyroscope action are found as follows: The x axis is the axis of spin, and the moment of inertia relative to this axis is

$$I_x = mr^2 = {}^{10}\!/_{386}(2)^2 = 0.1038 \text{ lb-sec}^2\text{-in.}$$

The precessional velocity is

$$\Omega_2 = 5\mathbf{K}' = 5\mathbf{K} \qquad \text{rad/sec}$$

because an angular-velocity vector is always a free vector. Equation (17-32) now applies where $I_s = I_x$, $\Omega^p = \Omega_2$, and $\Omega^s = \Omega_3$. Thus

$$\Omega_2 \times \Omega_3 = \begin{vmatrix} \mathbf{I} & \mathbf{J} & \mathbf{K} \\ 0 & 0 & 5 \\ 350 & 0 & 0 \end{vmatrix} = 1{,}750\mathbf{J}$$

so $\qquad\qquad \mathbf{T} = I_s \Omega_2 \times \Omega_3 = (0.1038)(1{,}750\mathbf{J}) = 181.5\mathbf{J} \qquad$ lb-in.

The position vector of B relative to A is $\mathbf{R}_{BA} = 6\mathbf{I}$. Then, taking moments about A,

$$\Sigma \mathbf{T}_A = \mathbf{T} + \mathbf{R}_{BA} \times \mathbf{F}_B = 0$$

or $\qquad\qquad 181.5\mathbf{J} + 6\mathbf{I} \times \mathbf{F}_B = 0$

so that $\qquad\qquad \mathbf{F}_B = 30.2\mathbf{K} \qquad$ lb $\qquad\qquad(3)$

Next, $\mathbf{R}_{AB} = -6\mathbf{I}$. Taking moments about B gives

$$\Sigma \mathbf{T}_B = \mathbf{T} + \mathbf{R}_{AB} \times \mathbf{F}_A = 0$$

or $\qquad\qquad 181.5\mathbf{J} + (-6\mathbf{I}) \times \mathbf{F}_A = 0$

so that $\qquad\qquad \mathbf{F}_A = -30.2\mathbf{K} \qquad$ lb $\qquad\qquad(4)$

Adding Eqs. (1) and (4) gives the total reaction at A as

At A $\qquad\qquad \mathbf{F}_{23} = 1.94\mathbf{I} + 2.24\mathbf{J} - 23.53\mathbf{K} \qquad$ lb $\qquad Ans.$

Similarly, Eqs. (2) and (3) are summed to give the reaction at B.

At B $\quad\quad\quad\quad \mathbf{F}_{23} = 1.12\mathbf{J} + 33.53\mathbf{K} \quad$ lb $\quad$ *Ans.*

We note that the effect of the gyroscopic couple is to lift the rear bearing off the plate and to push the front bearing against the plate.

PROBLEMS

17-1. The figure shows a two-throw opposed-crank crankshaft mounted in bearings at A and G. Each crank has an eccentric weight of 6 lb which may be considered as located at a radius of 2 in. from the axis of rotation and at the center of each throw. It is proposed to locate weights at B and F in order to reduce the bearing reactions, due to the rotating eccentric cranks, to zero. If these weights are to be mounted 3 in. from the axis of rotation, as shown, how much must they weigh?

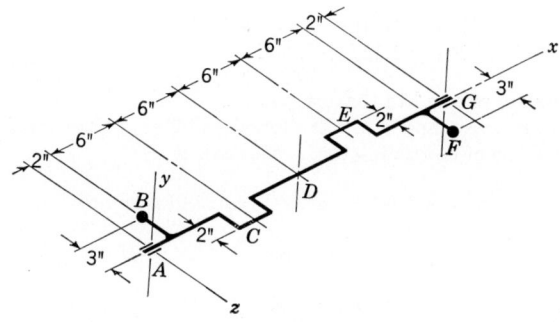

Prob. 17-1

17-2. The figure illustrates a two-throw crankshaft, mounted in bearings at A and F, with the cranks spaced 90° apart. Each crank may be considered to have an eccentric weight of 6 lb located at the center of the throw and 2 in. from the axis of crankshaft rotation. It is proposed to eliminate the rotating bearing reactions, which the cranks would cause, by mounting additional weights (called correction weights) on 3-in. arms at points B and E. Calculate the magnitude and the location of these weights.

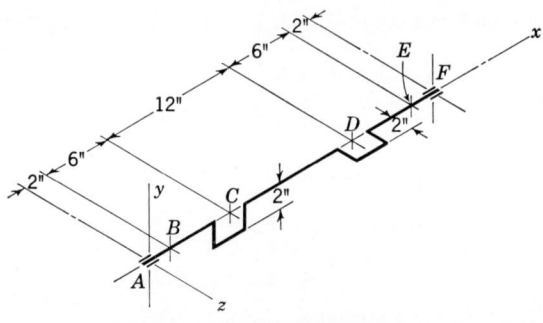

Prob. 17-2

17-3. Solve Prob. 17-2 if the angle between the two throws is reduced from 90 to 0°.

17-4. A connecting rod weighing 7.90 lb is pivoted on a knife-edge and caused to oscillate as a pendulum as shown in the figure. The mass center of the rod is 2¼ in. from the crankpin center. When caused to oscillate, the rod was observed to complete 64.5 oscillations in 1 min. Determine the moment of inertia of the rod about its own center of gravity.

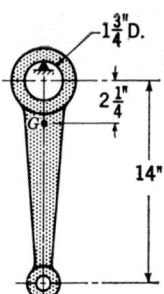

PROB. 17-4

17-5. The gear shown in the figure was suspended by a knife-edge at the rim and caused to oscillate. The period of oscillation was then observed and found to be 1.08 sec. If the weight of the gear is 41 lb, find the moment of inertia and the radius of gyration. Assume that the center of gravity of the gear and its axis of rotation are coincident.

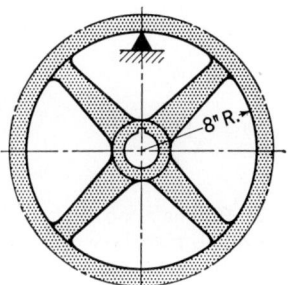

PROB. 17-5

17-6. The figure shows a wheel whose moment of inertia I is to be determined. The wheel is mounted on a shaft and placed in bearings whose frictional resistance to rotation must be kept very small. To the end of the shaft and on the outboard side of the bearings, in this example, is connected a rod having a weight W_b secured to the end. It is possible to measure the moment of inertia of the wheel by displacing the weight W_b from its position of static equilibrium and permitting the assembly to oscillate. If the weight of the pendulum arm, of length l, is neglected, show that the moment of inertia of the wheel can be obtained from the equation

$$I = W_b l \left(\frac{\tau^2}{4\pi^2} - \frac{l}{g} \right)$$

where τ is the period of oscillation in seconds.

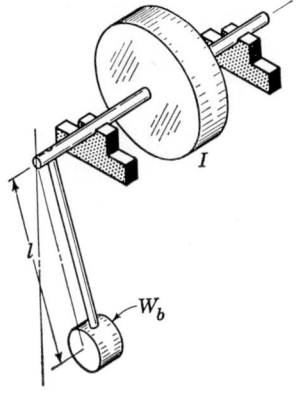

PROB. 17-6

17-7. If the weight of the pendulum arm is not neglected in Prob. 17-6 but is assumed to be uniformly distributed over the distance l, show that the moment of inertia can be calculated from the formula

$$I = l \left[\frac{\tau^2}{4\pi^2} \left(W_b + \frac{W_a}{2} \right) - \frac{l}{g} \left(W_b + \frac{W_a}{3} \right) \right]$$

where W_a is the weight of the arm.

17-8. Wheel 2 in the figure is a round disk which is arranged to rotate about a vertical axis z through its center. Wheel 2 carries a pin B at a distance R from the axis of rotation of the wheel, about which link 3 is free to rotate. Link 3 has its center of mass G located a distance r from the vertical axis through B, and it has a weight W_3 and a moment of inertia I about its own mass center. The wheel rotates at an angular velocity ω_2 with link 3 fully extended. Develop an expression for the angular velocity ω_3 that link 3 would acquire if the wheel is suddenly stopped.

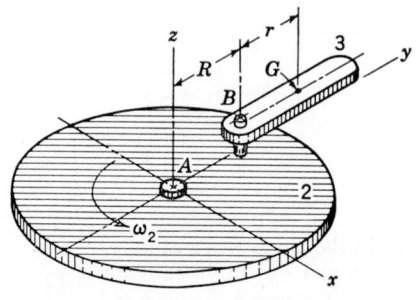

PROB. 17-8

17-9. The same as Prob. 17-8 except that the wheel rotates with link 3 radially inward. Under these conditions, is there a value of r for which the resulting ω_3 is zero?

17-10. Illustrated in the figure is a planetary gear reduction unit which utilizes 7-pitch spur gears cut on the 20° full-depth system. The input to the reducer is 25 hp at 600 rpm. Calculate the bearing reactions on the output and input shafts and on the planetary shafts. As a designer, what forces would you use in designing the mounting bolts? Why?

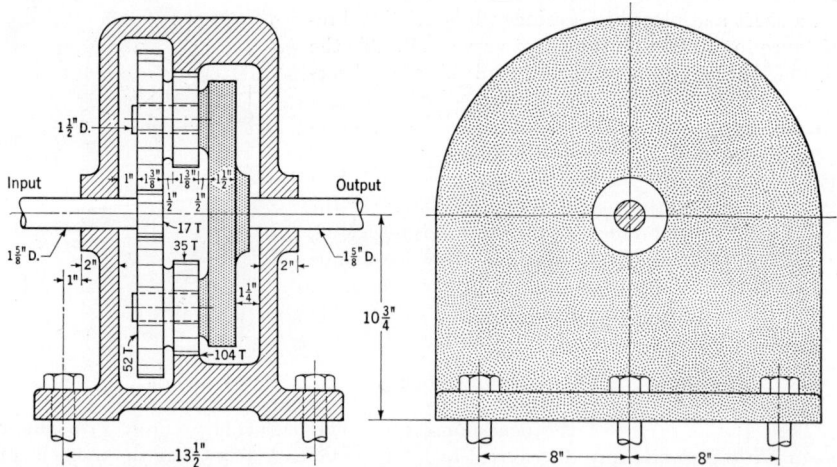

PROB. 17-10. All rotating parts are steel; $\rho = 0.282$ lb/in.³ The arm is rectangular and is 4 in. wide by 14 in. long with a central hub 4 in. in diameter and two planetary hubs each 3 in. in diameter. The segment separating the planet gears is a ½- by 4-in.-diameter cylinder. The inertia of the gears can be obtained by treating them as cylinders equal in diameter to the pitch circle.

17-11. It frequently happens in motor-driven machinery that the greatest torque is exerted when the motor is first turned on, because of the fact that some motors are capable of delivering more starting torque than running torque. Analyze the bearing reactions for Prob. 17-10 again, but this time use a starting torque equal to 250 per cent of the full load torque and a speed of zero. How does this condition affect the forces on the mounting bolts? The wr^2 for the motor is 2,250 lb-in.^{2}

17-12. The gear reduction unit of Prob. 17-10 is running at 600 rpm when the motor is suddenly turned off. Turning off the motor does not, however, change the resisting torque of the load. Solve Prob. 17-10 for this condition using $wr^2 = 2{,}250$ lb-in.^{2} for the rotor of the motor. Here, r is the radius of gyration of the rotor of the motor.

CHAPTER 18

VIBRATION ANALYSIS

The existence of vibrating elements in a machine produces unwanted noise, high stresses, wear, and, frequently, premature failure of one or more of the parts. The moving parts of all machines are inherently vibration producers, and for this reason the mechanical designer must expect vibrations to exist in the products he designs, but there is a great deal that he can do in anticipating a vibration problem and in minimizing its undesirable effects during the design of the machine. Even after these precautions have been taken, unexpected vibratory motions are often found after a machine has been designed and constructed. The engineer must then discover the source of the vibration and apply corrective measures.

Many excellent books are available on the subject of mechanical vibrations.[1] In the space of only one chapter we can cover but the most elementary aspects. This is not enough for the reader who expects to specialize in mechanical design or automatic control, and it is about the minimum for other readers.

18-1. Introduction. Any motion which exactly repeats itself after a certain interval of time may be called a *vibration*. Vibrations may be either *free* or *forced*. A machine element is said to have a *free vibration* if the periodic motion continues after the cause or the original disturbance is removed, but if a vibratory motion persists because of the existence of a disturbing force, then it is called a *forced vibration*. Any free vibration of a mechanical system will eventually cease because of loss of energy. In vibration analysis we usually account for these energy losses using a single factor called the *damping factor*. Thus a heavily damped system

---

[1] The author has found the following books most useful:
Lydik S. Jacobsen and Robert S. Ayre, "Engineering Vibrations," McGraw-Hill Book Company, Inc., New York, 1958.
J. P. Den Hartog, "Mechanical Vibrations," 4th ed., McGraw-Hill Book Company, Inc., New York, 1956.
Austin H. Church, "Mechanical Vibrations," John Wiley & Sons, Inc., New York, 1957.
John N. Macduff and John R. Curreri, "Vibration Control," McGraw-Hill Book Company, Inc., New York, 1958.

is one in which the vibration decays rapidly. The *period* of a vibration is the time of a single cycle; the *frequency* is the number of cycles occurring in unit time. The *natural frequency* is the frequency of a free vibration. If the forcing frequency should become equal to the natural frequency of a system, then *resonance* is said to occur.

We shall also use the terms *steady-state vibration*, to indicate that a motion is repeating itself exactly in each successive cycle, and *transient vibration*, to indicate a vibratory-type motion which is changing in character. If a periodic force operates on a mechanical system, the resulting motion will be transient in character when the force first begins to act, but after an interval of time the transient decays (owing to damping), and the resulting motion is termed a steady-state vibration.

The word *response* is frequently used in discussing vibratory systems. The words response, behavior, and performance have roughly the same meanings when used in vibration analysis. Thus we can apply an external force having a sine-wave relationship with time to a vibrating system in order to determine how the system "responds" or "behaves" when the frequency of the force is varied. A plot using the vibration amplitude as one axis and the forcing frequency as the other axis of the plot would then be described as a *performance* or *response* curve for the system. Sometimes it is desirable to apply arbitrary input disturbances or forces to a system. These may not resemble the force characteristics which a physical system would receive in use at all, yet the response of the system to these arbitrary disturbances can provide much useful information about the system. In general, then, the word response is used in the sense of describing the output characteristics or behavior of a vibrating system when subjected to specified input functions or forces.

In studying mechanical vibrations we shall also find it desirable to modify the assumption of a rigid body. Machine parts vibrate because they are elastic. When a rotating shaft has a torsional vibration, this means that a mark on the circumference at one end of the shaft is successively ahead of and then behind a corresponding mark on the other end of the shaft. In other words torsional vibration of a shaft is the alternate twisting and untwisting of the rotating material and requires elasticity for its existence. An inelastic material, such as a lump of putty, cannot support a vibration. In this chapter we shall usually, but not always, make the assumption that elastic parts have no mass and that masses are absolutely rigid. This is illustrated nicely by the idealized vibrating system of Fig. 18-1. Here a mass m is guided to move only in the x direction. The mass is connected to a fixed frame through the spring k and dashpot c. The assumptions used are:

1. The spring and dashpot are massless.
2. The mass is absolutely rigid.
3. All the damping is concentrated in the dashpot.

440   DYNAMIC ANALYSIS OF MACHINES

The results of many experiments show that a great number of mechanical systems can be analyzed with good accuracy using these assumptions.

The elasticity of the system of Fig. 18-1 is completely represented by the spring. The stiffness or scale of the spring is designated as k and defined by the equation

$$k = \frac{F}{x} \tag{18-1}$$

where F is the force in pounds required to deflect the spring a distance x in. Thus the units of k are pounds per inch.

Similarly, the friction or damping is assumed to be all viscous damping and is designated by c. Thus

$$c = \frac{F}{\dot{x}} \tag{18-2}$$

where the F of Eq. (18-2) is the force in pounds required to move the mass at a velocity of $\dot{x}$ in./sec. The units of c are therefore pound-seconds per inch.

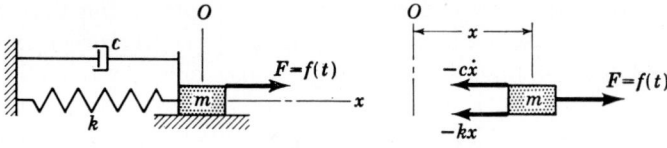

FIG. 18-1

The vibrating system of Fig. 18-1 has one degree of freedom because the position of the mass can be completely defined by a single coordinate. An external force $F = f(t)$ is shown acting upon the mass. Thus this system is classified as a forced, single-degree-of-freedom system with damping.

In order to write the equation of motion of the mass of Fig. 18-1 we choose as the origin of the coordinate system the position of the mass when the spring force is zero. Let x define the displacement of the mass from this position with the positive sense being taken to the right. It is also convenient to choose the same direction for positive values of velocity, acceleration, and force. Then, if the mass is displaced in the positive direction, the force of the spring on the mass is in the negative direction. If the velocity of the mass is positive, the damping force is negative. These forces are all shown on the free-body diagram of the mass after it has been displaced in the positive x direction (Fig. 18-1). Using Eq. (16-3) to write Newton's equation gives

$$-kx - c\dot{x} + f(t) + (-m\ddot{x}) = 0$$

or

$$\ddot{x} + \frac{c}{m}\dot{x} + \frac{k}{m}x = \frac{1}{m}f(t) \tag{18-3}$$

This is an important equation in vibration analysis, and in the next several sections we shall solve it for many specialized conditions.

Consider next the idealized torsional vibrating system of Fig. 18-2. Here a disk having a moment of inertia I is mounted upon the end of a weightless shaft having a torsional spring constant k_t as defined by

$$k_t = \frac{T}{\theta} \qquad (18\text{-}4)$$

where T is the torque in pound-inches necessary to produce an angular deflection of the disk of θ rad. The units of k_t are therefore pound-inches per radian. In a similar manner the torsional damping factor is defined by

$$c = \frac{T}{\dot{\theta}} \qquad (18\text{-}5)$$

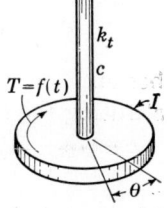

Fig. 18-2

and the units of c are pound-inch-seconds per radian. Designating the external torque applied to the disk as $T = f(t)$, the equation of motion of the torsional system is

$$-k_t\theta - c\dot{\theta} + f(\theta) + (-I\ddot{\theta}) = 0$$

or

$$\ddot{\theta} + \frac{c}{I}\dot{\theta} + \frac{k_t}{I}\theta = \frac{1}{I}f(t) \qquad (18\text{-}6)$$

which is the same form as Eq. (18-3). Thus, with appropriate substitutions, the solution of Eq. (18-6) will be the same as that of Eq. (18-3).

18-2. Free Vibration without Damping. Many cases of vibration occur in machines in which the amount of friction or damping present is so small that it can be neglected and the resulting analysis will still give very accurate results. Making the damping term and the external-force term of Eq. (18-3) zero gives the differential equation of motion for a free vibration:

$$\ddot{x} + \frac{k}{m}x = 0 \qquad (18\text{-}7)$$

The solution to this equation is well known and is

$$x = A \cos \omega_n t + B \sin \omega_n t \qquad (18\text{-}8)$$

where A and B are constants of integration and where

$$\omega_n = \pm \sqrt{\frac{k}{m}} \qquad (a)$$

If Eq. (18-7) is arranged in the form

$$\ddot{x} = -\omega_n^2 x \qquad (b)$$

then one can see that x must be a function whose second derivative is its own negative. Since both the sine and cosine functions satisfy this requirement, we could have derived Eq. (18-8) by simple inspection. The sine and cosine functions must both appear in Eq. (18-8) because of the requirement for two constants of integration. The reader can readily satisfy himself as to the validity of this equation by substituting it and its second derivative into Eq. (18-7).

The values of the constants of integration in Eq. (18-8) depend upon how the vibration originated. If we pull the mass out a distance $x = x_0$ and begin counting time at the instant we release it, the initial or starting conditions are

$$\text{When } t = 0 \quad x = x_0 \quad \text{and when } t = 0 \quad \dot{x} = 0$$

because the velocity of the mass is zero at the instant of release. Substituting the first of these conditions in Eq. (18-8) produces

$$x_0 = A \cos 0 + B \sin 0 \quad A = x_0 \qquad (c)$$

The first derivative of Eq. (18-8) is

$$\dot{x} = -A\omega_n \sin \omega_n t + B\omega_n \cos \omega_n t \qquad (d)$$

Substituting the second condition in Eq. (d) yields

$$0 = -A\omega_n \sin 0 + B\omega_n \cos 0 \quad B = 0 \qquad (e)$$

Therefore the equation

$$x = x_0 \cos \omega_n t \qquad (18\text{-}9)$$

describes the motion of the system for this method of starting.

One might also begin the motion by giving the mass an initial velocity v_0. The initial conditions are then

$$\text{When } t = 0 \quad x = 0 \quad \text{and when } t = 0 \quad \dot{x} = v_0$$

Substituting these conditions in Eq. (18-8), as before, gives for the constants of integration

$$A = 0 \quad \text{and} \quad B = \frac{v_0}{\omega_n}$$

and the solution now is

$$x = \frac{v_0}{\omega_n} \sin \omega_n t \qquad (18\text{-}10)$$

A more general method of originating the vibration is to let the mass have both displacement and velocity. Then the starting conditions are $x = x_0$ and $\dot{x} = v_0$ when $t = 0$. If the constants are again determined in the same manner, the solution will be found to be

$$x = x_0 \cos \omega_n t + \frac{v_0}{\omega_n} \sin \omega_n t \qquad (18\text{-}11)$$

These three solutions [Eqs. (18-9), (18-10), and 18-11)] are graphically represented in Fig. 18-3 using rotating vectors to generate the trigonometric functions. The ordinate of the graph is the displacement x, and the abscissa can be considered as the time axis or as the total angular displacement $\omega_n t$ of the rotating vectors for any given time after starting. The vectors x_0 and v_0/ω_n are shown in their initial positions, and as time passes, these vectors rotate counterclockwise with an angular velocity of ω_n rad/sec and generate the displacement curves. The drawing shows that $x_0 \cos \omega_n t$ starts from a maximum positive displacement and

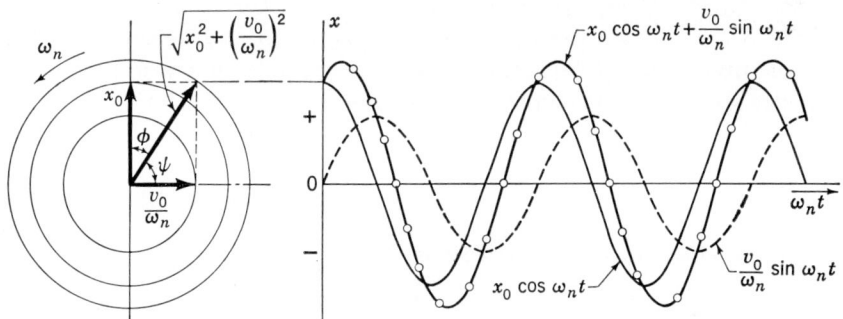

Fig. 18-3

$(v_0/\omega_n) \sin \omega_n t$ from a zero displacement. These, therefore, are very special, and the most general form of start is that given by Eq. (18-11), in which motion begins at some intermediate point.

The quantity

$$\omega_n = \sqrt{\frac{k}{m}} \qquad (18\text{-}12)$$

is called the *natural circular frequency* of the *undamped free vibration*, and its units are in radians per second. Note that this is *not* the same as the *natural frequency* defined earlier in this chapter which has the units of cycles per second. Nevertheless, we shall often describe ω_n as the natural frequency too, omitting the word "circular" for convenience, since its circular character follows from the units used. For most systems ω_n is a constant because the mass and spring constant do not vary. Since one cycle of motion is completed in an angle of 2π rad, the period of a vibration is given by the equation

$$\tau = \frac{2\pi}{\omega_n} = 2\pi \sqrt{\frac{m}{k}} \qquad (18\text{-}13)$$

where τ is in seconds. The frequency is the reciprocal of the period and is

$$f = \frac{\omega_n}{2\pi} = \frac{1}{2\pi} \sqrt{\frac{k}{m}} \qquad (18\text{-}14)$$

where the units of f are in cycles per second. This is abbreviated as cps.

Study of Fig. 18-3 suggests that one should also be able to express the motion by the equation[1]

$$x = X_0 \cos (\omega_n t - \phi) \qquad (18\text{-}15)$$

where X_0 and ϕ are the constants of integration and depend for their values upon the initial conditions. These constants are obtainable directly from the trigonometry of Fig. 18-3 and are

$$X_0 = \sqrt{x_0^2 + \left(\frac{v_0}{\omega_n}\right)^2} \qquad \phi = \tan^{-1} \frac{v_0}{\omega_n x_0}$$

Equation (18-15) can now be written in the form

$$x = \sqrt{x_0^2 + \left(\frac{v_0}{\omega_n}\right)^2} \cos (\omega_n t - \phi) \qquad (18\text{-}16)$$

This is a particularly convenient form of the equation because the coefficient is the *amplitude* of the vibration. The amplitude is the maximum displacement of the mass. The angle ϕ is called a *phase angle*, and it denotes the angular lag of the motion with respect to the cosine function.

The velocity and acceleration are obtained by successively differentiating the equation. Thus, from Eq. (18-15)

$$\dot{x} = -X_0 \omega_n \sin (\omega_n t - \phi) \qquad \ddot{x} = -X_0 \omega_n^2 \cos (\omega_n t - \phi)$$

[1] The motion can also be expressed in the form

$$x = X_0 \sin (\omega_n t + \psi) \qquad (18\text{-}17)$$

where X_0 and ψ are the constants of integration. This is probably not a very good way of expressing it, however, because, in the study of forced vibration, it might imply that the output can lead the input, which, of course, is not possible.

where the velocity amplitude is $X_0\omega_n$ and the amplitude of the acceleration is $X_0\omega_n^2$. The vector displacement, velocity, and acceleration are plotted in Fig. 18-4 to show their phase relationships. All three vectors maintain the fixed phase angles as they rotate at constant angular velocity ω_n. It is seen that the velocity leads the displacement by 90° and that the acceleration is 180° out of phase with the displacement.

18-3. Step-input Transient Forcing. Let us consider the vibrating system of Fig. 18-1 again, this time adding to the system a constant force F applied to the mass and acting in the positive x direction. As before, we consider the damping to be zero. For this condition Eq. (18-3) is written

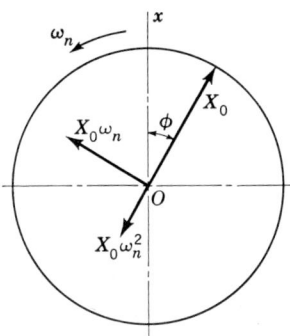

FIG. 18-4. Phase relationship of displacement, velocity, and acceleration.

$$\ddot{x} + \frac{k}{m} x = \frac{F}{m} \tag{18-18}$$

The solution to this equation is

$$x = A \cos \omega_n t + B \sin \omega_n t + \frac{F}{k} \tag{18-19}$$

where A and B are the constants of integration and where

$$\omega_n = \sqrt{\frac{k}{m}} \tag{a}$$

as before. It is noted that the dimensions of the quantity F/k are inches. The physical interpretation of this is that a force F applied to a spring of index k produces an elongation (or deformation) of the spring of magnitude F/k.

It will be interesting to start the motion when the system is at rest. Thus we apply a force F at the instant $t = 0$ and observe the behavior of the system. Since the system is motionless and in equilibrium the instant the force is applied, the starting conditions are $x = 0$, $\dot{x} = 0$, when $t = 0$. Substituting these conditions in Eq. (18-19) produces the following values for the constants of integration:

$$A = -\frac{F}{k} \qquad B = 0$$

The equation of motion is obtained by substituting these back into Eq. (18-19). This gives

$$x = \frac{F}{k} (1 - \cos \omega_n t) \tag{18-20}$$

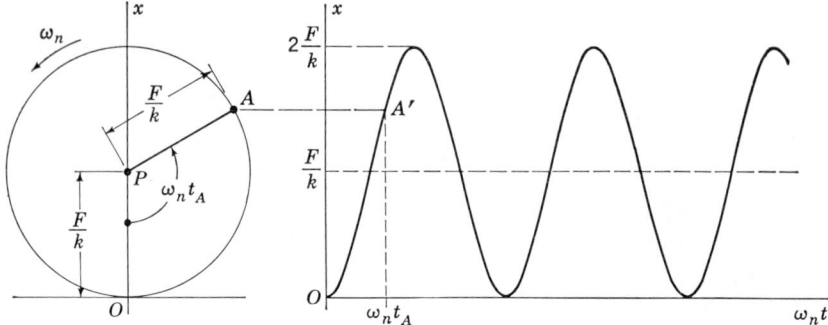

FIG. 18-5. Response of an undamped vibrating system to a constant force.

This equation is plotted in Fig. 18-5 to show how the system behaves. The figure shows that the application of a constant force F produces a vibration of amplitude F/k about a position of equilibrium displaced a distance F/k from the origin. This is evident from Eq. (18-20), since it contains a positive constant term

$$x_1 = \frac{F}{k}$$

and a negative trigonometric term

$$x_2 = -\frac{F}{k} \cos \omega_n t$$

For $t = 0$ these equations become

$$x_1 = \frac{F}{k} \qquad x_2 = -\frac{F}{k}$$

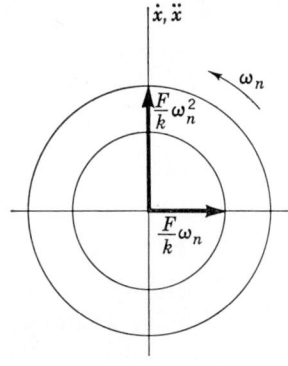

FIG. 18-6. Starting positions of the velocity and acceleration vectors.

On the rotating-vector diagram (Fig. 18-5) the motion is represented by a vector PA of length F/k rotating counterclockwise at ω_n rad/sec. This vector starts from the position PO when $t = 0$, rotates about P as a center, and generates the circle of radius F/k.

The velocity and acceleration are obtained by successive differentiation of Eq. (18-20) and are

$$\dot{x} = \frac{F}{k} \omega_n \sin \omega_n t \qquad (18\text{-}21)$$

$$\ddot{x} = \frac{F}{k} \omega_n^2 \cos \omega_n t \qquad (18\text{-}22)$$

These can be represented as rotating vectors too, as shown in Fig. 18-6. Their starting positions are found by calculating their values for $\omega_n t = 0$.

VIBRATION ANALYSIS

Velocity-time and acceleration-time graphs can also be plotted for these vectors employing the same methods used to obtain the displacement-time plot of Fig. 18-5.

Having produced a motion by virtue of the application of a constant force to a system originally at rest, it will now be of interest to observe the response of the same system when the force is removed. Thus at some instant in time later than $t = 0$, say $t = t_1$, we propose to make $F = 0$. The differential equation and its solution are now

$$\ddot{x} + \frac{k}{m} x = 0 \tag{b}$$

$$x = A \cos \omega_n t' + B \sin \omega_n t' \tag{c}$$

where A and B are new constants of integration and where $t' = t - t_1$. This means that when F becomes zero, a new era of the motion begins and the time is counted starting from zero at the beginning of this new era. The constants of integration depend for their values upon the state of the motion at the beginning of the second era. The ending conditions for the first era are

$$x = x_1 \qquad \dot{x} = \dot{x}_1 \qquad t = t_1$$

Substituting these conditions into Eqs. (18-20) and (18-21) gives

$$x_1 = \frac{F}{k} (1 - \cos \omega_n t_1) \qquad \dot{x}_1 = \frac{F}{k} \omega_n \sin \omega_n t_1 \tag{18-23}$$

The starting conditions for the second era are

$$x = x_1 \qquad \dot{x} = \dot{x}_1 \qquad t' = 0$$

Substituting these into Eq. (c) and solving for A and B produce

$$A = x_1 \qquad B = \frac{\dot{x}_1}{\omega_n}$$

and so Eq. (c) becomes

$$x = x_1 \cos \omega_n t' + \frac{\dot{x}_1}{\omega_n} \sin \omega_n t' \tag{d}$$

When this is transformed into a single trigonometric function and a phase angle, the result is

$$x = \sqrt{x_1^2 + \left(\frac{\dot{x}_1}{\omega_n}\right)^2} \cos (\omega_n t' - \phi) \tag{18-24}$$

$$\phi = \tan^{-1} \frac{\dot{x}_1}{\omega_n x_1}$$

The amplitude of this vibration is, from Eqs. (18-23),

$$X_0 = \sqrt{x_1^2 + \left(\frac{\dot{x}_1}{\omega_n}\right)^2} = \sqrt{\frac{F^2}{k^2}[(1 - \cos \omega_n t_1)^2 + \sin^2 \omega_n t_1]}$$

$$= 2\frac{F}{k}\left|\sin\frac{\omega_n t_1}{2}\right| \qquad (18\text{-}25)$$

The resulting motion has an amplitude of $2F/k$ whenever $\omega_n t_1 = \pi, 3\pi, 5\pi$, etc. On the other hand there is no motion at all, that is, the mass stops moving completely, whenever $\omega_n t_1 = 0, 2\pi, 4\pi$, etc. These are the limiting values of the amplitude, and so, regardless of when the force F is removed, the amplitude will never be greater than $2F/k$.

The motion resulting from removal of the force is plotted for several conditions in Fig. 18-7. This figure shows the forcing function F/k plotted along the $\omega_n t$ axis. During the period of application of the force the displacement is generated by a rotating vector of length F/k rotating about point P. As demonstrated previously the vector starts its motion from the position PO when the force is first applied. If the vector is permitted to generate $2\frac{1}{2}$ cycles of displacement, then, at the end of this era, it will occupy the position PB and the corresponding point on the displacement diagram will be B'. At this instant the total angle traversed is $\omega_n t_1 = \omega_n t_B = 5\pi$ rad. If the force is removed at this instant, Eq. (18-25) shows that the resulting motion will have an amplitude of $2F/k$. Thus the displacement for $t > t_B$ is generated by a rotating vector (not shown) of length $2F/k$ having its center of rotation at point O. Substituting x_1 and $\dot{x}_1$ from Eqs. (18-23) into (18-24) and solving for the phase angle give

$$\phi = \tan^{-1}\frac{\sin \omega_n t_1}{1 - \cos \omega_n t_1} = \frac{\pi}{2} - \frac{\omega_n t_1}{2} \qquad (18\text{-}26)$$

Thus, for $\omega_n t_B = 5\pi$, $\phi = 0$, and the motion continues in the same phase.

The result of making the forcing function zero at an earlier point in the motion, say at A, is shown in dashed lines. The first era occupies a period $t = t_A$ which, in this example, we have made such that $\omega_n t_1 = \omega_n t_A = 2\pi + \pi/3$ rad. Thus, substituting $\pi/3$ for $\omega_n t_1$ into Eq. (18-25) gives an amplitude $X_0 = F/k$. The resulting motion can be represented by a rotating vector of length F/k beginning from the position OA and rotating about O as a center. This means that for this particular set of conditions, the amplitude does not change at all. The phase relationship is obtained from Eq. (18-26) and is

$$\phi = \frac{\pi}{2} - \frac{\pi/3}{2} = \frac{\pi}{3}$$

Note that if the force is removed at some point O' on the displacement diagram, the resulting motion is zero.

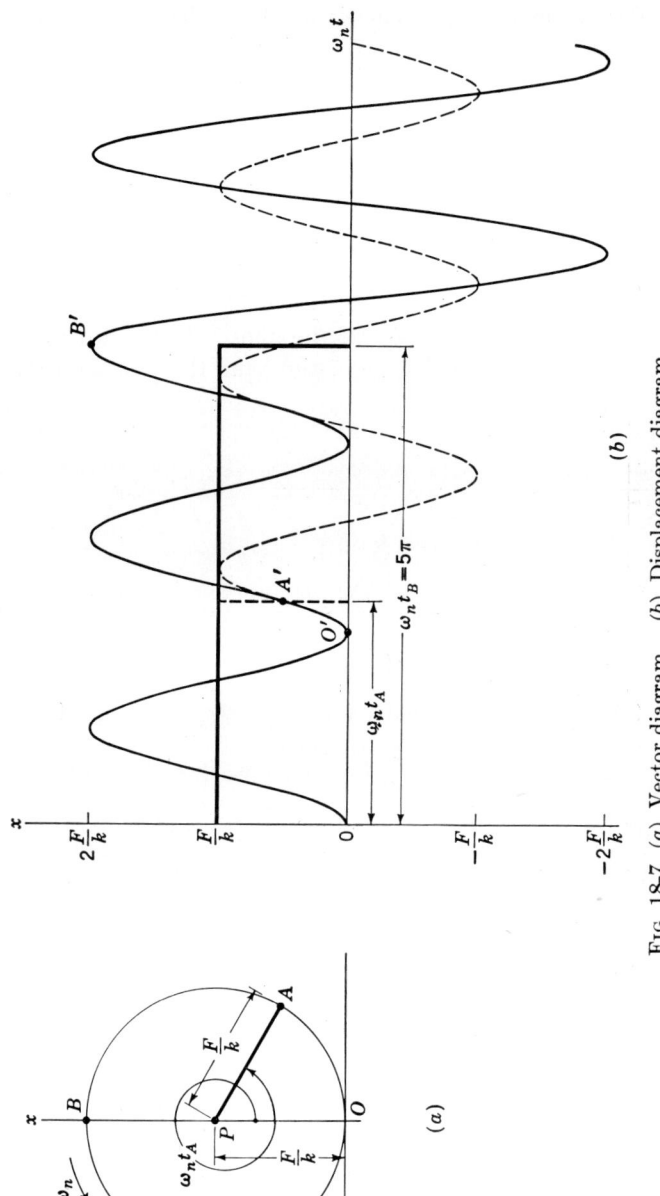

Fig. 18-7. (a) Vector diagram. (b) Displacement diagram.

450   DYNAMIC ANALYSIS OF MACHINES

18-4. Phase-plane Representation. The phase-plane method is a graphical means of solving transient vibration problems which is quite easy to understand and to use. The method eliminates the necessity for solving differential equations, some of which are very difficult, and even enables solutions to be obtained when the functions involved are not expressed in algebraic form. The engineer must concern himself as much with transient disturbances and motions of machine parts as with steady-state motions. The phase-plane method presents the physics of the problem with so much clarity that it will serve as an excellent vehicle for the study of mechanical transients.[1]

Before introducing the details of the phase-plane method it will be of value to show how the displacement-time and the velocity-time relations

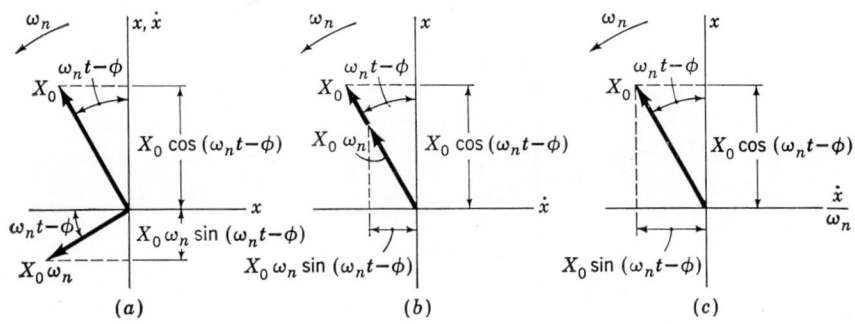

Fig. 18-8

can be generated by a single rotating vector. We have already observed that a free undamped vibrating system has an equation of motion which can be expressed in the form

$$x = X_0 \cos(\omega_n t - \phi) \tag{18-27}$$

and that its velocity is

$$\dot{x} = -X_0 \omega_n \sin(\omega_n t - \phi) \tag{18-28}$$

The displacement, as given by Eq. (18-27), can be represented by the projection on a vertical axis of a vector of length X_0 rotating at ω_n rad/sec in the counterclockwise direction (Fig. 18-8a). The angle $\omega_n t - \phi$, in this example, is measured from the vertical axis. Similarly, the velocity can be represented on the same vertical axis as the projection of another rotating vector of length $X_0 \omega_n$ rotating at the same angular velocity but leading X_0 by a phase angle of 90°, as shown in Fig. 18-8a. Therefore the angular location of the velocity vector is measured from the

---

[1] For additional information see Jacobsen and Ayre, *op. cit.* This book has an excellent bibliography on the phase-plane method and also on many other aspects of vibrations.

horizontal axis. If we take the coordinate system containing the velocity vector and rotate it backward (clockwise) through an angle of 90°, then the velocity and displacement vectors will be coincident and their angular locations can be measured from the same vertical axis. This step has been taken in Fig. 18-8b, where it is seen that the displacement is still measured along the same vertical axis. But, having rotated the coordinate system in which velocity is measured, we see that the velocity is obtained by projecting the $X_0\omega_n$ vector to the horizontal axis. Thus we can now measure velocities on a separate axis from displacements. Note, too, that the direction of positive velocities is to the right.

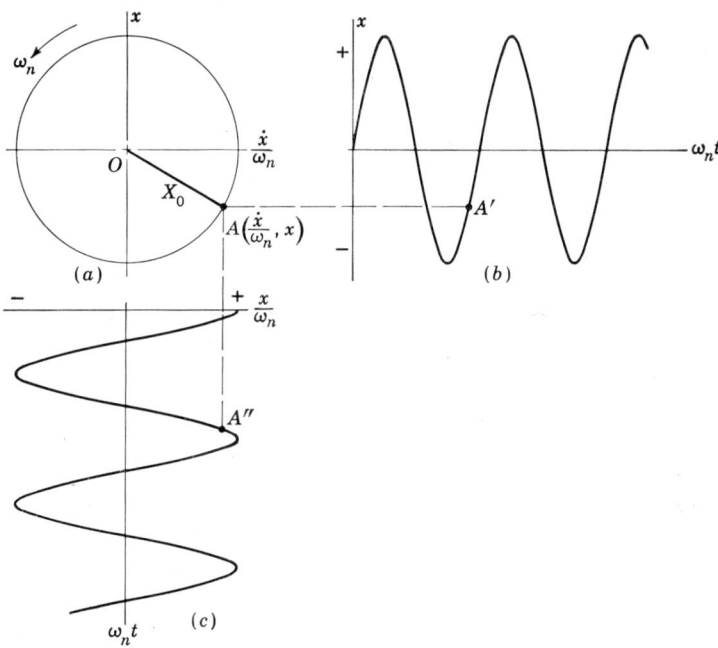

FIG. 18-9. (a) Vector diagram. (b) Displacement diagram. (c) Velocity diagram.

Our final step is taken by noting that the velocity vector differs in length from the displacement vector by the constant ω_n. Thus, instead of plotting velocities, if we plot the quantity $\dot{x}/\omega_n$, then we shall have a quantity which is proportional to velocity. This step has been taken in Fig. 18-8c, where the horizontal axis is designated as the $\dot{x}/\omega_n$ axis. With this change the projection of the rotating vector X_0 on the x axis gives the displacement and its projection on the $\dot{x}/\omega_n$ axis gives a quantity directly proportional to the velocity.

The displacement-time and velocity-time graphs of a free undamped vibration have been plotted in Fig. 18-9 to show how the vector diagram

452   DYNAMIC ANALYSIS OF MACHINES

is related to them. A point A on the vector diagram corresponds with A' on the displacement diagram and with A'' on the velocity diagram. Note that quantities obtained from the velocity plot must be multiplied by ω_n in order to obtain the actual velocity.

It is possible to arrive at these same conclusions in a different manner. If Eq. (18-27) is squared, we obtain

$$x^2 = X_0^2 \cos^2 (\omega_n t - \phi) \qquad (a)$$

Next, dividing Eq. (18-28) by ω_n and squaring it give

$$\left(\frac{\dot{x}}{\omega_n}\right)^2 = X_0^2 \sin^2 (\omega_n t - \phi) \qquad (b)$$

Then adding Eqs. (a) and (b) produces

$$x^2 + \left(\frac{\dot{x}}{\omega_n}\right)^2 = X_0^2 \qquad (18\text{-}29)$$

This is the equation of a circle having the amplitude X_0 as its radius and with its center at the origin of a coordinate system having the axes x and $\dot{x}/\omega_n$. Thus Eq. (18-29) describes the circle of Fig. 18-9a, where OA is the amplitude X_0 of the motion. The coordinate system x, $\dot{x}/\omega_n$ defines the positions of points in a region which is called the *phase plane*.[1]

As a first example of the use of the method we shall consider the vibrating system of Fig. 18-11. This is a spring-mass system having a spring attached to a frame which can be positioned at will. To begin the motion we might consider the system as initially at rest and then, at $t = 0$, suddenly move the frame a distance x_1 to the right. The effect of this sudden motion is to compress the spring an amount x_1 and to shift the equilibrium position of the mass a distance x_1 to the right.

---

[1] The phase plane is widely used in the solution of nonlinear differential equations. Such equations occur frequently in the study of vibrations and feedback control systems. Both of these subjects are included in this book, but the discussions are limited to linear systems. When the phase plane is used to solve nonlinear differential equations, it is customary to arrange the axes as shown in Fig. 18-10, with ω_n considered as positive in the clockwise direction instead of the counterclockwise direction as we are using it here. This arrangement of the axes seems more logical, since $\dot{x}/\omega_n$ is a function of x. However, for the analysis of transient disturbances to mechanical systems and for analyzing cam mechanisms, the arrangement of the axes as in Fig. 18-9 is more appropriate, and this is the one we shall employ throughout this book. The reason for this is that we shall employ the phase-plane method, not as an end in itself, as is done when it is used for the solution of nonlinear problems, but as a tool to obtain the transient response.

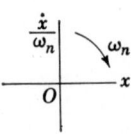

FIG. 18-10

VIBRATION ANALYSIS 453

For $t < 0$ the mass is in equilibrium at the position shown in the figure. For $t > 0$ the equilibrium position of the mass is a distance x_1 to the right of the position shown in the figure. If the motion of the frame occurs instantaneously, the initial conditions are

$$x = -x_1 \quad \dot{x} = 0 \quad t = 0$$

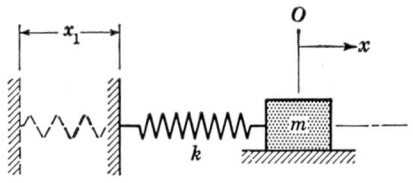

FIG. 18-11

where the motion is now measured from the shifted equilibrium position. If these initial conditions are used to evaluate the constants of integration of Eq. (18-15), there results

$$X_0 = -x_1 \quad \phi = 0$$

so that
$$x = -x_1 \cos \omega_n t \tag{c}$$

Next, note that the application of a constant force to an undamped spring-mass system gave as the equation of motion [Eq. (18-20)]

$$x = \frac{F}{k} - \frac{F}{k} \cos \omega_n t \tag{d}$$

and rearranging,

$$x - \frac{F}{k} = -\frac{F}{k} \cos \omega_n t \tag{e}$$

Letting $F/k = x_1$, Eq. (e) becomes

$$x - x_1 = -x_1 \cos \omega_n t \tag{f}$$

But if the origin of x is shifted a distance x_1, then Eq. (f) is the same as Eq. (c). Thus shifting of the frame of a vibrating system through a distance x_1 is equivalent to adding a constant force $F = kx_1$ acting upon the mass. This problem is illustrated in Fig. 18-12, where the old origin is O and the new origin O_1. These are separated, then, by the distance

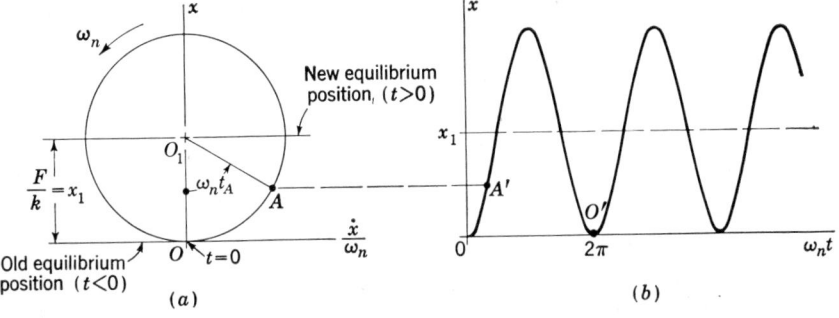

FIG. 18-12. (a) Phase-plane diagram. (b) Displacement-time diagram.

x_1. At the instant $t = 0$, the origin O is shifted to O_1, initiating the motion. A vector O_1A, equal in magnitude to the distance x_1, begins its rotation from the position O_1O. Rotating at ω_n rad/sec the projection of this vector on the x axis describes the displacement of the mass. This vector then continues to rotate until something else happens to the system. Arriving at point A it has traversed an angle $\omega_n t_A$, and the corresponding point on the displacement-time diagram is A'.

The phase-plane method permits us to move the origin in any manner we choose and at any instant in time that we may choose. In the example above the origin was shifted a distance x_1, and we have seen that this is equivalent to the sudden application of a force $F = kx_1$ applied to the mass in the positive direction. Suppose that we permit the vector to make one complete revolution of 360°. It then generates one complete cycle of displacement and returns to point O. If, at this instant, we shift the frame back to its original position, we might reasonably ask: What are the displacement and velocity at the instant of making this shift? The phase-plane diagram shows that both the displacement and velocity are zero at the end of a cycle. These, the displacement and velocity, are exactly the quantities we require in order to determine the motion of the system in the next era. Since these are both zero, there is no motion after returning the frame to its original position. It will be recalled that these are exactly the results which we obtained in analyzing square-wave forcing functions when the force endured for an integral number of cycles. Thus the rotating vector generates the displacement diagram through the angle $\omega_n t = 2\pi$ in this case, and at this point we return the frame to its original position. The initial conditions are now $x = 0$, $\dot{x} = 0$, and consequently, the mass stops its motion completely.

Let us now create a vibration by shifting the frame from $x = 0$ to $x = x_1$, waiting a period of time Δt, then shifting the frame back to $x = 0$. This constitutes a square-wave forcing function, or a step disturbance to the system, and the phase-plane and displacement diagrams for such a motion are illustrated in Fig. 18-13. The motion can be described in three eras. During the first, from t_0 to t_1, everything is at rest. At $t = t_1$ the frame suddenly moves to the right a distance $x = x_1$, compressing the spring. The duration of the second era is from t_1 to t_2 (Δt), and at time t_2 the frame suddenly returns to its original position. The third era represents time when $t \geq t_2$. The position of the frame during these three eras is shown in the displacement diagram. The displacement diagram also describes the motion of the mass, and the phase-plane diagram explains why it moves as it does. Between t_0 and t_1 the mass is at rest and nothing happens. At t_1 the frame shifts from O to O_1 on the phase-plane diagram, which is the distance x_1. If we designate the original frame position as the origin of the x, $\dot{x}/\omega_n$ system, then the conditions at $t = t_1$ are $x = +x_1$, $\dot{x} = 0$. However, the motion of the mass about O is equal to its motion about O_1 plus the distance from O to O_1. Therefore

$$x = -x_1 \cos \omega_n t' + x_1 = x_1(1 - \cos \omega_n t') \tag{g}$$

Also
$$\dot{x} = x_1 \omega_n \sin \omega_n t' \tag{h}$$

where t' is understood to begin at t_1. As shown in Fig. 18-13b, the mass vibrates about the position $x = x_1$ during this era. The vibration begins at point A on the phase-plane diagram and continues as a free vibration for the time Δt. During this period the line $O_1 A$ rotates counterclockwise with angular velocity ω_n rad/sec until at time t_2 it occupies the position $O_1 B$. At this instant the third era begins when the frame suddenly

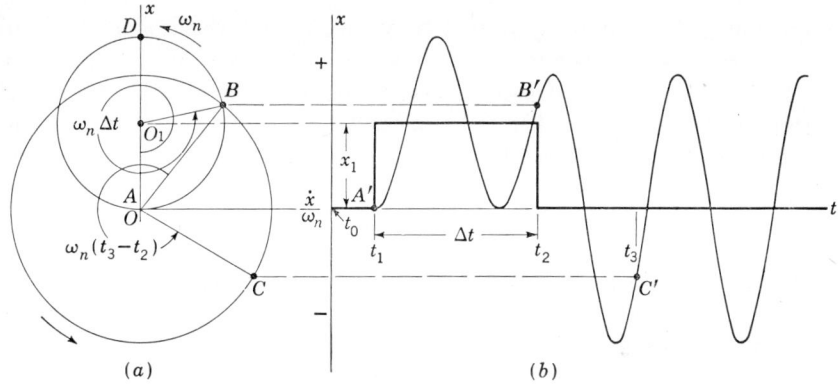

FIG. 18-13. A step disturbance.

returns through the distance x_1 to its original position. Thus the starting conditions for the third era are the same as the ending conditions for the second and are

$$x_2 = x_1(1 - \cos \omega_n \Delta t) \qquad \dot{x}_2 = x_1 \omega_n \sin \omega_n \Delta t \tag{i}$$

Substituting these conditions into Eq. (18-11) gives the equation of motion for the third era as

$$x = x_1(1 - \cos \omega_n \Delta t) \cos \omega_n t'' + x_1 \sin \omega_n \Delta t \sin \omega_n t'' \tag{j}$$

where t'' is the time measured from the start of the third era. Equation (j) can be transformed into an equivalent expression containing only a single trigonometric term and a phase angle as in Eq. (18-15), but we shall not do so here. The third era, we have seen, begins at point B on the phase-plane diagram (B' on the displacement diagram); the motion of point B as it moves about a circle with center at O in the counterclockwise direction describes the motion of the mass. Thus at instant t_3 the line OB will have rotated through the angle $\omega_n(t_3 - t_2)$ and be located at C. The corresponding point on the displacement diagram is C'.

The extension of the phase-plane method to any number of steps, taken in either or both the positive or negative x direction, should now be apparent. In each case the starting conditions for the next era are taken equal to the ending conditions for the previous era. The equations of motion should be written for each era with time counted from the start of that era. The reader can now understand that if a third era of Fig. 18-13 begins at point D on the phase-plane diagram, then the resulting motion will have an amplitude twice as large as that in the second era.

18-5. Transient Disturbances. Any action which destroys the static equilibrium of a vibrating system may be called a *disturbance* to that system. A *transient disturbance* is any action which endures for only a relatively short period of time. The analyses in the several preceding sections have dealt with transient disturbances having a stepwise relationship to time. Because all machine parts have elasticity and inertia, forces do not come into existence instantaneously. Consequently, we can usually expect to encounter forcing functions which vary smoothly with time. Although the step forcing function is not true to nature, it is our purpose in this section to demonstrate how the step function is used with the phase-plane method to obtain very good approximations of the vibration of systems excited by "natural" disturbances.

The procedure is to plot the disturbance as a function of time, to divide this into steps, and then to use the steps successively to make a phase-plane plot. The resulting displacement and velocity diagrams can then be obtained by graphically projecting points from the phase-plane diagram as previously explained. It turns out that very accurate results can frequently be obtained using only a small number of steps. Of course, as in any graphical solution, the best results are obtained when a large number of steps are employed and when the work is plotted to a large scale. It is difficult to set up general rules for selecting the size of the steps to be used. For slow vibrating systems and for relatively smooth forcing functions the step width can be quite large, but even a slow system will require narrow steps if the forcing function has numerous sharp peaks and valleys, that is, if it has a great deal of frequency content. For smooth forcing functions and slow vibrating systems a step width such that the rotating vector sweeps out an angle of 180° is probably about the largest that one should use. It is a good idea to check the step width during the construction of the phase-plane diagram. Too great a width will cause a large change in the slope of two curves at one of the points of discontinuity. If this occurs, then the step can immediately be narrowed and the procedure continued.

Figure 18-14 shows how to find the height of the steps. The first step for the forcing function of this figure has been given a width of Δt_1 and a height of h_1. This height is obtained by constructing a horizontal

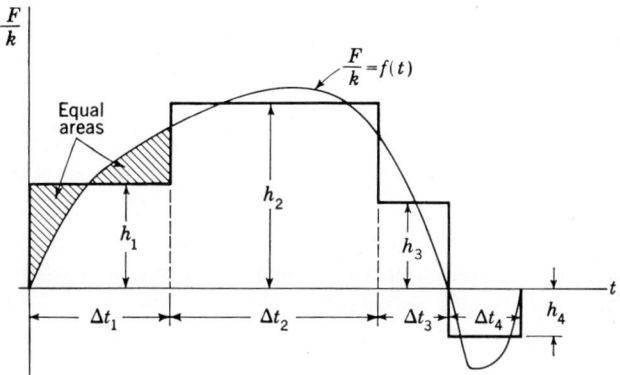

FIG. 18-14. Finding the height of the steps.

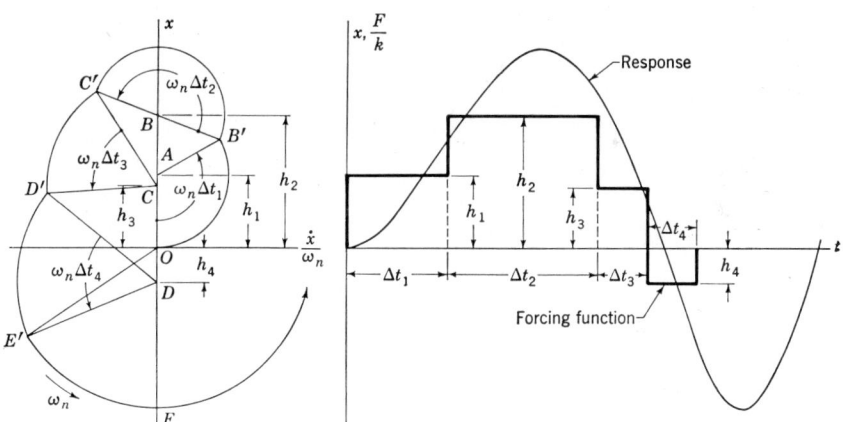

FIG. 18-15. Construction of the phase-plane and displacement diagrams for a four-step forcing function.

line across the force curve such that the areas of the two shaded triangles are equal. For a step width of Δt_1 sec the rotating vector sweeps out an angle $\omega_n \Delta t_1$ rad. Thus the angular rotation of the vector is obtained simply by multiplying the step width in seconds by the natural circular frequency of the system.

Figure 18-15 shows a four-step forcing function which we may assume has been deduced from a "natural" function. The width and height of the steps have been plotted to scale and are Δt_1 and h_1 for the first step, Δt_2 and h_2 for the second step, etc. Notice that the fourth step is a negative one. The motion begins at $t = 0$ when a vector of length h_1 starts rotating about A as a center from the initial position AO. This vector rotates through an angle $\omega_n \Delta t_1$ and generates the portion of the response curve contained in the period Δt_1. At the end of this period of time the second step begins with the center of rotation of the vector shifting from A to B. The length of the rotating vector for the duration of the second step is the distance BB'. This vector then rotates through the angle $\omega_n \Delta t_2$ about a center at B until it arrives at the position BC'. At this instant the third step begins. The center of rotation shifts to C, and the vector CC' rotates through the angle $\omega_n \Delta t_3$. At the end of the fourth step the vector has arrived at E'. The center of rotation now shifts to the origin O, and the motion continues as a free vibration of amplitude OE' until something else (not shown) happens to the system.

EXAMPLE 18-1. Measurements on a mechanical vibrating system show that the mass has a weight of 16.90 lb and that the springs can be combined to give an equivalent spring index of 30 lb/in. This system is observed to vibrate quite freely, and so damping can be neglected. A transient force resembling the first half cycle of a sine wave operates on the system. If the maximum value of the force is 10 lb, determine the response when the force is applied for 0.120 sec.

Solution. The circular frequency is

$$\omega_n = \sqrt{\frac{k}{m}} = \sqrt{\frac{(30)(386)}{16.90}} = 26.2 \text{ rad/sec}$$

The period and frequency are

$$\tau = \frac{2\pi}{\omega_n} = \frac{2\pi}{26.2} = 0.240 \text{ sec} \qquad f = \frac{1}{\tau} = \frac{1}{0.240} = 4.16 \text{ cps}$$

We shall first determine the response by replacing the entire force function by a single step disturbance. The height of the step should be the time average of the force function for the duration of the step. The average ordinate of a half cycle of the sine wave is

$$h_1 = \frac{1}{\pi} \int_0^\pi \frac{F_{\max}}{k} \sin bt \, d(bt) = 0.636 \frac{F_{\max}}{k} \tag{1}$$

or
$$h_1 = (0.636)(10/30) = 0.212 \text{ in.}$$

This is all we need to solve the problem. The graphical solution is shown in Fig. 18-16, together with the force and the step, all plotted to the same time scale. As shown, the amplitude of a vibration resulting from a single step is $X_0 = 0.424$ in.

In order to check the accuracy of a single step we shall solve the example again

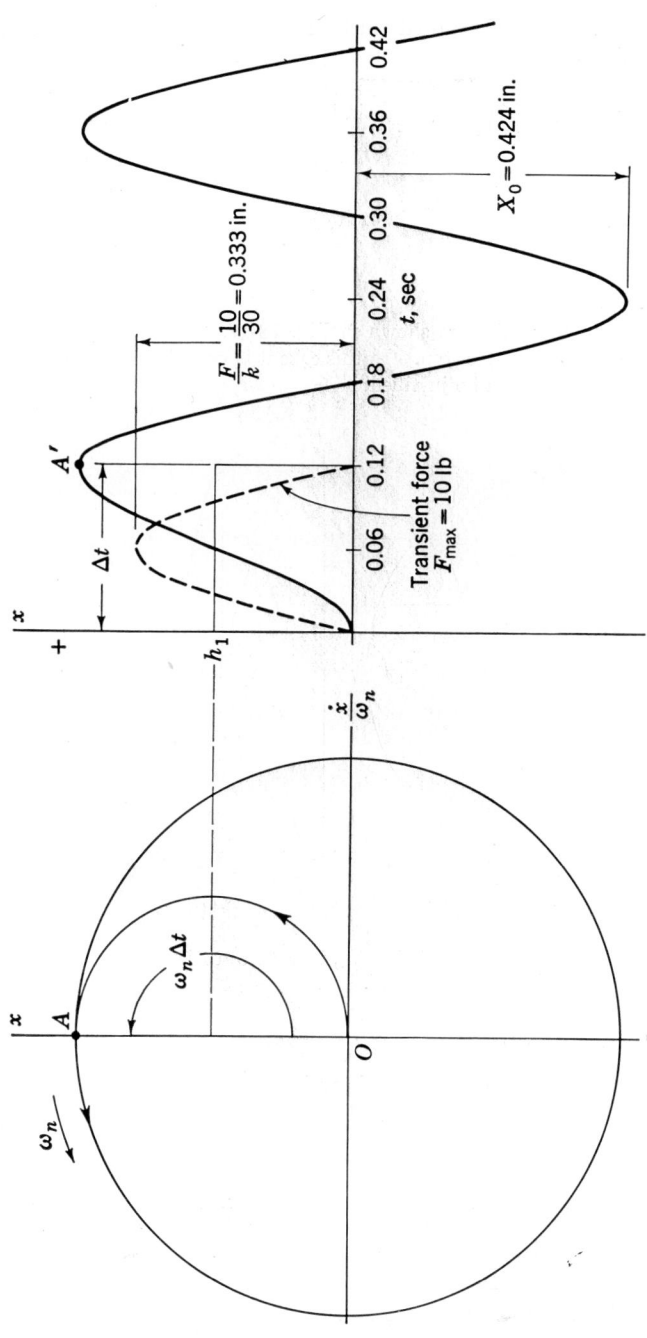

Fig. 18-16. Response determined with only a single step.

using three steps. These steps need not be equal, but it will be convenient in this example to make them so. For the three equal steps Eq. (1) is written

$$h_1 = h_3 = \frac{1}{\pi/3} \int_0^{\pi/3} \frac{F_{\max}}{k} \sin bt \, d(bt) = \frac{3}{2\pi} \frac{F_{\max}}{k}$$
$$= \frac{3}{2\pi} \frac{10}{30} = 0.159 \text{ in.}$$

$$h_2 = \frac{1}{\pi/3} \int_{\pi/3}^{2\pi/3} \frac{F_{\max}}{k} \sin bt \, d(bt) = \frac{3}{\pi} \frac{F_{\max}}{k}$$
$$= \frac{3}{\pi} \frac{10}{30} = 0.318 \text{ in.}$$

The construction and results are shown in Fig. 18-17. Note that the amplitude is slightly larger than for a single step. In this case it is very doubtful if the extra labor of a five-step solution would be justified.

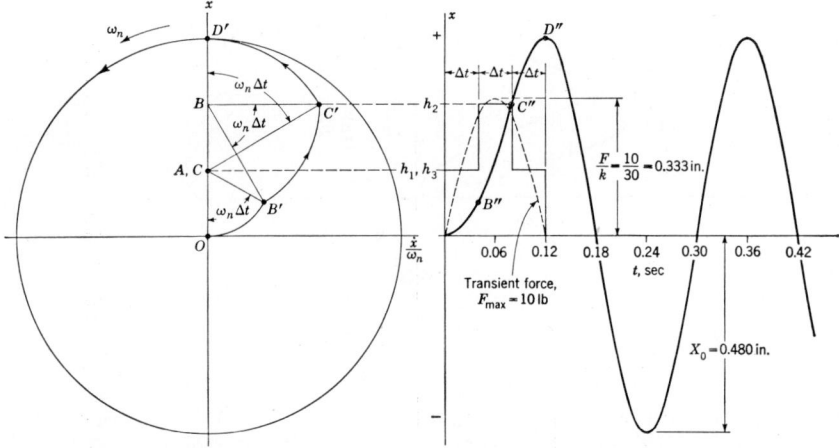

FIG. 18-17. Response determined in three steps. The first step has a height of h_1 and is taken from O to B' with a radius AO; the second has a height h_2, is from B' to C' with a radius BB'; the third has a height h_3, is from C' to D' with a radius CC'. The vibration is a free motion after point D'.

18-6. Free Vibration with Damping For a damped one-mass system with no external forces Eq. (18-3) is written

$$\ddot{x} + \frac{c}{m} \dot{x} + \frac{k}{m} x = 0 \tag{18-30}$$

Inspection of this equation shows that x and its derivatives must be alike in order that the relation can be satisfied. The exponential function satisfies this requirement, and so it is not unreasonable to guess that the solution might be in the form

$$x = A e^{st} \tag{a}$$

where A and s are constants still to be determined. The first and second derivatives of Eq. (a) are

$$\dot{x} = Ase^{st} \qquad \ddot{x} = As^2e^{st}$$

Substituting these in Eq. (18-30) and canceling give

$$s^2 + \frac{c}{m}s + \frac{k}{m} = 0 \tag{b}$$

Thus Eq. (a) is a solution provided the quantity s is selected to satisfy Eq. (b). The roots of Eq. (b) are

$$s = -\frac{c}{2m} \pm \sqrt{\left(\frac{c}{2m}\right)^2 - \frac{k}{m}} \tag{c}$$

and so Eq. (a) can be written

$$x = Ae^{s_1 t} + Be^{s_2 t} \tag{d}$$

where s_1 and s_2 are the two roots of Eq. (c) and A and B are the constants of integration.

The value of the damping which makes the radical of Eq. (c) zero has special significance, and we shall call it the *critical damping* and designate it by the symbol c_c. Substituting c_c for c under the radical gives

$$\frac{c_c}{2m} = \sqrt{\frac{k}{m}} = \omega_n \qquad \text{or} \qquad c_c = 2m\omega_n \tag{18-31}$$

It is also convenient to define a *damping ratio* ζ as the ratio of the actual to the critical damping. Thus

$$\zeta = \frac{c}{c_c} = \frac{c}{2m\omega_n} \tag{18-32}$$

After some algebraic manipulation Eq. (c) can be written

$$s = (-\zeta \pm \sqrt{\zeta^2 - 1})\omega_n \tag{e}$$

In the case where the actual damping is larger than the critical, $\zeta > 1$ and the radical is real. This is analogous to movement of the mass in a barrel of thick molasses, and there is no vibration. Although this case does have value in the analysis and design of machines, we shall not pursue it further in this book because of space limitations. It is not difficult to show, however, that if the mass is displaced and released, in a system with more than critical damping, it returns slowly to its position of equilibrium without overshooting.

When the damping is less than critical, $\zeta < 1$ and the radical of Eq. (e) is imaginary. It is then written

$$s = (-\zeta + j\sqrt{1 - \zeta^2})\omega_n \tag{f}$$

Substituting these roots into Eq. (d) gives

$$x = e^{-\zeta\omega_n t}(Ae^{j\sqrt{1-\zeta^2}\omega_n t} + Be^{-j\sqrt{1-\zeta^2}\omega_n t}) \tag{g}$$

which can be transformed to

$$x = e^{-\zeta\omega_n t}[(A + B)\cos\sqrt{1 - \zeta^2}\,\omega_n t + j(A + B)\sin\sqrt{1 - \zeta^2}\,\omega_n t] \tag{h}$$

It is expedient, as before, to transform Eq. (h) into a form having only a single trigonometric function. Making this transformation gives

$$x = X_0 e^{-\zeta\omega_n t}\cos(\sqrt{1 - \zeta^2}\,\omega_n t - \phi) \tag{18-33}$$

where X_0 and ϕ are the new constants of integration. It is apparent that Eq. (18-33) reduces to (18-15) if the damping is made zero. The

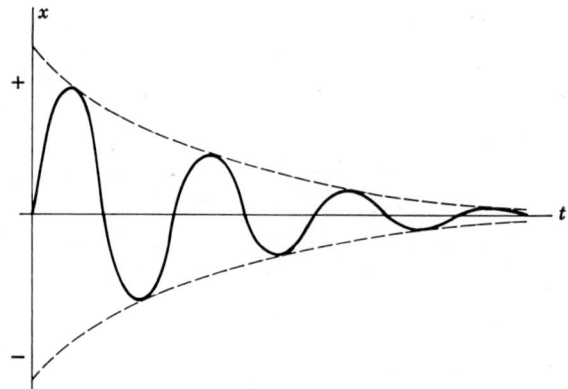

FIG. 18-18

constants of integration can also be determined in exactly the same manner.

Equation (18-33) is the product of a trigonometric function and a decreasing exponential function. It is therefore an oscillatory motion with an exponentially decreasing amplitude, as shown in Fig. 18-18. The frequency is less than that of an undamped system as given by the factor $\sqrt{1 - \zeta^2}$.

The rate of decay of a damped free vibration can be determined from the amplitudes of any two consecutive waves. These occur when the

VIBRATION ANALYSIS

cosine is approximately unity and are therefore

$$x_n = X_0 e^{-\zeta \omega_n t_n} \qquad x_{n+1} = X_0 e^{-\zeta \omega_n (t_n + \tau)}$$

where τ is the period of the vibration. The *logarithmic decrement* δ is defined as the natural logarithm of the ratio of these two amplitudes and is

$$\delta = \ln \frac{x_n}{x_n + 1} = \frac{X_0 e^{-\zeta \omega_n t_n}}{X_0 e^{-\zeta \omega_n (t_n + \tau)}} = \zeta \omega_n \tau \qquad (18\text{-}34)$$

The period of the vibration is

$$\tau = \frac{2\pi}{\omega_n \sqrt{1 - \zeta^2}} \qquad (18\text{-}35)$$

and so the decrement can also be written

$$\delta = \frac{2\pi \zeta}{\sqrt{1 - \zeta^2}} \qquad (18\text{-}36)$$

EXAMPLE 18-2. Let the vibrating system of Example 18-1 have a dashpot attached which exerts a force of 0.25 lb on the mass when the mass has a velocity of 1 in./sec. Find the critical damping constant, the logarithmic decrement, and the ratio of two consecutive maxima.

Solution. The data from Example 18-1 are repeated here for convenience.

$$W = 16.90 \text{ lb} \qquad k = 30 \text{ lb/in.} \qquad \omega_n = 26.2 \text{ rad/sec}$$

The critical damping constant is

$$c_c = 2m\omega_n = (2)\left(\frac{16.90}{386}\right)(26.2) = 2.29 \text{ lb-sec/in.} \qquad Ans.$$

The actual damping constant is $c = 0.25$ lb-sec-in. Therefore the damping ratio is

$$\zeta = \frac{c}{c_c} = \frac{0.25}{2.29} = 0.109$$

The logarithmic decrement is obtained from Eq. (18-36):

$$\delta = \frac{2\pi \zeta}{\sqrt{1 - \zeta^2}} = \frac{2\pi (0.109)}{\sqrt{1 - (0.109)^2}} = 0.690 \qquad Ans.$$

The ratio of two consecutive maxima is

$$\frac{x_n + 1}{x_n} = e^{-\delta} = e^{-0.690} = 0.501 \qquad Ans.$$

Therefore, for this amount of damping each amplitude is approximately 50 per cent of the previous one.

18-7. Phase-plane Representation of Damped Vibration. We have seen that when undamped mechanical systems are subjected to transient forces, the resulting motion (mathematically, at least) endures forever without decreasing in amplitude. We know, however, that energy losses always exist, though sometimes only in minute amounts, and that these losses will eventually cause the vibration to stop. In the case of vibrating systems known to be acted upon by transient forces, it is often desirable to introduce additional damping as one means of decreasing the number of cycles of vibration. For this reason the phase-plane solution of a damped vibration is particularly important.

We have seen that the phase-plane diagram of an undamped free vibration is generated by a vector of constant length rotating at a constant angular velocity. In the case of a damped free vibration the phase-plane diagram is generated by a rotating vector whose length is exponentially decreasing. Thus the trajectory for such a motion is a spiral instead of a circle.

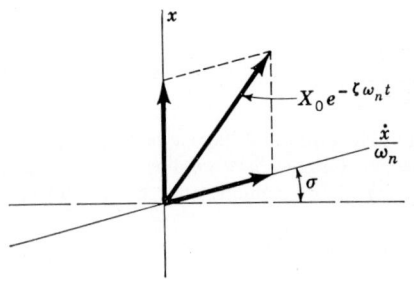

FIG. 18-19. The angle σ is the damping phase angle and is applied only to the velocity term.

Jacobsen and Ayre[1] point out that a simple spiral will give incorrect values for the velocity if plotted on the same x, $\dot{x}/\omega_n$ axes as used for an undamped free vibration. The reason for this is that the phase relationship between the velocity and displacement is not 90° as it is for an undamped system. Instead, for a damped system, the phase angle depends upon the amount of damping present.

In order to make a simple phase-plane diagram possible, Jacobsen and Ayre suggest tilting or rotating the $\dot{x}/\omega_n$ axis through an angle σ, as shown in Fig. 18-19. The angle σ is called the *damping phase angle*. The ensuing analysis follows that of Jacobsen and Ayre and shows how σ comes about.

Taking the time derivative of Eq. (18-33) gives

$$\dot{x} = -X_0 \zeta \omega_n e^{-\zeta \omega_n t} \cos \left(\sqrt{1-\zeta^2}\, \omega_n t - \phi \right) \\ - X_0 \sqrt{1-\zeta^2}\, \omega_n e^{-\zeta \omega_n t} \sin \left(\sqrt{1-\zeta^2}\, \omega_n t - \phi \right)$$

or

$$\frac{\dot{x}}{\omega_n} = -X_0 e^{-\zeta \omega_n t} [\zeta \cos \left(\sqrt{1-\zeta^2}\, \omega_n t - \phi \right) \\ + \sqrt{1-\zeta^2} \sin \left(\sqrt{1-\zeta^2}\, \omega_n t - \phi \right)] \quad (a)$$

Designating $\cos \sigma = \sqrt{1-\zeta^2}$ and $\sin \sigma = \zeta$ and noting the trigonometric identity

$$\sin (\alpha + \beta) = \sin \alpha \cos \beta + \cos \alpha \sin \beta$$

[1] *Ibid.*, p. 204.

we see that Eq. (a) can be written

$$\frac{\dot{x}}{\omega_n} = -X_0 e^{-\zeta\omega_n t} \sin(\sqrt{1-\zeta^2}\,\omega_n t - \phi + \sigma) \tag{18-37}$$

If the damping is zero, note that $\sqrt{1-\zeta^2} = 1$, $e^{-\zeta\omega_n t} = 1$, $\sigma = 0$, and Eqs. (18-33) and (18-37) reduce to the same set of equations which we employed to develop the phase-plane analysis of undamped motion. As shown in Fig. 18-19, the $\dot{x}/\omega_n$ axis should be rotated counterclockwise through the angle σ in plotting the phase-plane diagram. The velocity diagram is then obtained by projecting perpendicular to the $\dot{x}/\omega_n$ axis.

In analyzing a damped system which is acted upon by transient forces by the phase-plane method it is necessary to construct transparent spiral templates as shown in Fig. 18-20. These are made from a sheet of clear plastic which is placed over a drawing of the spiral so that the spiral and its center can be scribed on the plastic using the point of a pair of dividers. The scribed line can then be trimmed with a pair of manicurist's scissors or a sharp razor blade. Since only one template is needed for each damping ratio, they can be stored for use in future problems. Do not forget to label each template with the value of the damping ratio for which it was constructed.

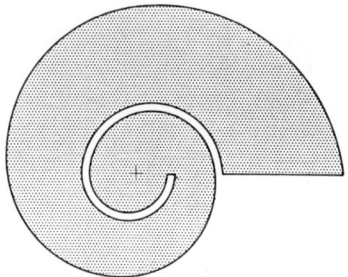

Fig. 18-20. A spiral template for $\zeta = 0.15$.

The spirals can be plotted directly from a table calculated using Eq. (18-33), but this is very tedious. A much more rapid method, which is accurate enough for graphical purposes, is to approximate each quadrant of the spiral with a circle arc. To do this it is necessary only to calculate the change in the length of the rotating vector in a quarter of a turn. From Eq. (18-33) the length of the vector is $x = X_0 e^{-\zeta\omega_n t}$ and its angular velocity is $\omega_n \sqrt{1-\zeta^2}$. Therefore, for 90° rotation,

$$\omega_n \sqrt{1-\zeta^2}\, t = \frac{\pi}{2} \quad \text{or} \quad \omega_n t = \frac{\pi}{2\sqrt{1-\zeta^2}} \tag{b}$$

Thus the length, after rotation through 90°, is

$$x_{90°} = X_0 e^{-\zeta\pi/2\sqrt{1-\zeta^2}} \tag{c}$$

To demonstrate the construction of the spirals let us employ a damping ratio $\zeta = 0.15$ and begin with a vector 2.50 in. long. Then, after 90° of rotation, the length, from Eq. (c), is

$$x_{90°} = (2.50)e^{-0.15\pi/2\sqrt{1-(0.15)^2}} = 1.97 \text{ in.}$$

The construction is shown in Fig. 18-21 and is explained as follows: Construct the axes 1 and 2 at right angles to each other and lay off 2.50 in. to A and 1.97 in. to B on axes 1 and 2, respectively. Draw two lines through the origin at 45° to the axes. The perpendicular bisector of AB intersects one of these lines at P_1; using P_1 as a center strike an arc from A to B. A line P_1B crosses another 45° line defining the center P_2 of arc BC. This process is continued in the same fashion with the point P_3 defining the center of the arc CD, and so on.

The template is used by placing its center at the same point from which a circle arc would be constructed for undamped motion. It is then turned until the template spiral coincides with the point from which the spiral diagram is to be started.

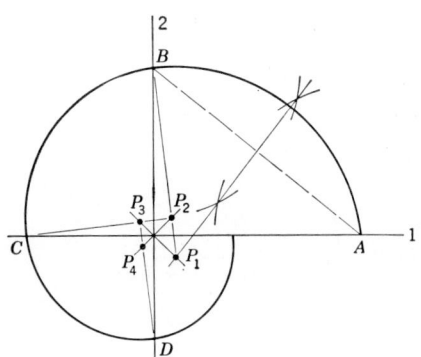

Fig. 18-21

EXAMPLE 18-3. A potential vibrating system has an 80-lb mass mounted upon springs with an equivalent spring constant of 360 lb/in. A dashpot is included in the system and is estimated to produce 15 per cent of critical damping. A transient force, which operates on the system, is assumed to act in three steps as follows: 720 lb for 0.0506 sec, -270 lb for 0.0634 sec, and 180 lb for 0.0380 sec. The negative sign on the second step force simply means that its direction is opposed to the direction of the first and third. Determine the response of the system assuming no motion when the force function begins to act.

Solution. The undamped natural frequency is

$$\omega_n = \sqrt{\frac{k}{m}} = \sqrt{\frac{(360)(386)}{80}} = 41.7 \text{ rad/sec}$$

and so the damped frequency is

$$\omega_n \sqrt{1 - \zeta^2} = 41.7 \sqrt{1 - (0.15)^2} = 41.4 \text{ rad/sec}$$

The period of the motion is $2\pi/\omega_n \sqrt{1 - \zeta^2}$, which gives 0.152 sec. The F/k values are

$$\frac{F_1}{k} = \frac{720}{360} = 2 \text{ in.} \qquad \frac{F_2}{k} = \frac{270}{360} = 0.75 \text{ in.} \qquad \frac{F_3}{k} = \frac{180}{360} = 0.50 \text{ in.}$$

These three steps are plotted in Fig. 18-22 to scale. The angular duration of each step for the phase-plane diagram is obtained by multiplying the damped circular frequency by the time of each step and converting to degrees. This gives

Step 1 $\qquad \omega_n \sqrt{1 - \zeta^2} \, \Delta t_1 \dfrac{180}{\pi} = (41.4)(0.0506) \dfrac{180}{\pi} = 120°$

Step 2 $\qquad \omega_n \sqrt{1 - \zeta^2} \, \Delta t_2 \dfrac{180}{\pi} = (41.4)(0.0634) \dfrac{180}{\pi} = 150°$

Step 3 $\qquad \omega_n \sqrt{1 - \zeta^2} \, \Delta t_3 \dfrac{180}{\pi} = (41.4)(0.0380) \dfrac{180}{\pi} = 90°$

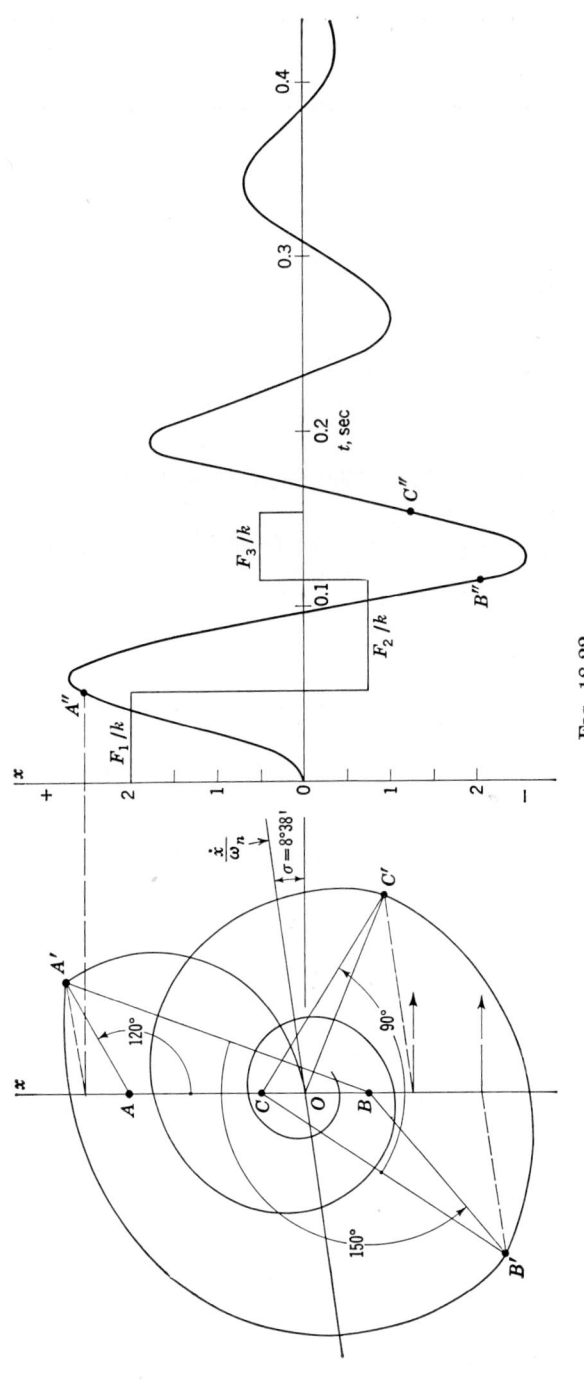

Fig. 18-22

468   DYNAMIC ANALYSIS OF MACHINES

The $\dot{x}/\omega_n$ axis must be turned $\sigma = \sin^{-1}(0.15) = 8°38'$ in the counterclockwise direction, as shown in the phase-plane diagram.

The construction of Fig. 18-22 is explained as follows: Project F_1/k to A on the x axis. A vector AO, with center at A, begins rotating at the instant $t = 0$. This vector rotates through an angle of 120° to AA' while exponentially decreasing its length. At A' the step function changes the origin of the vector to B, and it rotates 150° about this point to B', also exponentially decreasing. At B' the origin again shifts, this time to C, and the vector rotates 90° to C'. Now the origin shifts to O, and the vector continues its rotation with O as the origin until the motion becomes too small to follow. Note that various points on the spiral diagram are projected parallel to the $\dot{x}/\omega_n$ axis to the x axis and then horizontally to the displacement diagram. In a similar manner the velocity diagram, though not shown, would be projected perpendicular to the $\dot{x}/\omega_n$ axis. Points on the spiral would project parallel to the x axis until they intersect the $\dot{x}/\omega_n$ axis.

The displacement diagram shows the resulting response. The maximum amplitude occurs on the first positive half cycle and is 2.72 in.

18-8. Harmonic Forcing. We have seen that the action of any transient force function on a damped system is to create a vibration but that the vibration decays when the force is removed. A machine element which is connected to or is a part of any rotating or moving machinery is often subject to forces which vary periodically with time. Since all metal machine parts have both mass and elasticity, the opportunity for vibration exists. Many machines do operate at fairly constant speeds and constant output, and it is not difficult to see that vibratory forces may exist which have a fairly constant amplitude over a period of time. Of course, these varying forces will change in magnitude when the machine speed or output changes, but there is a rather broad class of vibration problems which can be analyzed and corrected using the assumption of a periodically varying force of constant amplitude. Sometimes these forces exhibit a time characteristic which is very similar to that of a sine wave. At other times they are quite complex and have to be analyzed as a Fourier series. Such a series is the sum of a number of sine and cosine waves, and the resultant motion is the sum of the responses to the individual terms. In this book we shall study only the motion resulting from the application of a single sinusoidal force.

Designating the amplitude of the force as F_0, we write Eq. (18-3) in the form

$$m\ddot{x} + c\dot{x} + kx = F_0 \sin \omega t \qquad (18\text{-}38)$$

where ω is the circular frequency of the force in radians per second. The solution to Eq. (18-38) is in two parts. The first is the transient term and is obtained by making the right side zero. We have already studied this solution in Sec. 18-6 and found that it decays with time. The second part is called the *steady-state* term and is a particular solution of the differential equation. The constants of integration appear only in the transient term, and therefore the steady-state term is a continued vibration with no beginning and no end. It is usually written in the form

$$x = X \sin(\omega t - \phi) \qquad (18\text{-}39)$$

where X is the amplitude and ϕ is the phase angle between the force and the displacement. The velocity and acceleration are obtained by differentiating. Thus

$$\dot{x} = \omega X \cos(\omega t - \phi) \qquad (a)$$
$$\ddot{x} = -\omega^2 X \sin(\omega t - \phi) \qquad (b)$$

It is convenient to write the velocity in the form

$$\dot{x} = \omega X \sin\left(\omega t - \phi + \frac{\pi}{2}\right) \qquad (c)$$

Substituting these values into Eq. (18-38) and moving all the terms to the right side give

$$m\omega^2 X \sin(\omega t - \phi) - c\omega X \sin\left(\omega t - \phi + \frac{\pi}{2}\right) - kX \sin(\omega t - \phi)$$
$$+ F_0 \sin \omega t = 0 \qquad (d)$$

The quantities in Eq. (d) can be plotted in vector form as shown in Fig. 18-23a. The five vectors all maintain a fixed angular relationship to one another and rotate counterclockwise at ω rad/sec. As shown, the displacement X and the inertia force $m\omega^2 X$ are in phase, but the spring force kX is 180° from X. The damping force is seen to lag the displace-

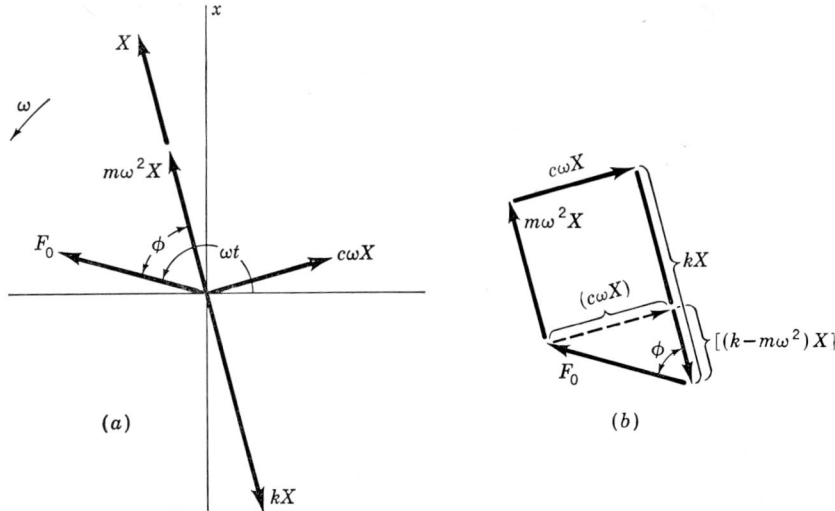

Fig. 18-23

ment by 90°. We shall see that the phase angle ϕ, between the force and the displacement, may, in general, have values between 0 and 180°.

Figure 18-23b shows the four forces arranged in polygon form. Subtracting the inertia force from the spring force makes it possible to evaluate X and ϕ by inspection. From the trigonometry it is seen that

$$X = \frac{F_0}{\sqrt{(k - m\omega^2)^2 + (c\omega)^2}} \qquad (e)$$

$$\phi = \tan^{-1} \frac{c\omega}{k - m\omega^2} \qquad (f)$$

These equations can be simplified by introducing the expressions

$$\omega_n = \sqrt{\frac{k}{m}} \qquad \zeta = \frac{c}{c_c} \qquad c_c = 2m\omega_n$$

They can then be written in the form

$$\frac{X}{F_0/k} = \frac{1}{\sqrt{(1 - \omega^2/\omega_n^2)^2 + (2\zeta\omega/\omega_n)^2}} \qquad (18\text{-}40)$$

$$\phi = \tan^{-1} \frac{2\zeta\omega/\omega_n}{1 - \omega^2/\omega_n^2} \qquad (18\text{-}41)$$

which is a dimensionless form more convenient for analysis. It is interesting to note that the quantity F_0/k is the deflection that a spring of scale k would experience if acted upon by the force F_0.

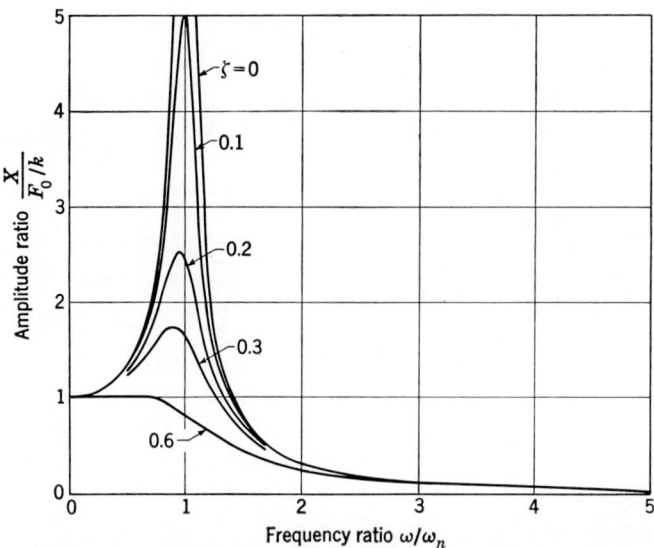

Fig. 18-24. Relative displacement of a damped forced system as a function of the damping and frequency ratios.

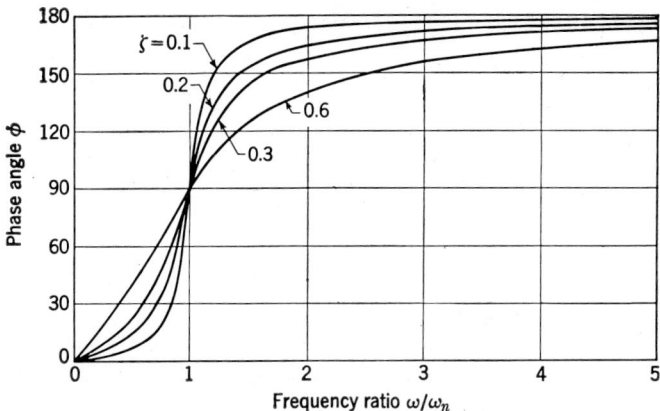

FIG. 18-25. Relationship of the phase angle to the damping and frequency ratios.

If various values are selected for the frequency and damping ratios, curves can be plotted showing the effect of varying these quantities upon the motion. This has been done in Figs. 18-24 and 18-25. Notice that the amplitude is very large near resonance (that is, where ω/ω_n is near unity) for small damping ratios. Notice, too, that the peak amplitudes occur slightly before the resonance point is reached. Figure 18-25 also indicates that the phase angle changes very rapidly near the resonant condition.

18-9. Forcing Due to Unbalance. The type of forcing which is due to unbalance occurs so frequently in machines that it deserves our special attention. The results of this kind of forcing are particularly interesting because the magnitude of the force depends upon the speed of the machinery. Such forcing may be due to the existence of reciprocating parts or to the fact that gears, wheels, and other rotating parts turn about an axis not coincident with the center of mass. The solutions for reciprocating and for rotating unbalance are identical and can be obtained by reference to the schematic drawing of Fig. 18-26.

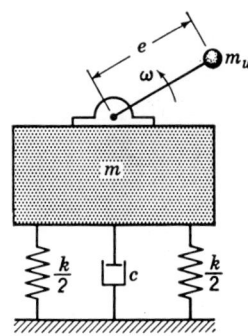

FIG. 18-26. Schematic representation of rotating unbalance.

An unbalanced mass m_u having an eccentricity e is rotating at ω rad/sec in Fig. 18-26. The machine is connected to ground through a dashpot and springs and has a total mass m which includes the unbalanced mass. The magnitude of the force is, from Sec. 16-4, $m_u e \omega^2$, and this force vector rotates at the angular velocity ω. Both the horizontal and vertical components of this force are important, but here we shall assume a single-degree-of-freedom system and, because of this assumption, deal only with the vertical component.[1] Equation (18-3) is written

$$m\ddot{x} + c\dot{x} + kx = m_u e\omega^2 \sin \omega t \quad (18\text{-}42)$$

Equation (18-42) is identical with (18-38) except that F_0 has been replaced

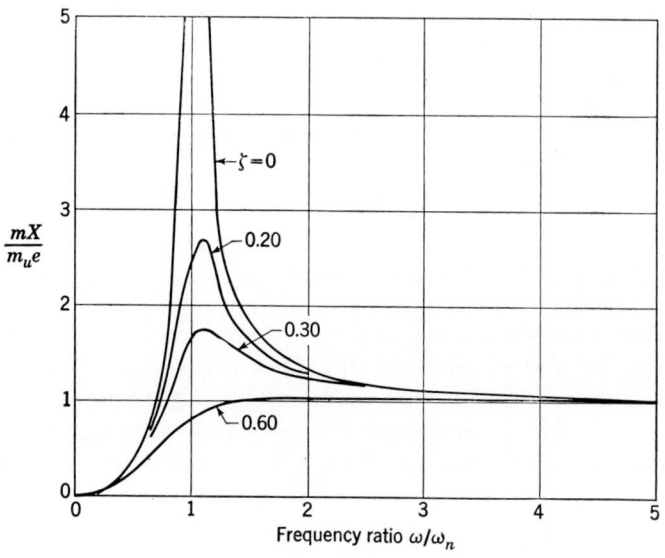

FIG. 18-27

by the quantity $m_u e\omega^2$. Therefore the solution is identical with Eq. (18-39), where

$$X = \frac{m_u e\omega^2}{\sqrt{(k - m\omega^2)^2 + (c\omega)^2}} \quad (a)$$

$$\phi = \tan^{-1} \frac{c\omega}{k - m\omega^2} \quad (b)$$

Substituting $\omega_n = \sqrt{k/m}$, $\zeta = c/c_c$, and $c_c = 2m\omega_n$ in these expressions produces

$$\frac{mX}{m_u e} = \frac{(\omega/\omega_n)^2}{\sqrt{(1 - \omega^2/\omega_n^2)^2 + (2\zeta\omega/\omega_n)^2}} \quad (18\text{-}43)$$

$$\phi = \tan^{-1} \frac{2\zeta\omega/\omega_n}{1 - \omega^2/\omega_n^2} \quad (18\text{-}44)$$

Equation (18-43) is dimensionless and is plotted in Fig. 18-27 for several damping ratios. Notice that the curves will begin at zero amplitude because there can be no exciting force until an exciting frequency begins.

[1] See Charles E. Crede, "Vibration and Shock Isolation," pp. 32–40, John Wiley & Sons, Inc., New York, 1951.

For small damping ratios the amplitude becomes very large near resonance. Thus if machinery must be operated near resonance, a great deal of damping must be purposefully introduced to avoid dangerous amplitudes. On the other hand the curves show an amplitude ratio near unity when the operating speed is three or more times the natural frequency and the effect of damping is negligible. Equation (18-44) is identical with Eq. (18-41) and so is described by Fig. 18-25.

18-10. Vibration Isolation. The investigations of this chapter thus far will enable the engineer to design mechanical equipment so as to minimize vibration, but in many cases it will be impractical to eliminate all of it because machines are inherently vibration generators. When this is the case, a more practical and economical approach is that of reducing as much as possible the annoyances which it causes. This may take any one of several directions depending upon the nature of the

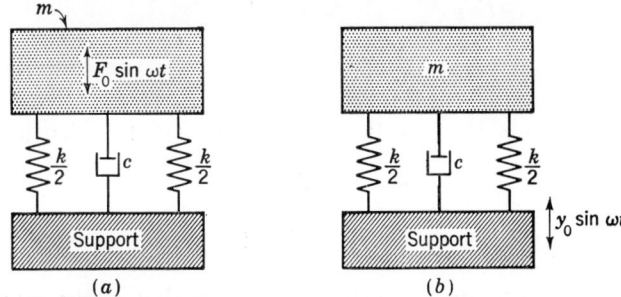

FIG. 18-28. Problems in vibration isolation. (a) The vibrating forces originating in the machine are to be isolated from the support. (b) The equipment is to be isolated from the vibrating support.

problem. For example, everything that is economically feasible may have been done toward the elimination of vibrations in a machine but the residuals are still strong and cause objectionable noise by transmitting these vibrations to the base structure. Another type of problem is that in which an item of equipment, such as a radio receiver, which is not of itself a vibration generator receives objectionable vibrations from another source. Both of these problems can be solved by isolating the equipment from the support. In solving these problems we may be interested in the *isolation of forces* or in the *isolation of motions*, as shown in Fig. 18-28.

Let us consider the system of Fig. 18-28a first. The forces acting against the support must be transmitted through the springs and dashpot because these are the only connections. In Sec. 18-8 it was shown that the spring and damping forces are vectors always at right angles to each other. This enables us to write the equation

$$F_{TR} = \sqrt{(kX)^2 + (c\omega X)^2} = kX\sqrt{1 + \left(\frac{2\zeta\omega}{\omega_n}\right)^2} \quad (a)$$

where X is the amplitude of vibration of the mass and F_{TR} is called the *transmitted force*. The system is identical with the one described in Sec. 18-8, and the amplitude is given by the equation

$$X = \frac{F_0/k}{\sqrt{(1 - \omega^2/\omega_n^2)^2 + (2\zeta\omega/\omega_n)^2}} \tag{b}$$

Therefore the transmitted force is

$$F_{TR} = \frac{F_0 \sqrt{1 + (2\zeta\omega/\omega_n)^2}}{\sqrt{(1 - \omega^2/\omega_n^2)^2 + (2\zeta\omega/\omega_n)^2}} \tag{c}$$

Notice that Eq. (c) contains both the transmitted force and the exciting force. The ratio of these two terms is called the *transmissibility* and is

$$tr = \frac{F_{TR}}{F_0} = \frac{\sqrt{1 + (2\zeta\omega/\omega_n)^2}}{\sqrt{(1 - \omega^2/\omega_n^2)^2 + (2\zeta\omega/\omega_n)^2}} \tag{18-45}$$

which again is in dimensionless form.

Before discussing the significance of Eq. (18-45), let us consider the system of Fig. 18-28b. In this case the motion of the support is given by the equation

$$y = y_0 \sin \omega t \tag{d}$$

Reserving the coordinate x for the absolute motion of the mass, we see that three forces act upon it. These are the inertia force, which is proportional to the absolute motion of the mass, and the spring and damping forces, which are proportional to the motion of the mass relative to the moving support. Thus the equation of motion is written

$$m\ddot{x} + c(\dot{x} - \dot{y}) + k(x - y) = 0 \tag{e}$$

Substituting Eq. (d) and its derivative gives

$$m\ddot{x} + c\dot{x} + kx = c\omega y_0 \cos \omega t + k y_0 \sin \omega t \tag{18-46}$$

As in Sec. 18-8, we assume the solution in the form

$$x = X \sin (\omega t - \phi) \tag{f}$$

Then the velocity and acceleration are

$$\dot{x} = \omega X \sin \left(\omega t - \phi + \frac{\pi}{2}\right) \qquad \ddot{x} = -\omega^2 X \sin (\omega t - \phi) \tag{g}$$

Substituting these terms into Eq. (18-46) and moving all terms to the right side give

$$m\omega^2 X \sin(\omega t - \phi) - c\omega X \sin\left(\omega t - \phi + \frac{\pi}{2}\right) - kX \sin(\omega t - \phi)$$
$$+ c\omega y_0 \sin\left(\omega t + \frac{\pi}{2}\right) + ky_0 \sin \omega t = 0 \quad (h)$$

where we have replaced $c\omega y_0 \cos \omega t$ by the term $c\omega y_0 \sin(\omega t + \pi/2)$. The vectors represented by the terms of Eq. (h) are now plotted in Fig. 18-29. For fixed values of the quantities these vectors rotate at an

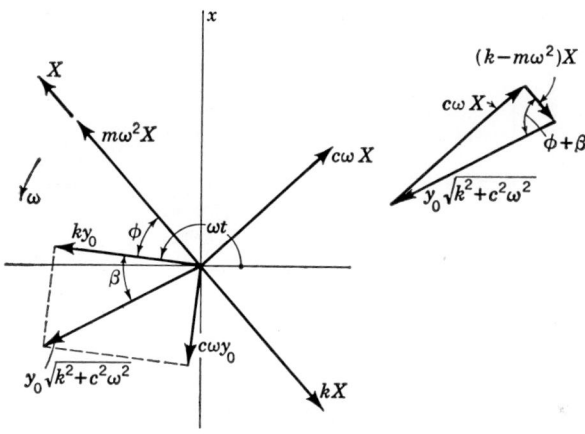

Fig. 18-29

angular velocity ω and maintain a fixed phase relationship. The following relationships can be obtained directly by inspection of the vector polygon:

$$X = \frac{y_0 \sqrt{k^2 + c^2\omega^2}}{\sqrt{(k - m\omega^2)^2 + c^2\omega^2}} \quad (i)$$

$$\phi + \beta = \tan^{-1} \frac{c\omega}{k - m\omega^2} \quad (j)$$

$$\beta = \tan^{-1} \frac{c\omega}{k} \quad (k)$$

Transforming to dimensionless form gives

$$\frac{X}{y_0} = \frac{\sqrt{1 + (2\zeta\omega/\omega_n)^2}}{\sqrt{(1 - \omega^2/\omega_n^2)^2 + (2\zeta\omega/\omega_n)^2}} \quad (18\text{-}47)$$

$$\phi + \beta = \tan^{-1} \frac{2\zeta\omega/\omega_n}{1 - \omega^2/\omega_n^2} \qquad \beta = \tan^{-1} 2\zeta \frac{\omega}{\omega_n} \quad (18\text{-}48)$$

Thus Eq. (18-45) for the ratio of the force transmitted and Eq. (18-47) for the ratio of the motion transmitted are identical and the two cases can be studied together.

Figure 18-30 is a plot of the transmissibility equation for the force transmitted from a vibrating machine to its support and of the relative

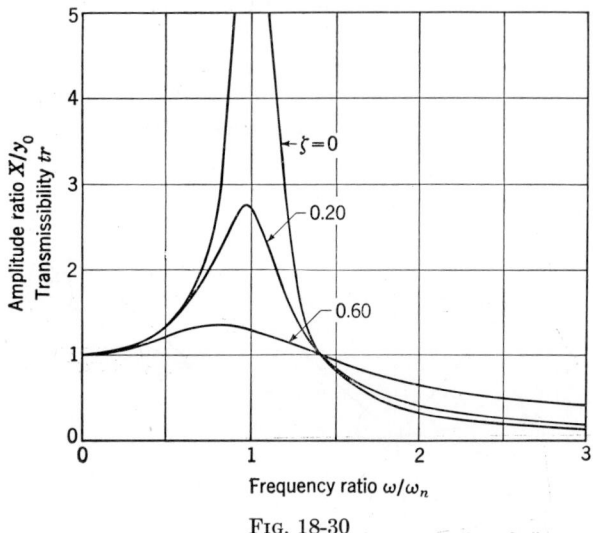

Fig. 18-30

motion of a mass spring-mounted on a vibrating support. The curves show an amplitude ratio and a transmissibility of unity when the exciting force is $\sqrt{2}$ times the natural frequency. Both ratios are less than unity for frequency ratios greater than $\sqrt{2}$. The mounting springs should be designed to have a low spring constant, that is, they should be soft, in order that the natural frequency be a fraction of the disturbing frequency. The curves also show that the damping actually increases the amplitude and transmissibility when $\omega/\omega_n > 1.414$. Thus damping is a desirable quantity only when the machine is starting up and is passing through resonance. After the machine reaches its operating speed, damping is no longer helpful. It is desirable to design the mounting with a low natural frequency. An appreciable time is required for vibration at the resonant frequency to build up. If the natural frequency of the mounting system is low compared with the operating speed, then the chances are good that the dangerous frequencies will be traversed quickly and the vibration will not have an opportunity to build up.

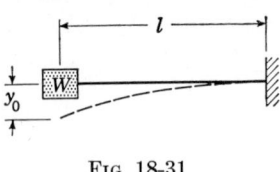

Fig. 18-31

18-11. Natural Frequencies of Beams and Rotating Shafts. Figure 18-31 illustrates a vibrating system which consists of a weight W mounted upon the end of a cantilever spring. The equation for the static deflection of the spring, due to the weight on the end, is obtained from any book on strength of materials and is

$$y_0 = \frac{Wl^3}{3EI} \qquad (a)$$

where l is the length of the spring, E its modulus of elasticity, and I its flexural moment of inertia. Since the spring constant is the force required to produce unit deflection, Eq. (a) can be rearranged to give

$$k = \frac{W}{y_0} = \frac{3EI}{l^3} \qquad (b)$$

Substituting this value for k in the equation for natural frequency produces

$$\omega_n = \sqrt{\frac{k}{m}} = \sqrt{\frac{3EI/l^3}{W/g}} = \sqrt{\frac{3EIg}{Wl^3}} \qquad (c)$$

so that

$$f = \frac{1}{2\pi}\sqrt{\frac{3EIg}{Wl^3}} \quad \text{cps} \qquad (d)$$

Next, note, for any single-degree-of-freedom system, that the quantity W/k represents the static deflection of a spring due to the weight W acting upon it. Rearranging the natural-frequency equation by substituting $m = W/g$ gives

$$\omega_n = \sqrt{\frac{k}{m}} = \sqrt{\frac{g}{W/k}} = \sqrt{\frac{g}{y_0}} \qquad (18\text{-}49)$$

where y_0 is the static deflection due to the weight W acting upon the spring. If we substitute in Eq. (18-49) the value of y_0 from Eq. (a), there results

$$\omega_n = \sqrt{\frac{g}{y_0}} = \sqrt{\frac{g}{Wl^3/3EI}} = \sqrt{\frac{3EIg}{Wl^3}} \qquad (e)$$

which is the same obtained in Eq. (c). Equation (18-49) is, incidentally, a very useful one for determining natural frequencies of mechanical systems because static deflections can usually be measured quite easily.

In our investigation of the mechanics of vibration we have been concerned with systems whose motions can be described using a single coordinate. In the cases of vibrating beams and rotating shafts there may be many masses involved or the mass may be distributed. A separate differential equation must be written for each mass or element of mass of any system and these equations solved simultaneously if we are to obtain the equations of motion of any multimass system. It is clear that the mathematics of such an approach would be overwhelming. While there are a number of optional approaches available, here we shall present an energy method, which is due to Lord Rayleigh,[1] because of its importance in the study of vibration.

It is probable that Eq. (18-49) first suggested to Rayleigh the idea of employing the static deflection to find the natural frequency of a system. If we consider a freely vibrating system without damping, then, during motion, no energy is added to the system nor is any taken away. Yet when the mass has velocity, kinetic energy exists, and when the spring is compressed or extended, potential energy exists. Since no energy is added or taken away, the maximum kinetic energy of a system must be the same as the maximum potential energy. This is the basis of Rayleigh's method; a mathematical statement of it is

$$u_{K,\max} = u_{P,\max} \tag{18-50}$$

In order to see how it works let us apply the method to the simple system of Fig. 18-31. We begin by assuming that the motion is harmonic and of the form

$$y = y_0 \sin \omega_n t \tag{f}$$

The potential energy is a maximum when the spring is fully extended or compressed and occurs when $\sin \omega_n t = 1$. It is

$$u_{P,\max} = \frac{W y_0}{2} \tag{g}$$

The velocity of the weight is given by

$$\dot{y} = y_0 \omega_n \cos \omega_n t \tag{h}$$

The kinetic energy reaches a maximum when the velocity is a maximum, that is, when $\cos \omega_n t = 1$. The kinetic energy is

$$u_{K,\max} = \frac{1}{2} m v_{\max}^2 = \frac{W}{2g} (y_0 \omega_n)^2 \tag{i}$$

Applying Eq. (18-50),

$$\frac{W}{2g} (y_0 \omega_n)^2 = \frac{W y_0}{2} \quad \text{or} \quad \omega_n = \sqrt{\frac{g}{y_0}} \tag{j}$$

which is identical with Eq. (18-49).

It is true in multimass systems that the dynamic deflection curves are not the same as the static deflection curves. The importance of the

---

[1] John William Strutt (Baron Rayleigh), "Theory of Sound," republished by Dover Publications, New York, 1945. This book is in two volumes and is the classic treatise on the theory of vibrations. It was originally published in 1877–1878.

method, however, is that any *reasonable* deflection curve can be used in the process. The static deflection curve *is* a reasonable one, and hence it gives a good approximation. Rayleigh also shows that the correct curve will always give the lowest value for the natural frequency, though we shall not demonstrate this fact here. It is sufficient to know that if many deflection curves are assumed, the one giving the lowest natural frequency is the best.

Rayleigh's method is applied to a multimass system composed of weights W_1, W_2, W_3, etc., by assuming, as before, a deflection of each mass according to the equation

$$y_n = y_{0n} \sin \omega_n t \qquad (k)$$

The maximum deflections are therefore y_{01}, y_{02}, y_{03}, etc., and the maximum velocities are $y_{01}\omega_n$, $y_{02}\omega_n$, $y_{03}\omega_n$, etc. The maximum potential energy for the system is

$$\begin{aligned} u_{P,\max} &= \tfrac{1}{2} W_1 y_{01} + \tfrac{1}{2} W_2 y_{02} + \tfrac{1}{2} W_3 y_{03} + \cdots \\ &= \tfrac{1}{2} \Sigma W_n y_{0n} \end{aligned} \qquad (l)$$

The maximum kinetic energy is

$$\begin{aligned} u_{K,\max} &= \frac{1}{2g} W_1 y_{01}{}^2 \omega_n{}^2 + \frac{1}{2g} W_2 y_{02}{}^2 \omega_n{}^2 + \frac{1}{2g} W_3 y_{03}{}^2 \omega_n{}^2 + \cdots \\ &= \frac{\omega_n{}^2}{2g} \Sigma W_n y_{0n}{}^2 \end{aligned} \qquad (m)$$

Applying Rayleigh's principle and solving for the frequency yield

$$\omega_n = \sqrt{\frac{g \Sigma W_n y_{0n}}{\Sigma W_n y_{0n}{}^2}} \quad \text{or} \quad f = \frac{1}{2\pi} \sqrt{\frac{g \Sigma W_n y_{0n}}{\Sigma W_n y_{0n}{}^2}} \qquad (18\text{-}51)$$

Equation (18-51) can be applied to a beam consisting of several masses or to a rotating shaft having mounted upon it masses in the form of gears, pulleys, flywheels, and the like. If the speed of the rotating shaft should become equal to the natural frequency given by Eq. (18-51), then violent vibrations will occur. For this reason it is common practice to designate this frequency as the *critical speed*.

In the investigation of torsional vibrations to follow, we shall show that there are as many natural frequencies in a multimass vibrating system as there are degrees of freedom. Similarly, a three-mass lateral system will have three degrees of freedom and, consequently, three natural frequencies. Equation (18-51) gives only the first or lowest of these frequencies.[1]

---

[1] Methods are available, however, for finding these frequencies. See Den Hartog, *op. cit.*, pp. 162–165.

It is probable that the critical speed can be determined by Eq. (18-51), for most cases, within about 5 per cent. This is because of the assumption that the static and dynamic deflection curves are identical. Bearings, couplings, belts, etc., will all have an effect upon the spring constant and the damping. Shaft vibration usually occurs over an appreciable range of shaft rpm, and for this reason Eq. (18-51) is accurate enough for many engineering purposes.

EXAMPLE 18-4. Figure 18-32 shows a rotating shaft mounted in bearings A and B carrying three masses which may be gears, pulleys, or the like. If the shaft is steel and has a modulus of elasticity of $E = 28{,}500{,}000$ psi, find the critical speed using Rayleigh's method.

Solution. For many problems in the calculation of critical speeds the shaft can be assumed to be simply supported in its bearings.[1] Thus the problem reduces to that

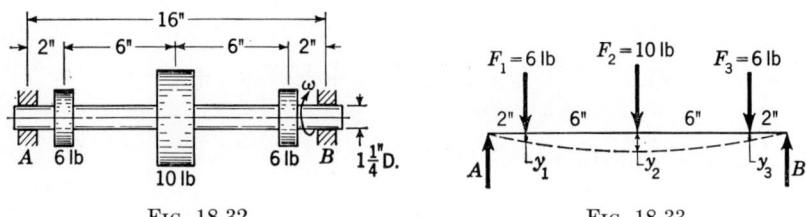

FIG. 18-32   FIG. 18-33

of finding the deflection y_1, y_2, and y_3 of the simply supported beam shown in Fig. 18-33. Formulas for obtaining these deflections can be found in any book on strength of materials and will not be given here. The moment of inertia is

$$I = \frac{\pi d^4}{64} = \frac{\pi (1.25)^4}{64} = 0.120 \text{ in.}^4$$

The deflections can then be calculated and are found to be

$$y_1 = 0.000137 \text{ in.} \qquad y_2 = 0.000350 \text{ in.} \qquad y_3 = 0.000137 \text{ in.}$$

Substituting in Eq. (18-51) gives

$$f = \frac{60}{2\pi} \sqrt{\frac{386[(6)(0.000137) + (10)(0.000350) + (6)(0.000137)]}{(6)(0.000137)^2 + (10)(0.000350)^2 + (6)(0.000137)^2}}$$
$$= 11{,}200 \text{ rpm} \qquad Ans.$$

Note that the weight of the shaft has been neglected in this example.

As a general rule shafts are designed with shoulders against which bearings, gears, and other shaft-mounted masses are located. Also, the

---

[1] Additional constraint may be present if double-row bearings are used or if the bearings are preloaded. See Joseph E. Shigley, "Machine Design," pp. 266, 267, McGraw-Hill Book Company, Inc., New York, 1956.

weight of the shaft should usually be considered in the analysis. Such a problem is most difficult when handled by the usual mathematical techniques, and the method of graphical integration as described in Sec. 4-12 is recommended.

EXAMPLE 18-5. Using graphical integration to obtain the deflection curve, find the critical speed in rpm of the shaft and gear assembly shown in Fig. 18-34. The shaft is made of steel having a modulus of elasticity of 28,500,000 psi.

Solution. First, construct the loading diagram as shown in Fig. 18-35a and find the bearing reactions. Divide the shaft into parts, as shown, and find the weight and moment of inertia of each part. Note that these divisions are not necessarily equal ones. It is convenient to designate the parts in some manner for later identification, and this is done using the letters of the alphabet.

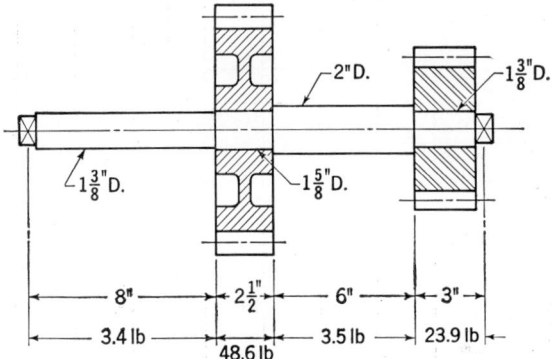

FIG. 18-34. The weights given include the shaft and gears and are based on steel at 0.282 lb/in.³

The shear force diagram is calculated and drawn to scale next, directly under and in line with the loading diagram, as shown in Fig. 18-35b. It is noted that the shear force diagram could have been obtained by graphically integrating the loading diagram, but it is convenient not to do so here.

The next step is to integrate the shear force diagram graphically by choosing a pole distance h_1 as explained in Sec. 4-12. This produces the moment diagram shown in Fig. 18-35c; the scale calculation is also shown.

Now, since the moment of inertia changes along the shaft, the value of the moments must be divided by the moments of inertia. The modulus of elasticity E will also have to be introduced somewhere in the analysis, and it is expedient to do so here. Thus we scale the moments directly from the moment diagram at the end of each division and divide them by EI. If the beginning or end of a division coincides with a change in the moment of inertia, notice that the moment will have to be divided by EI for both sides, producing two values of M/EI. These calculations are not shown, but the reader can easily verify them using the values shown on the moment diagram. The resulting M/EI diagram is shown in Fig. 18-35d. The scale is not a function of the scale of the moment diagram but can be selected to produce a diagram of convenient size.

The integration is carried out successively now to produce the slope diagram and the deflection curve, as shown in Fig. 18-35e and f. The closing chord of the deflection curve is seldom horizontal in this process because it depends upon guessing the position of zero slope correctly. Slight deviations do no harm to the accuracy of scaling the deflection. Notice, though, that the measurements are made parallel to the y axis and *not* perpendicular to the closing chord. Inch measurements taken at mid-space are recorded on the diagram for convenience. The actual deflection is obtained by multiplying these by the scale.

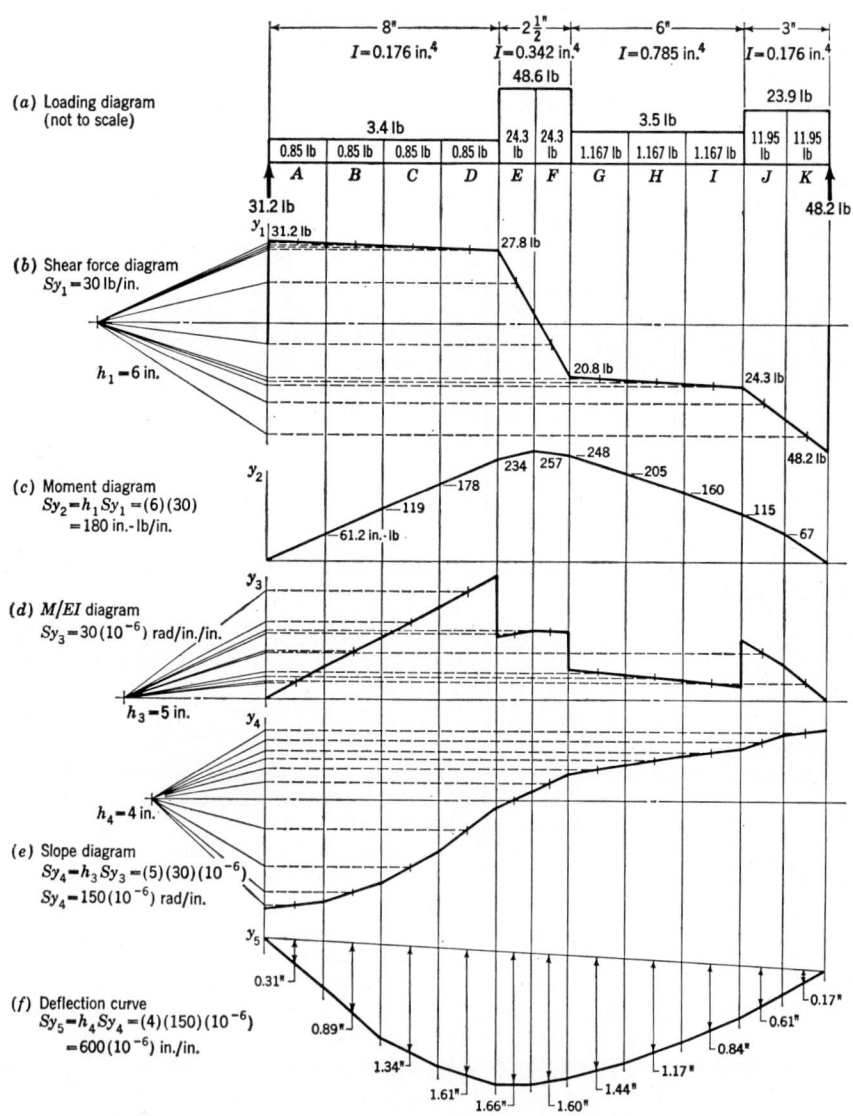

Fig. 18-35

TABLE 18-1

Division	Weight W_n, lb	y_5, in.	Deflection y_{0n}, 10^{-6} in.	$W_n y_{0n}$, 10^{-6} lb-in.	$W_n y_{0n}^2$, 10^{-9} lb-in.^{2}
A	0.85	0.31	186	158	29.4
B	0.85	0.89	534	453	242
C	0.85	1.34	804	682	548
D	0.85	1.61	965	820	791
E	24.3	1.66	996	24,200	25,100
F	24.3	1.60	960	23,400	22,400
G	1.167	1.44	863	1,007	868
H	1.167	1.17	701	817	573
I	1.167	0.84	504	586	295
J	11.95	0.61	366	4,370	1,600
K	11.95	0.17	102	1,220	124.5
Totals.....	......	....	...	57,713	52,570.9

The data can now be arranged in tabular form as shown in Table 18-1, and the last two columns summed. The critical speed is calculated from Eq. (18-51) and is

$$f = \frac{60}{2\pi}\sqrt{\frac{g\Sigma W_n y_{0n}}{\Sigma W_n y_{0n}^2}} = \frac{60}{2\pi}\sqrt{\frac{(386)(57,713)(10)^{-6}}{(52,570.9)(10)^{-9}}} = 6{,}210 \text{ rpm} \quad Ans.$$

18-12. Torsional Systems. Figure 18-36 shows a shaft supported on bearings at A and B with two masses connected at the ends. The masses represent any rotating machine parts, an engine and its flywheel, for example. We wish to study the possibilities of free vibration of the system when it rotates at constant angular velocity. In order to investigate the motion of each mass it is necessary to picture a reference system fixed to the shaft and rotating at the same angular velocity. Then we can measure the angular displacement of either mass by finding the instantaneous angular location of a mark on the mass relative to one of the rotating axes. Thus we define θ_1 and θ_2 as the angular displacement, respectively, with respect to the rotating axes. Now, assuming no damping, Eq. (18-6) is written for each mass:

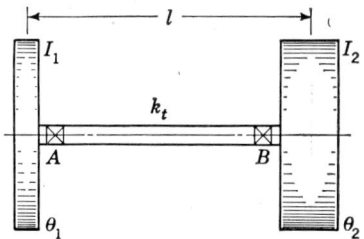

Fig. 18-36. A two-mass torsional system; A and B are bearings.

$$I_1 \ddot{\theta}_1 + k_t(\theta_1 - \theta_2) = 0 \\ I_2 \ddot{\theta}_2 + k_t(\theta_2 - \theta_1) = 0 \quad (a)$$

where the angle $(\theta_1 - \theta_2)$ represents the total twist of the shaft. A solution is assumed in the form

$$\theta_n = \gamma_n \sin \omega_n t \qquad (b)$$

The acceleration is

$$\ddot{\theta}_n = -\gamma_n \omega_n^2 \sin \omega_n t \qquad (c)$$

If these equations are substituted into Eqs. (a), there results

$$(k_t - I_1\omega_n^2)\gamma_1 - k_t\gamma_2 = 0 \\ -k_t\gamma_1 + (k_t - I_2\omega_n^2)\gamma_2 = 0 \qquad (d)$$

The determinant of these equations must be zero. Therefore

$$(k_t - I_1\omega_n^2)(k_t - I_2\omega_n^2) - k_t^2 = 0$$

so that

$$\omega_n^2 \left(\omega_n^2 - k_t \frac{I_1 + I_2}{I_1 I_2} \right) = 0$$

and

$$\omega_{n1} = 0 \qquad \omega_{n2} = \sqrt{k_t \frac{I_1 + I_2}{I_1 I_2}} \qquad (18\text{-}52)$$

Substituting $\omega_{n1} = 0$ into Eqs. (d) we find that $\gamma_1 = \gamma_2$. Therefore the masses rotate together without relative displacement and there is no vibration.

Substituting the value of ω_{n2} into Eqs. (d) it is found that

$$\gamma_1 = -\frac{I_2}{I_1}\gamma_2 \qquad (e)$$

indicating that the motion of the second mass is opposite to that of the first and proportional to I_1/I_2. Figure 18-37 is a graphical representation of the vibration. Here l is the distance between the two masses and γ_1 and γ_2 the instantaneous angular displacements plotted to scale. If a line is drawn connecting the ends of the angular displacements, this line crosses the axis at a *node* which is a section of the shaft having zero angular displacement. It is convenient to designate the configuration of the vibrating system as the *mode* of vibration. This system has two modes of vibration, corresponding to the two frequencies, although we have seen that one of them is degenerate. A system of n masses would have n modes corresponding to n different frequencies.

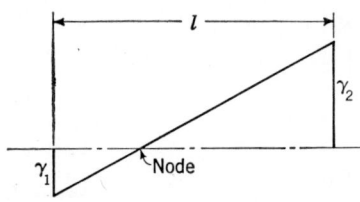

FIG. 18-37

18-13. Holzer's Tabulation Method. In the design and analysis of machines it is desirable to have a method of estimating the speeds at which torsional vibrations will occur. If they are estimated to exist near the operating speeds, then the design can be changed to eliminate the vibration or a damper installed to decrease its amplitude. Such a method will also provide useful information in the analysis of vibrations in existing machines. In these cases it is the frequency that is of primary concern rather than the amplitude. It should also be noted that we have already investigated the factors influencing amplitude and that these same factors generally apply for multimass systems. The *Holzer tabulation method* is a means of obtaining the natural frequencies of torsional vibration in a multimass system. It is widely used for this purpose, and a number of refinements have been made since the method originated. We shall present here only the elementary method.

Figure 18-38 shows a torsional system consisting of the inertias I_1, I_2, I_3, and I_4 separated by the torsional spring constants k_{t1}, k_{t2}, and k_{t3}. The deflection of each mass from a reference system rotating with the shaft at constant angular velocity is assumed to be such that $\theta_1 < \theta_2 < \theta_3 < \theta_4$. Then the differential equations of motion for each mass can be written

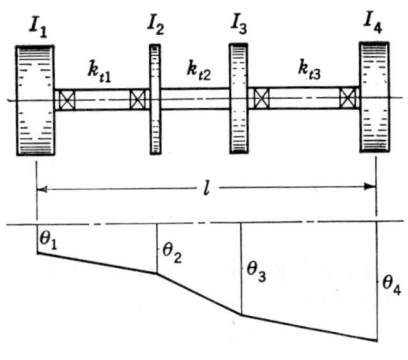

Fig. 18-38

$$\begin{aligned} I_1\ddot{\theta}_1 &= k_{t1}(\theta_2 - \theta_1) \\ I_2\ddot{\theta}_2 &= -k_{t1}(\theta_2 - \theta_1) + k_{t2}(\theta_3 - \theta_2) \\ I_3\ddot{\theta}_3 &= -k_{t2}(\theta_3 - \theta_2) + k_{t3}(\theta_4 - \theta_3) \\ I_4\ddot{\theta}_4 &= -k_{t3}(\theta_4 - \theta_3) \end{aligned} \quad (a)$$

Assuming a solution $\theta_n = \gamma_n \sin \omega_n t$ and substituting it and its derivatives into Eqs. (a) yield

$$\begin{aligned} -I_1\omega_n^2\gamma_1 &= k_{t1}(\gamma_2 - \gamma_1) \\ -I_2\omega_n^2\gamma_2 &= -k_{t1}(\gamma_2 - \gamma_1) + k_{t2}(\gamma_3 - \gamma_2) \\ -I_3\omega_n^2\gamma_3 &= -k_{t2}(\gamma_3 - \gamma_2) + k_{t3}(\gamma_4 - \gamma_3) \\ -I_4\omega_n^2\gamma_4 &= -k_{t3}(\gamma_4 - \gamma_3) \end{aligned} \quad (b)$$

Amplitude and frequency are not dependent upon each other, and furthermore, we are principally interested in the frequency. Therefore assume $\gamma_1 = 1$ rad. Solving the first of Eqs. (b) for γ_2 gives

$$\gamma_2 = \gamma_1 - \frac{I_1\omega_n^2\gamma_1}{k_{t1}} \quad (c)$$

With γ_2 known, γ_3 can be found. Adding the first two equations and solving for γ_3,

$$\gamma_3 = \gamma_2 - \frac{I_1\omega_n^2\gamma_1 + I_2\omega_n^2\gamma_2}{k_{t2}} \quad (d)$$

Next γ_4 is obtained by adding the first three equations. Thus

$$\gamma_4 = \gamma_3 - \frac{I_1\omega_n{}^2\gamma_1 + I_2\omega_n{}^2\gamma_2 + I_3\omega_n{}^2\gamma_3}{k_{t3}} \tag{e}$$

In general, then,

$$\gamma_n = \gamma_{n-1} - \frac{\sum_{m=1}^{m=n-1} I_m\omega_n{}^2\gamma_m}{k_{t(n-1)}} \tag{18-53}$$

Also, adding all four of Eqs. (b) gives zero; so, for any system of n masses,

$$\sum_{m=1}^{m=n} I_m\omega_n{}^2\gamma_m = 0 \tag{18-54}$$

which states that the sum of the inertia torques must equal zero.

In using the Holzer method values are assumed for ω_n until one is found which makes the summation of Eq. (18-54) zero. If the amplitude of the first mass is made unity, the relative amplitudes of the other masses are found using Eq. (18-53). The approximations can be made in an orderly manner if a table is arranged as follows:

No.	I	$I\omega_n{}^2$	γ	$I\omega_n{}^2\gamma$	$\Sigma I\omega_n{}^2\gamma$	k_t	$\Sigma I\omega_n{}^2\gamma/k_t$
1	I_1	$I_1\omega_n{}^2$	1.00	$I_1\omega_n{}^2$	$I_1\omega_n{}^2$	k_{t1}	$I_1\omega_n{}^2/k_{t1}$
2	I_2	$I_2\omega_n{}^2$	γ_2	$I_2\omega_n{}^2\gamma_2$	$I_1\omega_n{}^2 + I_2\omega_n{}^2\gamma^2$	k_{t2}	$(I_1\omega_n{}^2 + I_2\omega_n{}^2\gamma_2)/k_{t2}$
3	I_3						

A separate table is constructed for each assumed value of ω_n with the object of finding a natural frequency such that the last figure of the summation of column 6 is zero.

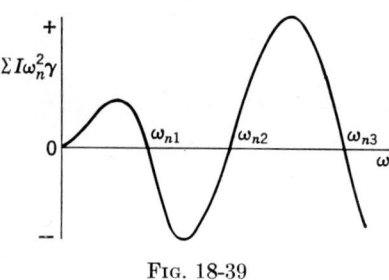

Fig. 18-39

Several trials of the method and observation of the values of γ in column 4 will assist the reader to assume values for ω_n which are in the correct direction. Figure 18-39 is a graph showing how the summation $\Sigma I\omega_n{}^2\gamma$ varies for changing values of ω_n. In this example a system with three natural frequencies ω_{n1}, ω_{n2}, and ω_{n3} is shown. If a value for ω_{n2} is assumed which is slightly greater than its true value, then the remainder in column 6 is positive, as shown on the graph, but if the assumed value is less than ω_{n2}, then the remainder is negative. By bracketing a natural frequency in this manner one can usually estimate the true value closely by plotting

VIBRATION ANALYSIS

a straight-line graph of the two assumptions and finding its intersection with another line representing the zero remainder.

EXAMPLE 18-6. Find the critical speeds for the torsional system shown in Fig. 18-40.[1]

Solution. The calculations are shown in Table 18-2. The first guess is $\omega_n = 3$, and this gives a remainder of -18.5, the negative sign indicating that the guess is too high. Therefore a second guess is made at $\omega_n = 2.3$, which yields a positive remainder. The critical frequency lies between these two values. If a line graph is plotted, it will show a zero remainder near $\omega_n = 2.5$; so $\omega_n = 2.5$ is used for a third trial, and the resulting remainder shows that this value is sufficiently close.

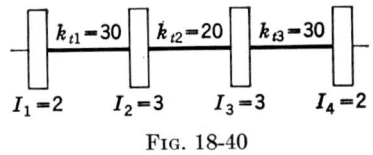

FIG. 18-40

For the second critical speed $\omega_n = 5$ exactly, as shown in the table. In general, however, the second critical speed is not twice the first.

TABLE 18-2

Assumed speed	No.	I	$I\omega_n^2$	γ	$I\omega_n^2\gamma$	$\Sigma I\omega_n^2\gamma$	k_t	$\dfrac{\Sigma I\omega_n^2\gamma}{k_t}$
$\omega_n = 3$	1	2	18	1.00	18	18	30	0.60
	2	3	27	0.40	10.8	28.8	20	1.44
	3	3	27	-1.04	-28.1	0.7	30	0.02
	4	2	18	-1.06	-19.2	-18.5		
$\omega_n = 2.3$	1	2	10.6	1.00	10.6	10.6	30	0.35
	2	3	15.9	0.64	10.3	20.9	20	1.04
	3	3	15.9	-0.40	-6.3	14.6	30	-0.49
	4	2	10.6	0.09	0.9	15.5		
$\omega_n = 2.5$	1	2	12.5	1.00	12.5	12.5	30	0.42
	2	3	18.75	0.58	10.9	23.4	20	1.17
	3	3	18.75	-0.59	-11.1	12.3	30	0.41
	4	2	12.5	-1.00	-12.5	-0.2		
$\omega_n = 5$	1	2	50	1.00	50	50	30	1.67
	2	3	75	-0.67	-50	0	20	0
	3	3	75	-0.67	-50	-50	30	-1.67
	4	2	50	1.00	50	0		
$\omega_n = 5.63$	1	2	63.5	1.00	63.5	63.5	30	2.12
	2	3	95.2	-1.12	-106.8	-43.3	20	-2.16
	3	3	95.2	1.14	108.7	65.4	30	2.18
	4	2	63.5	-1.04	-66.0	-0.6		

[1] Note that this system is not a realistic one because the springs are quite soft and the masses heavy.

After several trials, which are not shown, the third critical speed is found to be $\omega_n = 5.63$, as shown in the table.

The relationship between the amplitudes of column 4 is especially interesting. As shown in Fig. 18-41, the character of the vibration for each critical speed can readily be determined.

18-14. Equivalent Systems. For the solution of all problems in mechanical engineering it is necessary to construct abstract models of the systems. The models are analyzed and the results interpreted in terms of the real system. The translation of the real system into a model often requires real ingenuity, judgment, and a considerable amount of practical experience. These models are termed abstract because of the simplifications and assumptions which are necessary to create them. They exist only to make the problems mathematically solvable.

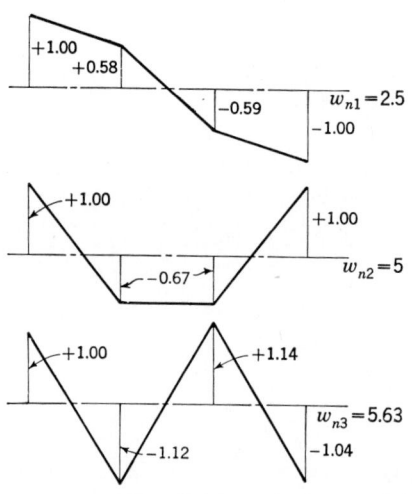

FIG. 18-41. Amplitude relationships for the critical speeds.

As an example of the need for vibration models consider the crankshaft and camshaft of an automotive engine. These shafts run at different speeds because they are connected by gears; yet torsional vibrations generated in one may cause trouble in the other. A method is needed, therefore, to convert them into masses connected to an equivalent shaft so that they are all rotating at the same equivalent speed. Or consider the cam, follower, rocker arm, and valve of an automotive engine. To analyze the vibrations of this system one requires that it be converted into an equivalent mass-spring system so that dangerous frequencies and amplitudes can be calculated.

The *constant* or *scale* of a spring is defined as the ratio of the force exerted to the deflection produced by that force. Therefore, if an equation is available which gives the relation between the force and the deflection, then we can use the equation to determine the spring scale. As an example, the deflection of a helical compression or tension spring is given by the equation[1]

$$y = \frac{8FD^3N}{d^4G} \tag{a}$$

---

[1] Shigley, *op. cit.*, p. 227.

where y = spring deflection, in.
F = force, lb
D = mean diameter of spring, in.
N = number of active coils
d = wire diameter, in.
G = modulus of rigidity, psi

Rearranging Eq. (a) gives for the scale

$$k = \frac{F}{y} = \frac{d^4 G}{8 D^3 N} \tag{18-55}$$

where k is in pounds per inch as before.

The torsional spring constant of a solid or hollow round bar is determined in a similar manner. The angular deflection of such a bar is given by the equation

$$\theta = \frac{Tl}{GJ} \tag{b}$$

where θ = angular deflection, rad
T = torque, lb-in.
l = length of bar, in.
J = polar moment of inertia, in.^{4}

Then the torsional spring constant is

$$k_t = \frac{T}{\theta} = \frac{GJ}{l} \tag{18-56}$$

where k_t is in pound-inches per radian.

The elongation of a bar of any cross section when loaded in pure tension or compression is

$$\delta = \frac{Fl}{AE} \tag{c}$$

where δ = deformation, in.
A = cross-sectional area, in.^{2}
E = modulus of elasticity, psi

The spring constant is

$$k = \frac{F}{\delta} = \frac{AE}{l} \tag{18-57}$$

In each case the procedure is the same; note that we require only a relation between the force and the deflection. Beams, for example, also act as springs, since they experience deflection when acted upon by a force.

When the abstract model is created, the interconnected machine parts are separately analyzed for the spring scales using the equations given

490   DYNAMIC ANALYSIS OF MACHINES

above or similar ones. Following this it is necessary to put together the various constants to obtain an *equivalent* spring constant for the system. When this is done, it is found that the elements occur in series and parallel combinations of springs.

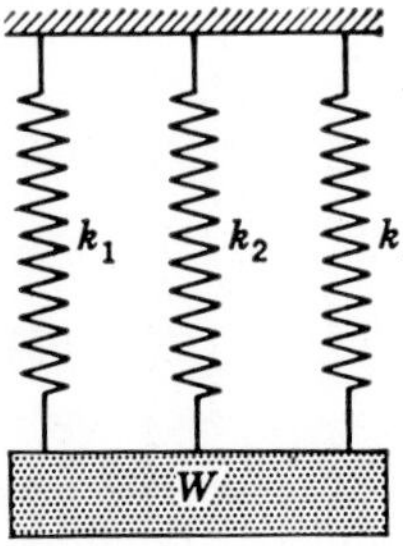

Parallel Springs (Fig. 18-42). The force required to deflect the springs a unit distance is

$$F = k_1 + k_2 + k_3$$

Therefore the equivalent spring constant is

$$k_e = \frac{F}{x} = k_1 + k_2 + k_3 \qquad (18\text{-}58)$$

FIG. 18-42. Parallel springs.

Series Springs. If a weight is placed on the system of Fig. 18-43a, it will move through a distance which is the sum of the deflection of each spring. The equivalent spring will have to move through the same distance. Therefore

$$\frac{W}{k_e} = \frac{W}{k_1} + \frac{W}{k_2} + \frac{W}{k_3} \qquad (d)$$

or

$$k_e = \frac{1}{1/k_1 + 1/k_2 + 1/k_3} \qquad (18\text{-}59)$$

Equation (18-59) is valid for the torsional system of Fig. 18-43b too.

Geared Shafts. The arrangement of parts consisting of a motor, a pair of gears, and a mass shown in Fig. 18-44a may vibrate torsionally. If the gears are quite rigid, then this system can be reduced to the equivalent system shown in Fig. 18-44b. The criterion of equivalency is that the kinetic energy of the two systems must remain the same. Since the kinetic energy of a rotating mass is proportional to the square of the speed, we can write

$$I_e \omega_e^2 = I \omega^2$$

or

$$I_e = \left(\frac{\omega}{\omega_e}\right)^2 I \qquad (18\text{-}60)$$

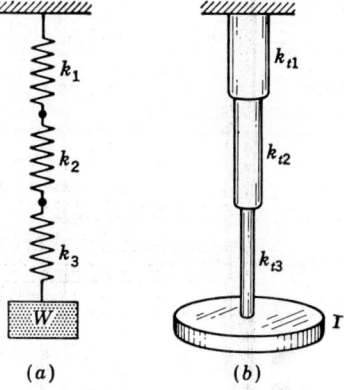

FIG. 18-43. Series springs. (a) Linear. (b) Torsional.

There are two inertias connected to the motor shaft of Fig. 18-44a. These are the inertia of the rotor and of the pinion. Equation (18-60) is applied for each of these masses, where ω is the speed of the motor shaft and ω_e is the speed of the driven shaft. Note that we are making the speed of the driven shaft equal to that of the equivalent shaft. The equivalent moment of inertia of the pinion is added to the moment of inertia of the gear forming I_{2e}. The equivalent moment of inertia of the rotor then becomes I_{3e}.

The equivalent spring constant of the motor shaft will also have to be obtained. It is

$$k_{te} = \left(\frac{\omega}{\omega_e}\right)^2 k_t \tag{18-61}$$

Distributed Mass. In the investigations of vibration in this chapter our model has been a weightless spring and a rigid mass. Now let us

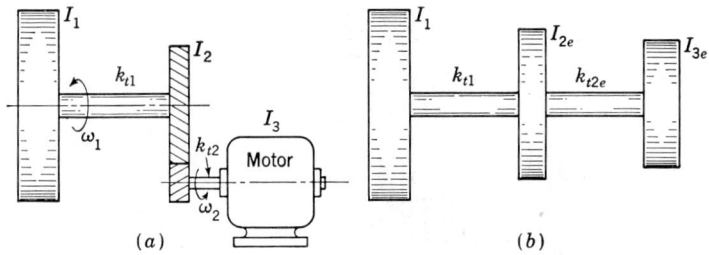

FIG. 18-44. (a) Real system. (b) Equivalent system.

investigate the validity of this assumption so that, when it is made, we shall be aware of its effect upon the results.

We begin with a spring k_1 which has one end fastened to a support and the other end to a concentrated mass. Let the mass of k_1 be m_s. In Fig. 18-45 we give the spring a length l, at any instant in time, and an instantaneous velocity v to the end connected to the concentrated mass. The end connected to the support will then have zero velocity, and the velocity of various points on the spring will vary linearly with their distance from the support. At some distance x from the support the velocity is $v_x = (x/l)v$. The kinetic energy of a vibrating mass is $mv^2/2$, and hence the kinetic energy of a portion of the spring of length dx is

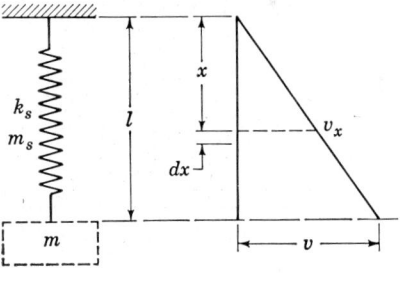

FIG. 18-45

$$du = \frac{dm_s}{2} v_x^2 = \frac{1}{2}\frac{m_s}{l} dx \left(\frac{x}{l}\right)^2 v^2 \tag{e}$$

so the total kinetic energy for the spring is

$$u = \frac{m_s v^2}{2l^3} \int_0^l x^2\, dx = \frac{1}{2}\frac{m_s v^2}{3} \tag{18-62}$$

Equation (18-62) shows that one-third of the spring mass should be assigned to the concentrated mass; the remaining two-thirds, being concentrated at the support, will then have no effect on the motion. The same procedure can be used to show that one-third of the mass moment of inertia of a round shaft vibrating torsionally should be added to the concentrated moment of inertia. The remaining two-thirds is then assigned to the fixed end, if one exists, or to the node (because it has no motion). Since the node always exists between two disks, the assignments should be made beginning at the node and proceeding first to one disk and then to the other.

The deflection curve for a bar in tension or compression and for a helical spring is linear, and this is the justification for concentrating an equivalent mass according to the one-third rule. The deflection curve

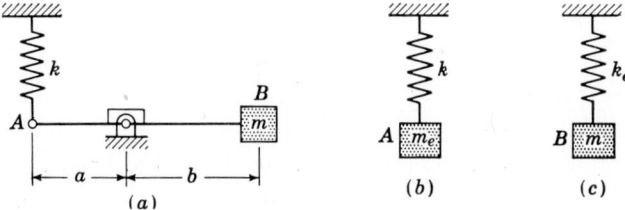

FIG. 18-46

of a beam acting as a spring is *not* a straight line, and consequently we cannot use the same rule. Den Hartog[1] shows that a cantilever beam should have 23 per cent of its distributed mass concentrated at the end; similarly, a simply supported beam should have one-half its mass concentrated at the center.

The system shown in Fig. 18-46a can be reduced to that of (b) having an equivalent mass moving at the speed of the spring or to that of (c) having an equivalent spring whose end moves at the speed of the mass. The reduction is based on the fact that the energy of the equivalent system must be the same as that of the original system.

Suppose that we reduce Fig. 18-46a to (b) by finding an equivalent mass m_e to be placed at A, thus eliminating the lever. Writing that the kinetic energy at A must be the same as that at B gives

$$\tfrac{1}{2} m v_B{}^2 = \tfrac{1}{2} m_e v_A{}^2 \tag{f}$$

Since $v_A = (a/b)v_B$,

$$m v_B{}^2 = m_e \frac{a^2}{b^2} v_B{}^2 \tag{g}$$

so that

$$m_e = \frac{b^2}{a^2} m \tag{18-63}$$

[1] *Op. cit.*, p. 430.

VIBRATION ANALYSIS   493

Since we are usually more interested in the motion of the mass than we are in the motion of the end of the spring, reduction of Fig. 18-46a to (c) is preferred. In this case we wish to replace the present spring with an equivalent spring k_e acting at B. The potential energy stored in a spring is the average force times the distance, $kx^2/2$, where x is the spring deflection. If we equate the potential energy of the spring at A to that of an equivalent spring at B, we have

$$\tfrac{1}{2} k x_A^2 = \tfrac{1}{2} k_e x_B^2 \qquad (h)$$

but the deflection at B is b/a times as much as at A. Therefore

$$k x_A^2 = k_e \frac{b^2}{a^2} x_A^2 \qquad (i)$$

or

$$k_e = \frac{a^2}{b^2} k \qquad (18\text{-}64)$$

Note that Eqs. (18-63) and (18-64) are really identical; the only difference is that in the first case we are studying the motion at A and in the second case that at B.

PROBLEMS

18-1. The vibrating system shown in the figure has $k_1 = k_3 = 5$ lb/in., $k_2 = 10$ lb/in., and $W = 8.33$ lb. What is the natural frequency in cycles per second? *Ans.:* 3.14 cps.

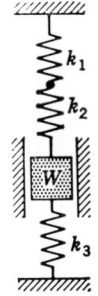

PROB. 18-1

18-2. The figure shows a weight $W = 15$ lb connected to a pivoted rod which is assumed to be weightless. A spring connected to the center of the rod holds the system in static equilibrium at the position shown. Assuming that the system can vibrate with a small deflection, calculate the period in seconds.

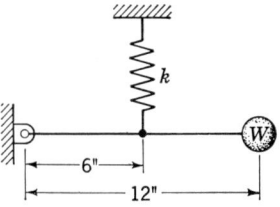

PROB. 18-2. $k_2 = 60$ lb/in.

18-3. The figure shows an upside-down pendulum of length l retained by two springs connected a distance a from the pivot. The springs are mounted so that the pendulum is in static equilibrium when it is in the vertical position. (a) Assuming only small-amplitude oscillations, calculate the natural frequency in cycles per second. (b) If the length of the pendulum can be varied, at what value of l will the system become unstable?

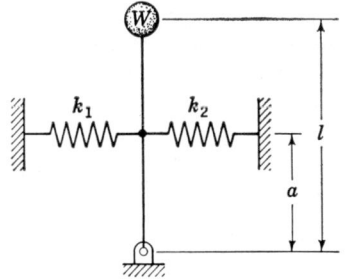

PROB. 18-3. $W = 1.00$ lb; $k_1 = 5$ lb/in.; $k_2 = 10$ lb/in.; $a = 5$ in.; $l = 15$ in.

18-4. The figure illustrates a method which might be used to determine the coefficient of friction between two surfaces. A bar of total weight W rests upon two wheels which rotate in opposite directions. If the bar is not centered when it is placed on the wheels, then more of its weight will rest upon one wheel than upon the other. Consequently, the frictional forces will not be equal and the bar will move in the direction of the largest of the two forces. But the inertia of the bar carries it too far and then the other wheel exerts the largest frictional force. Thus the bar moves back and forth at a frequency which is dependent upon the coefficient of friction. Develop a formula for calculating the coefficient of friction when the frequency of oscillation is known. *Ans.:* $\mu = \pi f^2 l/g$.

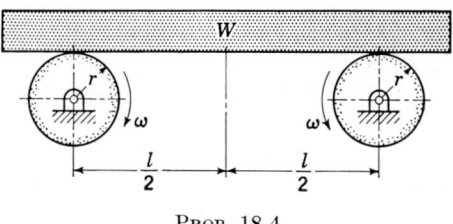

PROB. 18-4

18-5. Show that the equation $x = A \sin \omega_n t + B \cos \omega_n t$ can also be expressed in the form $x = C \sin (\omega_n t + \psi)$. Next, using $x = x_0$ and $\dot{x} = v_0$ as starting conditions, find C and ψ.

18-6. An undamped vibration system consists of a spring whose scale is 200 lb/in. and a weight of 2.60 lb. A square-wave force $F = 10$ lb is exerted on the mass for 0.041 sec. Write the equations of motion of the system for the period in which the force acts and for the period after the force has become zero. Plot the displacement-time diagram. What is the amplitude during the period in which the force is acting? What is the amplitude after the force has returned to zero? What is the phase angle between the motion of the second era and that of the first era?

18-7. The figure shows a round shaft, whose torsional spring constant is k_t lb-in./rad, connecting two wheels having the mass moments of inertia I_1 and I_2. Show that the system is capable of vibrating torsionally with a frequency of

$$\omega_n = \sqrt{\frac{k_t(I_1 + I_2)}{I_1 I_2}}$$

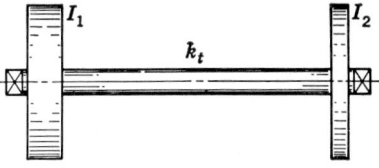

PROB. 18-7

18-8. A motor is connected to a flywheel by a ⅝-in.-diameter steel shaft 36 in. long, as shown. Using the methods of this chapter it can be demonstrated that the torsional spring constant of the shaft is 4,790 lb-in./rad. The mass moments of inertia of the motor and flywheel are 24.0 and 56.0 lb-sec²-in., respectively. The motor is turned on for 2 sec, and during this period it exerts a constant torque of 200 lb-in. on the shaft. (a) What speed in rpm does the shaft attain? (b) What is the natural circular frequency of vibration of the system? (c) Assuming no damping, what is the amplitude of vibration of the system in degrees during the first era? During the second era?

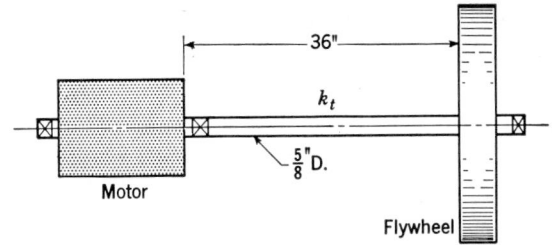

PROB. 18-8

18-9. A power-transmission train consists of a motor and pinion driving a gear which is mounted outboard of a pair of bearings. The figure illustrates the motor and pinion in dashed lines and the gear and its shaft in solid lines. Since the gear is mounted on the outboard end of the shaft, it is possible to regard it as a cantilever beam having a length of 8 in. (as shown) with a weight on the end equal to the weight of the gear. If assumptions of no damping and of a weightless shaft are made,

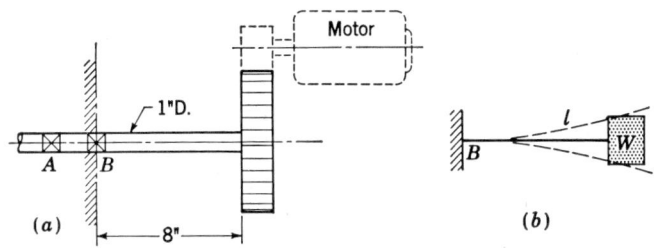

PROB. 18-9. (a) The power-transmission train. (b) The vibrating model.

then it is possible to treat the system as an undamped single-degree-of-freedom vibrating system. Such a system will have a spring constant (linear), as given by Eq. (b) of Sec. 18-11. Using steel as the shaft material ($E = 28,500,000$ psi) and recalling that the moment of inertia for a solid round shaft is $I = \pi d^4/64$ make it possible to calculate k. The weight of the gear is 7.5 lb. When the motor is turned on, the pinion exerts a force on the gear of 180 lb. (a) Calculate the amplitude and frequency of the vibration which results when the motor is turned on. (b) If the motor is turned off after 3 sec, what is the amplitude of the resulting vibration? (c) Plot a portion of the response for both eras including the end point of the first era.

18-10. The weight of the mass of a vibrating system is 10 lb, and it has a natural frequency of 1 cps. Using the phase-plane method plot the response of the system to the force function shown in the figure. What is the final amplitude of the motion?

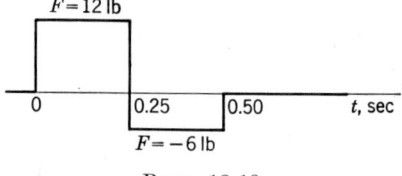

PROB. 18-10

18-11. Plot the response and find the final amplitude for Prob. 18-10 if the natural frequency is 0.5 cps.

18-12. The same as Prob. 18-10 except that $f = 2$ cps.

18-13. An undamped vibrating system has a spring scale of 200 lb/in. and a weight of 50 lb. Find the response and the final amplitude of vibration of the system if it is acted upon by the forcing function shown in the figure. Use the phase-plane method.

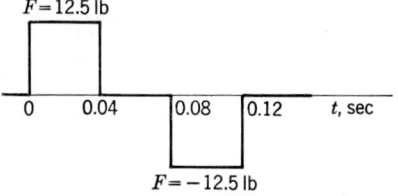

PROB. 18-13

18-14. A vibrating system has a spring $k = 400$ lb/in. and a weight $W = 80$ lb. Plot the response of this system to the forcing function shown in the figure (a) using three steps and (b) using six steps.

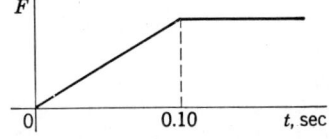

PROB. 18-14. $F_{max} = 200$ lb.

18-15. What is the value of the coefficient of critical damping for a spring-mass-damper system in which $k = 320$ lb/in. and $W = 80$ lb? If the actual damping is 20 per cent of critical, what is the natural circular frequency of the system? How does this value compare with the undamped natural frequency? What is the period of vibration of the damped system? What is the value of the logarithmic decrement?

18-16. A vibrating system has a spring $k = 20$ lb/in. and a weight $W = 30$ lb. When disturbed, it was observed that the amplitude decayed to one-fourth of its original value in 4.80 sec. What is the damping coefficient? What is the damping factor?

18-17. The same as Prob. 18-14 except that the system has damping equal to 15 per cent of critical.

18-18. A vibrating system has damping $\zeta = 0.15$, a spring constant $k = 600$ lb/in., and a weight $W = 20$ lb. When in static equilibrium it is subjected to the forcing function shown in the figure. Using the phase-plane method, plot the displacement-time and velocity-time curves of the response.

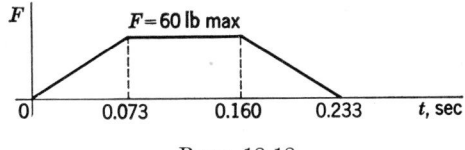

Prob. 18-18

18-19. A damped vibrating system has an undamped natural frequency of 10 cps and a weight of 800 lb. The damping ratio is 0.15. Using the phase-plane method determine the response of the system to the forcing function shown in the figure.

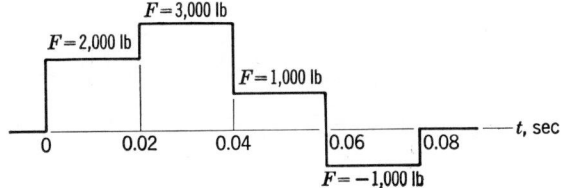

Prob. 18-19

18-20. The same as Prob. 18-19 except that the natural undamped frequency is 100 cps.

18-21. A vibrating system has 15 per cent of critical damping, a spring constant of 600 lb/in., and a weight of 36 lb. It is acted upon by the forcing function shown in the figure. Plot the phase-plane diagram and the displacement-time diagram for the system.

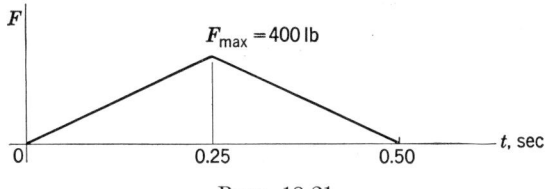

Prob. 18-21

18-22. A vibrating system has a spring scale of 3,000 lb/in., a damping factor of 55 lb-sec/in., and a weight of 800 lb. It is excited by a harmonically varying force $F_0 = 100$ lb at a frequency of 435 cycles/min. (a) Calculate the amplitude of the forced vibration and the phase angle between the vibration and the force. (b) Plot several cycles of the displacement-time and force-time diagrams.

18-23. The same as Prob. 18-22 except that the frequency of the force is 290 cycles/min.

18-24. An aircraft radio transmitter weighing 28 lb is mounted on four springs which deflect $\frac{1}{4}$ in. when the transmitter is placed upon them. Calculate the percentage of engine vibration received by the transmitter for engine speeds of 1,000, 2,000, and 3,000 rpm. Assume no damping.

18-25. When a 6,000-lb press is mounted upon structural-steel floor beams, it causes them to deflect ¾ in. If the press has a reciprocating unbalance of 420 lb and it operates at a speed of 80 rpm, how much of the force will be transmitted from the floor beams to other parts of the building? Assume no damping. Can this mounting be improved?

18-26. Four vibration mounts are used to support a 900-lb machine which has a rotating unbalance of 30 lb-in. and runs at 300 rpm. The vibration mounts have damping equal to 30 per cent of critical. What must the spring constant of the mounting be if 20 per cent of the exciting force is transmitted to the foundation? What is the resulting amplitude of motion of the machine?

18-27. A round solid steel shaft is supported in two radial bearings at A and D and carries gears at B and C whose weights are 11 and 40 lb, respectively. The shaft material has a modulus of elasticity of 28,500,000 psi and weighs 0.282 lb/in.³ Using graphical integration and Rayleigh's method calculate the critical speed of the shaft in rpm. The weight of each gear may be replaced by a single concentrated force if desired.

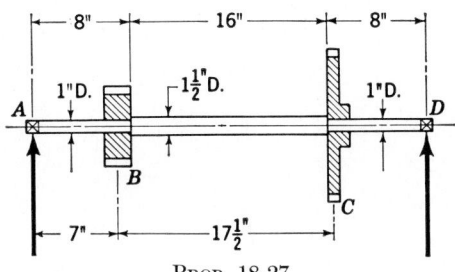

Prob. 18-27

18-28. A solid round steel shaft is supported in bearings at A and B, as shown in the figure, and carries two wheels. Using Rayleigh's method calculate the critical speed of the shaft in rpm. Use $E = 28,500,000$ psi.

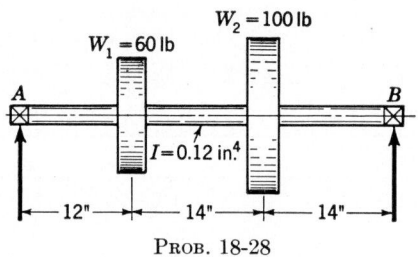

Prob. 18-28

18-29. The torsional system shown is a model of an engine and flywheel. Using Holzer's method find the natural frequencies of the system.

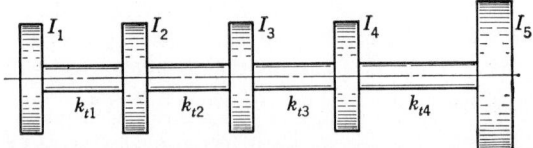

Prob. 18-29. $I_1 = I_2 = I_3 = I_4 = 0.18$ lb-sec²-ft.; $I_5 = 8.2$ lb-sec²-ft.; $k_{t1} = k_{t2} = k_{t3} = 1.48(10)^6$ lb-in./rad; $k_{t4} = 1.07(10)^6$ lb-in./rad.

18-30. Find the natural torsional frequencies of vibration for the system shown in the figure. Use Holzer's tabulation method.

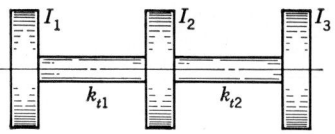

PROB. 18-30. $I_1 = I_2 = I_3 = 40.0$ lb-sec^2-in.; $k_{t1} = k_{t2} = 0.50(10)^6$ lb-in./rad.

18-31. The figure shows two cantilever springs between which a helical spring $k = 50$ lb/in. is assembled. The cantilever springs are of steel ($E = 28,500,000$ psi) and have the same moments of inertia $I = 0.00163$ in.^{4} (a) If the helical spring was originally 5 in. long, what are the deflections δ_1 and δ_2 of the cantilever springs after assembly? (b) A weight $W = 30$ lb is suspended at A. What is the equivalent spring constant of the system? (c) Calculate the natural frequency. (d) Suppose the helical spring closes to its solid height during a part of the vibration cycle. What effect would this have upon the equivalent spring constant and upon the natural frequency?

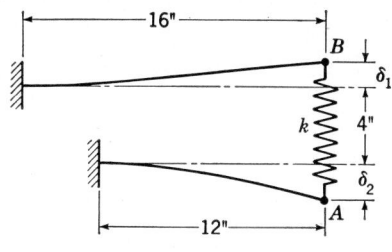

PROB. 18-31

18-32. A vibrating system consists of a cantilever spring in series with a coil spring as illustrated, together with a stop or bumper which prevents the cantilever spring from moving upward. When the system is motionless and in static equilibrium, the cantilever spring just touches the stop but without exerting any force upon it. The cantilever spring is steel ($E = 28,500,000$ psi) and has a moment of inertia of 0.002 in.^{4} The helical spring has a scale of 300 lb/in., and the weight is 150 lb. (a) What are the two natural frequencies of the system? (b) The weight is pulled down 1 in. and released to start the motion. Using the phase-plane method plot several cycles of the displacement-time diagram.

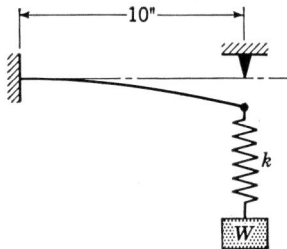

PROB. 18-32

CHAPTER 19

BALANCING

Balancing is the technique of correcting or eliminating unwanted inertia forces. In previous chapters we have seen that such forces cause vibrations which at times may reach dangerous amplitudes. Even if not dangerous, vibrations do increase the material stresses and subject bearings to repeated loads which cause parts to fail prematurely by fatigue. Thus it is not sufficient in the design of machinery merely to avoid operation near the critical speeds, but one must also eliminate, or at least reduce, the inertia forces which produce these vibrations in the first place.

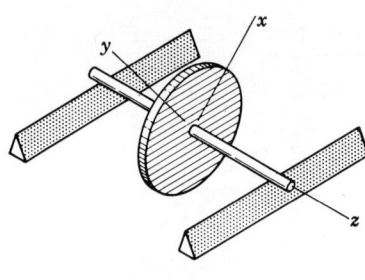

Fig. 19-1

Production tolerances used in the manufacture of machinery are adjusted as closely as possible without running up the cost of manufacture prohibitively. In general it is more economical to produce parts which are not quite true and then to subject them to a balancing procedure than it is to produce them so perfectly that no correction is needed. Because of this each part produced is an individual case in that no two parts can normally be expected to require the same corrective measures. This, the determination of the unbalance and the application of corrections, is the principal problem in the study of balancing.

19-1. Static Unbalance. The arrangement shown in Fig. 19-1 consists of a disk and shaft combination resting on hard rigid rails so that the shaft, which is assumed to be perfectly straight, can roll without friction. A reference system xyz is attached to the disk and moves with it. Simple experiments to determine if the disk is statically unbalanced can be conducted as follows: Roll the disk gently by hand and permit it to coast until it comes to rest. Then mark with chalk the lowest point of the periphery of the disk. Repeat four or five times. Now if the chalk marks are scattered at different places around the periphery, then the disk is in static balance. If all the chalk marks are coincident, then the disk is statically unbalanced, which means that the axis of the shaft and the center of mass of the disk are not coincident. The position of the chalk marks with respect to the xy system indicates the angular location, but *not* the amount, of unbalance.

It is unlikely that any of the marks will be located 180° from the remaining ones even though it is theoretically possible to obtain static equilibrium with the unbalance above the shaft axis.

If static unbalance is found to exist, it can be corrected by drilling out material at the chalk mark or by adding mass to the periphery 180° from the mark. Since the amount of unbalance is unknown, these corrections must be made by trial and error.

If the unbalanced disk and shaft is mounted in bearings and caused to rotate (Fig. 19-2), then the centrifugal force $mr_G\omega^2$ exists, as demonstrated in Sec. 16-4. The centrifugal force acting upon the shaft produces the rotating bearing reactions shown in the figure. For convenience Eq. (18-43) is repeated:

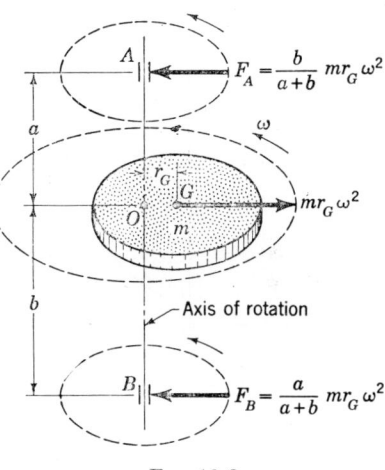

FIG. 19-2

$$\frac{mX}{m_u e} = \frac{(\omega/\omega_n)^2}{\sqrt{(1 - \omega^2/\omega_n^2)^2 + (2\zeta\omega/\omega_n)^2}} \quad (a)$$

This is the equation for the amplitude ratio of the vibration of a rotating disk and shaft combination. Here m is the total mass being vibrated and m_u is the unbalanced mass. The damping of rotating parts usually is found to be between 5 and 15 per cent of critical, but in this example, let us assume it can be neglected. If we neglect the weight of the shaft too, then, for Fig. 19-2, $m = m_u$, $e = r_G$, and Eq. (a) becomes

$$X = r_G \frac{(\omega/\omega_n)^2}{1 - (\omega/\omega_n)^2} \quad (19\text{-}1)$$

where r_G is the eccentricity and X is the amplitude of the vibration corresponding to any frequency ratio ω/ω_n. Now if, in Fig. 19-2, we designate O as the center of the shaft at the disk and G as the mass center of the disk, then we can draw some interesting conclusions by plotting Eq. (19-1). This is done in Fig. 19-3, where the amplitude is plotted on the vertical axis and the frequency ratio along the abscissa. The natural frequency is ω_n, which corresponds to the critical speed, while ω is the actual speed of the shaft. When rotation is just beginning, ω is much less than ω_n and the graph shows that the amplitude of the vibration is very small. As the shaft speed increases, the amplitude also increases and becomes infinite at the critical speed. As the shaft goes through the critical, the amplitude changes over to a negative value and decreases as the shaft speed increases. The graph shows that the amplitude never returns to zero no matter how much the shaft speed is increased but

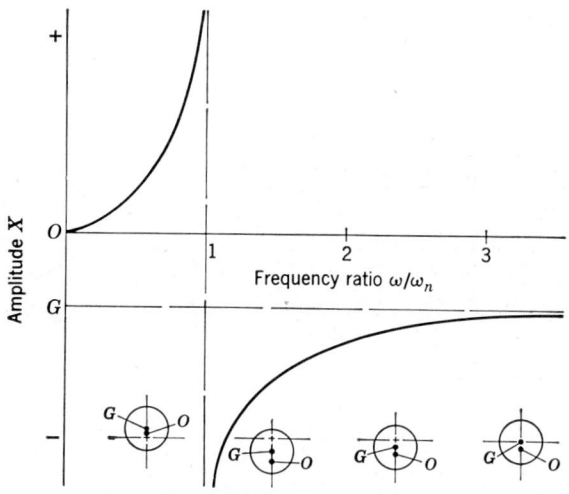

FIG. 19-3. The small figures below the graph indicate the relative position of three points for various frequency ratios. The mass center of the disk is at G, the center of the shaft at O, and the axis of rotation at the intersecting center lines. Thus, this figure shows both amplitude and phase relationships.

reaches a limiting value of $-r_G$. Note, in this range, that the disk is rotating about its own center of gravity, which is then coincident with the bearing center line.[1]

The preceding discussion demonstrates that statically unbalanced rotating systems produce undesirable vibrations and rotating bearing reactions. Using static balancing equipment the eccentricity r_G can be reduced, but it is impossible to make it zero. Therefore, no matter how small r_G is made, trouble can always be expected whenever $\omega = \omega_n$. When the operating frequency is higher than the natural frequency, then the machine should be designed to pass through the natural frequency as rapidly as possible in order to prevent dangerous vibrations from building up.

[1] For an excellent discussion of this motion see E. B. Cole, "The Theory of Vibrations for Engineers," 3d ed., pp. 315–324, The Macmillan Company, New York, 1957.

19-2. Static Balancing Machines. The purpose of a balancing machine is first to indicate whether a part is in balance. If it is out of balance, then the machine must measure the unbalance by indicating its *magnitude* and *location*.

Static balancing machines are used only for parts whose axial dimensions are small, such as gears, fans, and impellers, and the machines are often called *single-plane balancers* because the mass must practically

FIG. 19-4. A helicopter-rotor assembly balancer. (*Courtesy Micro-Poise Engineering and Sales Company, Detroit, Mich.*)

lie in a single plane. In the sections to follow we shall discuss balancing in several planes, but it is important to note here that, if several wheels are to be mounted upon a shaft which is to rotate, then the parts should be individually statically balanced prior to mounting. While it is possible to balance the assembly in two planes after the parts are mounted, we shall find that additional bending moments inevitably come into existence when this is done.

Static balancing is essentially a weighing process in which the part is acted upon by either a gravity force or a centrifugal force. We have seen that the disk and shaft of the preceding section could be balanced by placing it on two parallel rails, rocking it, and permitting it to seek equilibrium. In this case the location of the unbalance is found through the aid of the force of gravity. Another method of balancing the disk would be to rotate it at a predetermined speed. Then the bearing reactions could be measured, and their magnitudes used to indicate the amount of unbalance. Since the part is rotating while the measurements are taken, a stroboscope is used to indicate the location of the required correction.

When machine parts are manufactured in large quantities, a balancer is required which will measure both the amount and location of the unbalance and give the correction directly and quickly. Time can also be saved if it is not necessary to rotate the part. Such a balancing machine is shown in Fig. 19-4. This machine is essentially a pendulum

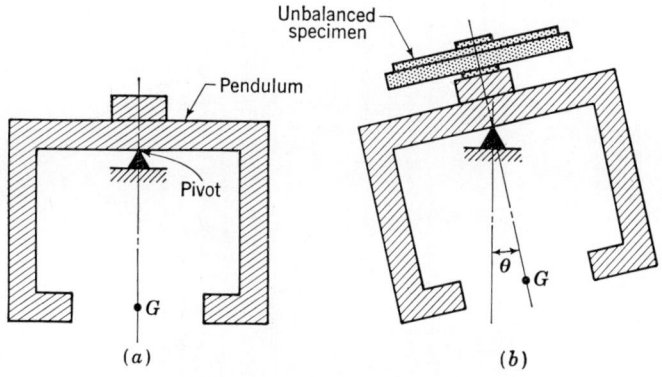

FIG. 19-5. Operation of a static balancing machine.

which can tilt in any direction, as illustrated by the schematic drawing of Fig. 19-5a. When an unbalanced specimen is mounted upon the platform of the machine, the pendulum tilts. The direction of the tilt gives the location of the unbalance, while the angle θ (Fig. 19-5b) indicates the magnitude of the unbalance. Damping is employed to eliminate oscillations of the pendulum. Figure 19-6 shows a universal level which is mounted on the platform of the balancer. In balancing, the position of the bubble gives both the location and the magnitude of the correction.

19-3. Dynamic Unbalance. Figure 19-7 shows a long rotor which is to be mounted in bearings at A and B. We might suppose that two equal masses m_1 and m_2 are placed at opposite ends of the rotor and at equal distances r_1 and r_2 from the axis of rotation. Since the masses are equal and on opposite sides of the rotational axis, the rotor can be placed on rails as described earlier to show that it is statically balanced in all angular positions.

If the rotor of Fig. 19-7 is placed in bearings and caused to rotate at an angular velocity ω rad/sec, then the centrifugal forces $m_1 r_1 \omega^2$ and $m_2 r_2 \omega^2$ act, respectively, at m_1 and m_2 on the rotor ends. These centrifugal forces produce the unequal bearing reactions $\mathbf{F}_A$ and $\mathbf{F}_B$, and the entire system of forces rotates with the rotor at the angular velocity ω. Thus a

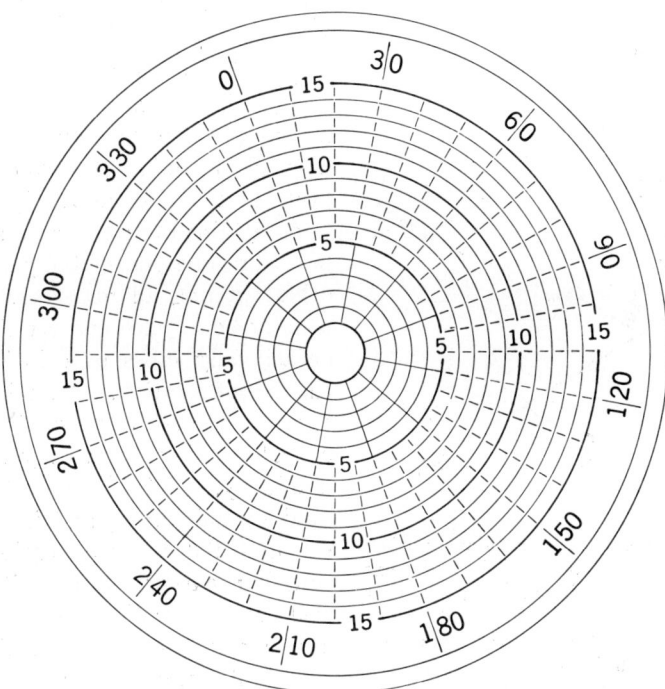

FIG. 19-6. Drawing of the universal level used on the Micro-Poise balancer. The numbers on the periphery are degrees; the radial distances are calibrated in units proportional to ounce-inches. The position of the bubble indicates both the location and the magnitude of the unbalance. (*Micro-Poise Engineering and Sales Company, Detroit, Mich.*)

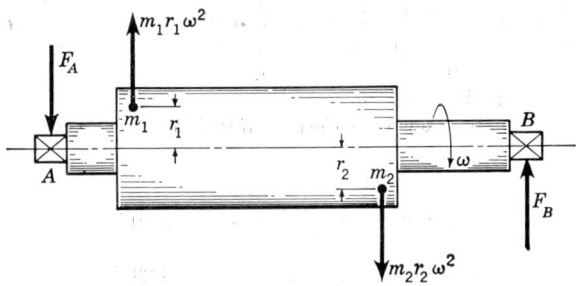

FIG. 19-7. The rotor is statically balanced if $m_1 = m_2$ and $r_1 = r_2$ but dynamically unbalanced.

part may be statically balanced and, at the same time, dynamically unbalanced (Fig. 19-8).

In the general case distribution of the mass along the axis of the part depends upon the configuration of the part, but errors occur in machining and also in casting and forging. Other errors or unbalance may be caused by improper boring, by keys, and by assembly. It is the designer's responsibility to design so that a line joining all mass centers will be a straight line coinciding with the axis of rotation. However, perfect parts and perfect assembly are seldom attained, and consequently a line from one end of the part to the other, joining all mass centers, will usually be a space curve which may occasionally cross or coincide with the axis of rotation. An unbalanced part, therefore, will usually be out of balance both statically and dynamically. This is the most general kind of unbalance, and if the part is supported by two bearings, one can then expect the magnitudes as well as the directions of these rotating bearing reactions to be different. This is illustrated in the example to follow, which is a case of combined static and dynamic unbalance.

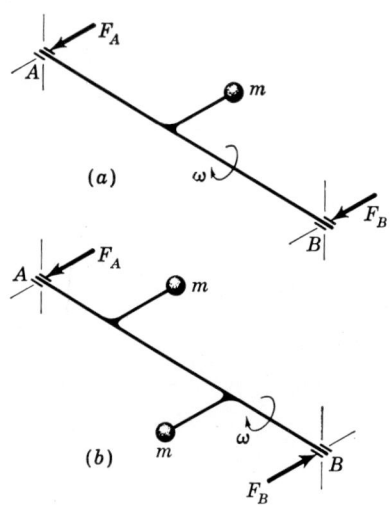

FIG. 19-8. (a) Static unbalance; when the shaft rotates, both bearing reactions are in the same plane and in the same direction. (b) Dynamic unbalance; when the shaft rotates, the unbalance creates a couple tending to turn the shaft end over end. The shaft is in equilibrium because of the opposite couple formed by the bearing reactions. Note that the bearing reactions are still in the same plane but opposite in direction.

EXAMPLE 19-1. Figure 19-9 represents a rotating system which has been idealized for illustrative purposes. It consists of a weightless shaft supported in bearings at A and B and rotating at $\Omega = 100I$ rad/sec. The shaft has three weights, w_1, w_2, and w_3, connected to it and rotating with it. We are to calculate the magnitude and direction of the bearing reactions at A and B for the particular position shown.

Solution. We begin by calculating the centrifugal force due to each rotating weight. These are

$$m_1 r_1 \omega^2 = \frac{2}{(386)(16)} (3)(100)^2 = 9.70 \text{ lb}$$

$$m_2 r_2 \omega^2 = \frac{1}{(386)(16)} (2)(100)^2 = 3.23 \text{ lb}$$

$$m_3 r_3 \omega^2 = \frac{1.5}{(386)(16)} (2.5)(100)^2 = 6.06 \text{ lb}$$

BALANCING 507

These three forces are parallel to the yz plane, and we can write them in vector form by inspection:

$$\mathbf{F}_1 = m_1 r_1 \omega^2 / \theta = 9.70\mathbf{J}$$
$$\mathbf{F}_2 = m_2 r_2 \omega^2 / \theta = -1.62\mathbf{J} + 2.80\mathbf{K}$$
$$\mathbf{F}_3 = m_3 r_3 \omega^2 / \theta = -5.85\mathbf{J} - 1.57\mathbf{K}$$

where θ, in this example, is the angular direction of the vectors when viewed from the positive end of the x axis. The moments of these forces taken about the bearing

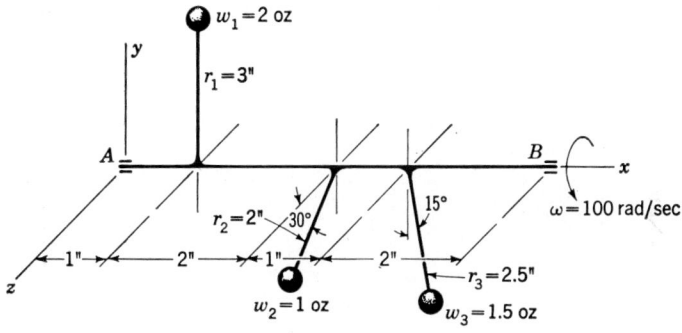

FIG. 19-9

at A must be balanced by the moment of the bearing reaction at B. Therefore

$$\Sigma \mathbf{T}_A = (1\mathbf{I}) \times (9.70\mathbf{J}) + (3\mathbf{I}) \times (-1.62\mathbf{J} + 2.80\mathbf{K}) + (4\mathbf{I}) \times (-5.85\mathbf{J} - 1.57\mathbf{K}) + (6\mathbf{I}) \times \mathbf{F}_B = 0$$

so that $\mathbf{F}_B = 3.10\mathbf{J} - 0.352\mathbf{K}$ lb

To find the reaction at A we repeat the analysis. Taking moments about B gives

$$\Sigma \mathbf{T}_B = (-2\mathbf{I}) \times (-5.85\mathbf{J} - 1.57\mathbf{K}) + (-3\mathbf{I}) \times (-1.62\mathbf{J} + 2.80\mathbf{K}) + (-5\mathbf{I}) \times (9.70\mathbf{J}) + (-6\mathbf{I}) \times \mathbf{F}_A = 0$$

so that $\mathbf{F}_A = -5.32\mathbf{J} + 0.877\mathbf{K}$ lb

The magnitude of the two bearing reactions is found to be 5.40 lb at A and 3.22 lb at B. These reactions, their directions, and the original centrifugal forces are shown in Fig. 19-10.

The reader should note that these are the rotating bearing reactions and that they do not include the static or stationary components due to the gravity force acting upon the masses themselves.

19-4. Graphical Analysis of Unbalance. Using the two equations

$$\Sigma F = 0 \qquad \Sigma T = 0 \qquad (a)$$

it is possible to develop graphical methods of analyzing unbalance to determine the amount and location of the corrections. We begin by noting that centrifugal force is proportional to the product wr of a rotating eccentric mass. Thus vector quantities, proportional to the centrifugal force of each of the three masses w_1R_1, w_2R_2, and w_3R_3 of Fig. 19-11a, will act in radial directions as shown. The first of Eqs. (a) is applied by constructing a force polygon (Fig. 19-11b). Since this polygon requires another vector w_cR_c for closure, the magnitude of the correction is

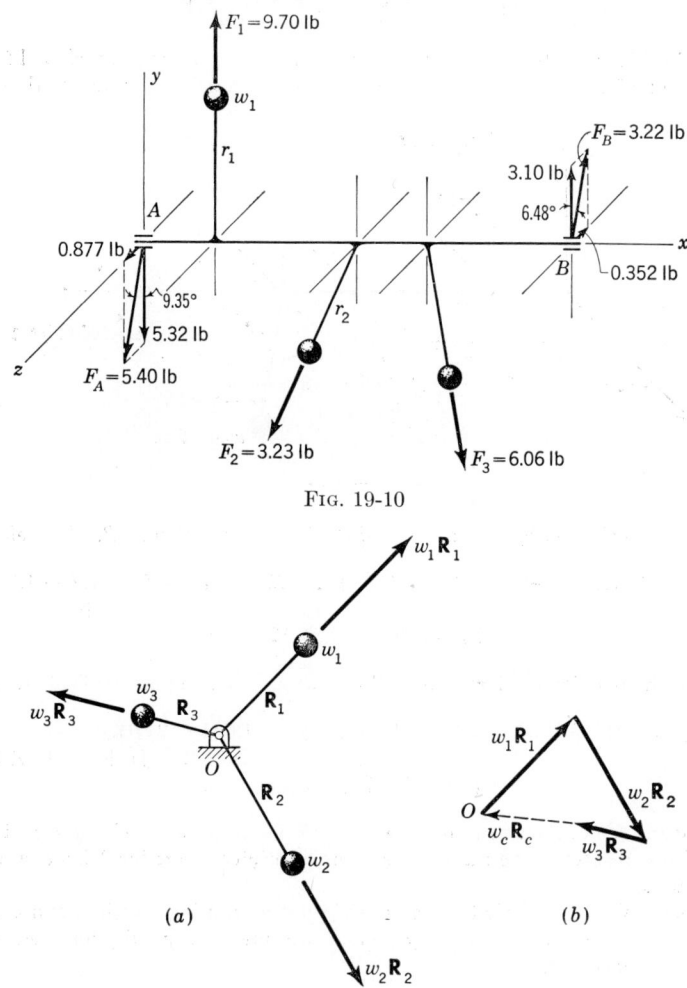

Fig. 19-10

Fig. 19-11. (a) A three-mass system rotating in a single plane. (b) Centrifugal force polygon gives w_cR_c as the required correction.

w_cR_c and its direction is parallel to R_c. The three masses of Fig. 19-11 are assumed to rotate in a single plane, and so this is a case of static unbalance.

When the rotating masses are in different planes, both of Eqs. (a) must be used. Figure 19-12a is an end view of a shaft having mounted upon it the three masses w_1, w_2, and w_3 at radial distances R_1, R_2, and R_3, respectively. Figure 19-12b is a side view of the same shaft showing left and right correction planes and the distances to the three masses. It is desired to determine the magnitude and angular location of the corrections for each plane.

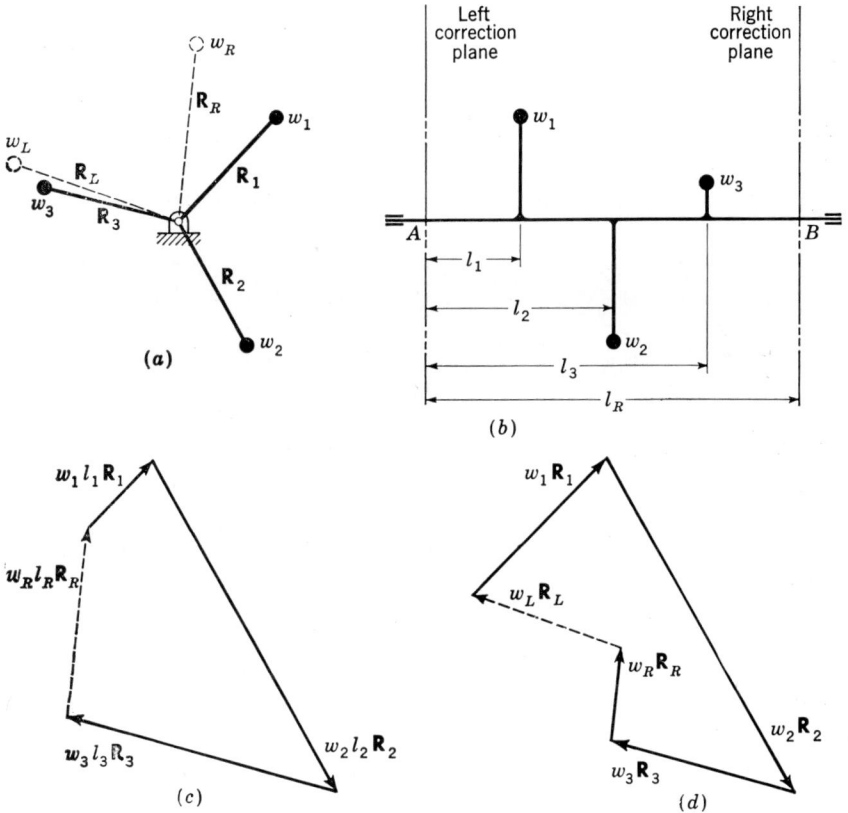

FIG. 19-12. Graphical analysis of unbalance.

The first step in the solution is to take a summation of the moments of the centrifugal forces, including the corrections, about some point. We choose to take this summation about A in the left correction plane in order to eliminate the moment of the left correction weight. Thus, applying the second of Eqs. (a) gives

$$\Sigma T_A = w_1 l_1 R_1 + w_2 l_2 R_2 + w_3 l_3 R_3 + w_R l_R R_R = 0 \qquad (b)$$

This is a vector equation in which the directions of the vectors are parallel, respectively, to the vectors R_N in Fig. 19-12a. Consequently, the moment polygon of c can be constructed. The closing vector $w_R l_R \mathbf{R}_R$ gives the magnitude and direction of the correction required for the right-hand plane. The quantities w_R and $\mathbf{R}_R$ are now known. Therefore the equation

$$\Sigma \mathbf{F} = w_1 \mathbf{R}_1 + w_2 \mathbf{R}_2 + w_3 \mathbf{R}_3 + w_R \mathbf{R}_R + w_L \mathbf{R}_L = 0 \quad (c)$$

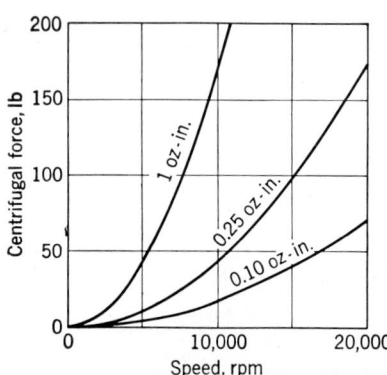

FIG. 19-13. The graph shows the effect of various amounts of unbalance on the centrifugal force for different speeds.

can be written. This equation is solved for the left-hand correction $w_L \mathbf{R}_L$ by constructing the force polygon of Fig. 19-12d.

19-5. Dynamic Balancing. The units in which unbalance is measured are the ounce-inch (oz-in.), the gram-centimeter (g-cm), and the gram-inch (g-in.). There are 72 g-cm in 1 oz-in. In this book we shall employ the ounce-inch as the unit of measurement, although the gram-inch is now coming into use in this country too.

It is desirable to adapt the equation for centrifugal force to use the unbalanced weight in ounces and the shaft speed in rpm. The equation then becomes

$$F = mr\omega^2 = \frac{wr}{(386)(16)} \left(\frac{2\pi n}{60}\right)^2 = (1.775)(10)^{-6} wrn^2 \quad (19\text{-}2)$$

where w = weight, oz
r = radius, in.
n = speed, rpm
F = centrifugal force, lb

As an example of the use of this equation consider a rotating part weighing 1,000 oz (62.5 lb) with an eccentricity of 0.001 in. The unbalance is $wr = 1$ oz-in. If this part is caused to rotate at 8,000 rpm, Eq. (19-2) gives the resulting centrifugal force as 114 lb. Figure 19-13 is a graph showing the relationship between centrifugal force and speed for various amounts of unbalance.

We have seen that static balancing is sufficient for rotating disks, wheels, gears, and the like when the mass can be assumed to exist in a single rotating plane. In the case of longer machine elements, such as turbine rotors or motor armatures, the unbalanced centrifugal forces result in couples whose effect is to tend to cause the rotor to turn end over end. The purpose of balancing is to measure the unbalanced couple and to substitute a new couple in the opposite direction and of the same magnitude. The new couple is introduced by the addition of weights on two preselected correction planes or by the subtracting of weights (drilling out) on these two planes. A rotor to be balanced will usually have both static and dynamic unbalance, and consequently the correction weights, their radial location, or both will not be the same for the two correction planes. This also means that the angular separation of the correction weights on the two correction planes will usually not be 180°. Thus, to balance a rotor, one must measure the magnitude and angular location of the correction weight for each of the two correction planes.

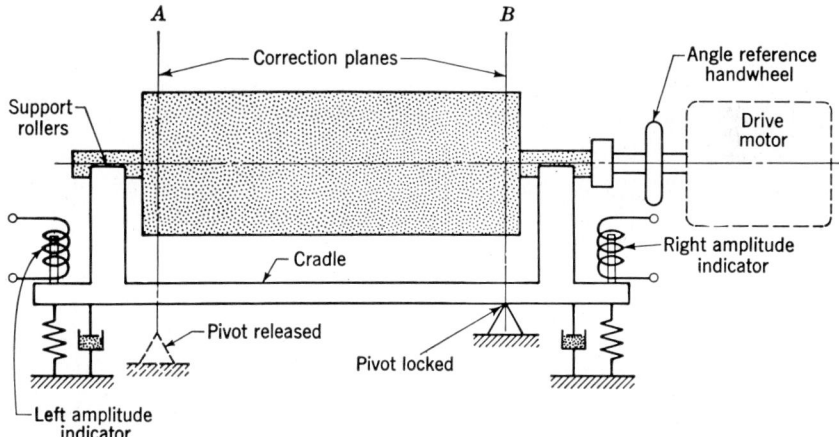

Fig. 19-14. Schematic drawing of pivoted-cradle balancing machine.

Three methods of measuring the corrections for two planes are in general use. These are the *pivoted-cradle*, the *nodal-point*, and the *mechanical-compensation* methods. We proceed next to a detailed exploration of each of these methods.

19-6. Pivoted-cradle Balancing. Figure 19-14 shows a specimen to be balanced mounted on half bearings or rollers attached to a cradle. The right end of the specimen is connected to a drive motor through a universal joint. The cradle can be rocked about either of two points which are adjusted to coincide with the correction planes on the specimen to be balanced. In the figure the left pivot is shown in the released position and the cradle and specimen are free to rock or oscillate about the right pivot, which is shown in the locked position. Springs and dashpots are secured at each end of the cradle so as to provide a single-degree-of-freedom vibrating system. Often these are made adjustable so that the natural frequency can be tuned to the motor speed. Also shown are amplitude indicators at each end of the cradle. These are velocity-type vibration pickups. A permanent magnet mounted on the cradle moves relative to a stationary coil and generates a voltage proportional to the unbalance. This voltage is amplified, and the result read directly from a voltmeter calibrated in ounce-inches.

With the pivots located in the two correction planes, one can lock either pivot and take readings of the amount and angle of location of the correction. The readings obtained will be completely independent of measurements taken in the other correction plane because an unbalance in the plane of the locked pivot will have no moment about that pivot. With the right-hand pivot locked, an unbalance correctable in the left correction plane will cause vibration whose amplitude is measured by the left amplitude indicator. When this correction is made (or measured), the right-hand pivot is released, the left pivot locked, and another set of measurements made for the right-hand correction plane using the right-hand amplitude indicator.

The relation between the *amount* of the unbalance and the measured amplitude is given by Eq. (18-43). Rearranging and substituting $wr/16g$ for $m_u e$ give

$$X = \frac{wr(\omega/\omega_n)^2}{16mg\sqrt{(1 - \omega^2/\omega_n^2)^2 + (2\zeta\omega/\omega_n)^2}} \quad (19\text{-}3)$$

where wr = unbalance, oz-in.
m = mass of cradle and specimen, lb-sec^2/in.
ω = angular velocity of specimen, rad/sec
ω_n = natural frequency of cradle and specimen, rad/sec
ζ = damping ratio
X = amplitude of vibration, in.

Equation (19-3) shows that the amplitude of the vibration is directly proportional to the amount of the unbalance. This equation is plotted in Fig. 19-15 for one damping ratio. The figure shows that the machine will be most sensitive near resonance, since in this region the greatest amplitude is recorded for a given unbalance. Damping is deliberately introduced in balancing machines to filter noise and other vibrations which might affect the results. This damping is also beneficial in that it helps to maintain a constant calibration regardless of variations in temperature or other environmental conditions.

Not shown in Fig. 19-14 is a sine-wave signal generator which is attached to the drive shaft. Because the signal generator is attached to the drive shaft, the circular frequency of the generated sine wave and of the drive shaft are identical. If this sine wave is compared on an oscilloscope with the wave generated by one of the amplitude indicators, a phase difference will be found. This angular phase difference is the angular location of the unbalance. In a balancing machine an electronic phasemeter measures the phase angle and gives the result on another meter calibrated in degrees. To locate the correction on the specimen (Fig. 19-14) the angular reference handwheel is turned by hand until the indicated angle is in line with a reference pointer. This places the heavy side of the specimen in any preselected position and permits the correction to be made.

The phase angle is given by Eq. (18-44) and is repeated here for convenience:

$$\phi = \tan^{-1} \frac{2\zeta\omega/\omega_n}{1 - \omega^2/\omega_n^2} \qquad (19\text{-}4)$$

A plot of this equation for a single damping ratio and for various frequency ratios is shown in Fig. 19-16. This curve shows that, at resonance, when the speed of the shaft and the frequency of the vibration

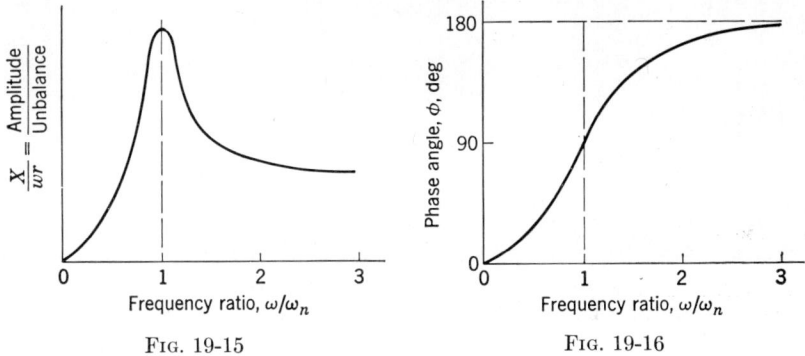

Fig. 19-15   Fig. 19-16

are the same, the displacement lags the unbalance by 90°. Thus, if the top of the specimen is turning away from the operator, then the unbalance will be horizontal and directly in front of the operator when the displacement is maximum downward. The figure also shows that the angular location approaches 180° as the frequency ratio increases above unity.

A pivoted-cradle balancing machine designed for high-speed production is illustrated in Fig. 19-17. These machines can be obtained singly in many sizes or in batteries for use in automatic production lines. The Tinius Olsen Company, for example, manufactures batteries consisting of three balancing machines in which automotive crankshafts are loaded from a conveyor, the unbalance measured, the proper correction applied, the crankshaft retested for balance and then returned to the conveyor, all automatically and under the supervision of a single operator.[1] This

---

[1] Warren M. Gruber, In-line Production Balancing, *Automation*, vol. 5, no. 2, pp. 80–83, 1958.

FIG. 19-17. A Tinius Olsen static-dynamic pivoted-cradle balancing machine with specimen mounted for balancing. (*Tinius Olsen Testing Machine Company, Willow Grove, Pa.*)

battery checks up to 90 crankshafts per hour and balances them to an accuracy of 0.30 oz-in.

A block diagram of the electronic circuits in a Tinius Olsen machine is shown in Fig. 19-18. The two electronic channels, one for *amount* and one for *location*, are shown. The polarity switch in the location channel rotates the angle through 180°. Sometimes it is necessary to shift this switch when the angle is near 0 or 360°; otherwise small variations may cause the pointer to jump from one end of the scale to the other.

19-7. Nodal-point Balancing. Plane separation using a point of zero or minimum vibration is called the *nodal-point* method of balancing. To see how this method is used for balancing let us examine Fig. 19-19. Here the specimen to be balanced is shown mounted on bearings which are fastened to a nodal bar. We assume that the specimen is already balanced in the left-hand correction plane and that an unbalance still exists in the right-hand plane as shown. Because of this unbalance a vibration of the entire assembly takes place, causing the nodal bar to oscillate about some point O, occupying, first, position CC, then DD. Point O is easily located by sliding a dial indicator along the nodal bar until a point of zero motion is found. This is the null or nodal point. Its location is the center of oscillation for a center of percussion in the right-hand correction plane.

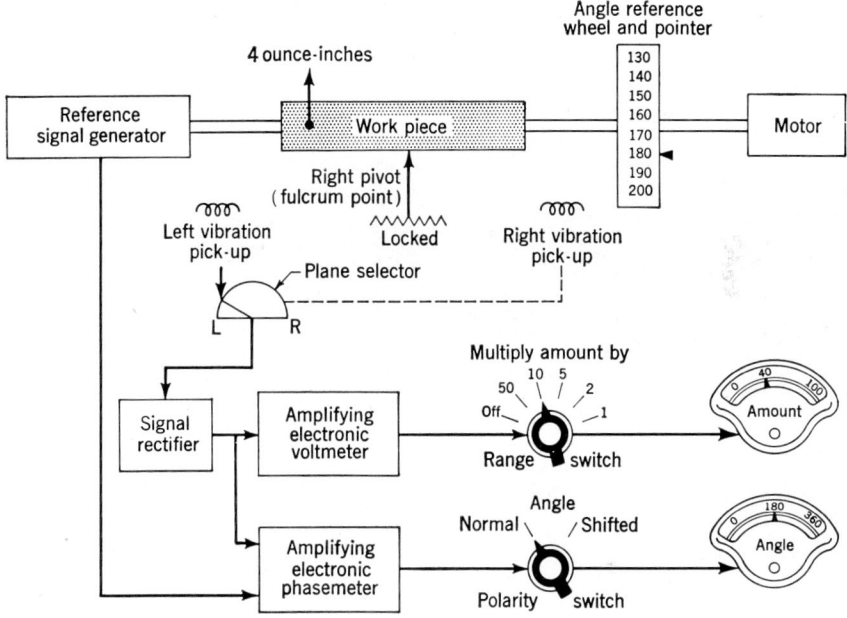

FIG. 19-18. Functional diagram of a Tinius Olsen balancing machine with the circuits set for the left plane of correction. (*Tinius Olsen Testing Machine Company, Willow Grove, Pa.*)

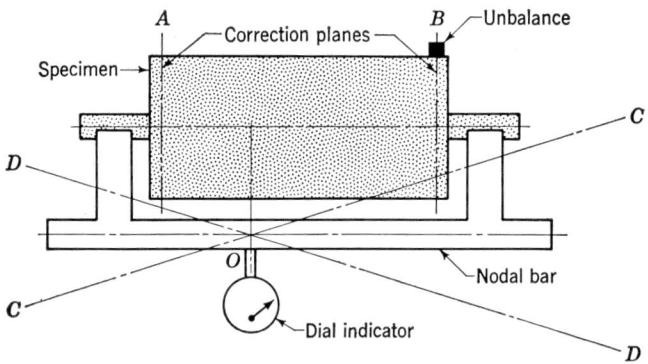

FIG. 19-19. Plane separation by the nodal-point method. The nodal bar experiences the same vibration as the specimen.

We assumed, at the beginning of this discussion, that no unbalance existed in the left-hand correction plane. However, if unbalance *is* present, its magnitude will be given by the dial indicator located at the nodal point just found. Thus, by locating the dial indicator at this nodal point, we measure the unbalance in the left-hand plane without any interference from that in the right-hand plane. In a similar manner another nodal point can be found which will measure only the unbalance in the right-hand correction plane.

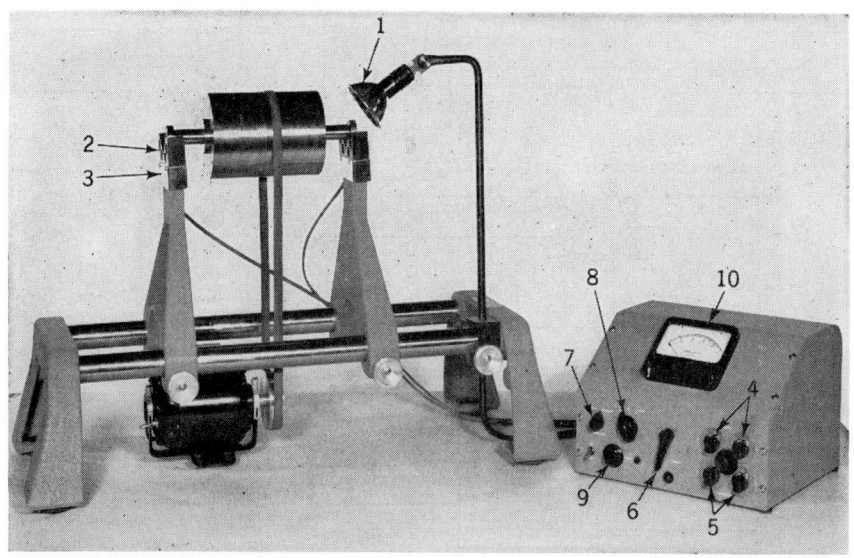

FIG. 19-20. The Micro SU-6 Dynamic Balancer. Rotor weight, 0.25 to 100 lb; balancing speed, 900 to 3,000 rpm; sensitivity, 0.00005 in. displacement; maximum swing diameter, 20 in. The arrows identify the following items: (1) strobe light, (2) vee bearing block, (3) left-hand pickup, (4) correction-plane selector, (5) sensitivity control, (6) left-right switch, (7) position-amount switch, (8) scale-selector switch, (9) synchronism control, (10) amount meter. (*Micro Balancing, Inc., Garden City Park, Long Island, N.Y.*)

In commercial balancing machines employing the nodal-point principle the plane separation is accomplished in electrical networks.[1] Typical of many of these is the Micro Dynamic Balancer shown in Fig. 19-20. A rotor mounted in this machine will vibrate in a horizontal direction and actuate two transducers or pickups connected to the vee blocks. The two transducers convert the motion to electrical signals which are fed to an electronic computer and mixed. A switching knob selects either correction plane and displays the unbalance on a voltmeter which

[1] For a more detailed description see Steve Elonka, Balancing Rotating Machinery, *Power*, vol. 103, no. 6, pp. 213–236, 1959.

is calibrated in ounce-inches. This arrangement is shown schematically in Fig. 19-21.

The computer of Fig. 19-21 contains a filter which eliminates bearing noise and other frequencies not related to the unbalance. A multiplying network is used to give any sensitivity desired and to cause the meter to read in preselected balancing units. The strobe light is driven by an oscillator which is synchronized to the rotor speed.

The rotor is driven at a speed which is much greater than the natural frequency of the system, and since the damping is quite small, Fig. 19-16 shows that the phase angle will be approximately 180°. Marked on the right-hand end of the rotor of Fig. 19-20 are degrees or numbers which are readable and stationary under the strobe light during rotation of the rotor. Thus it is only necessary to observe the particular station number

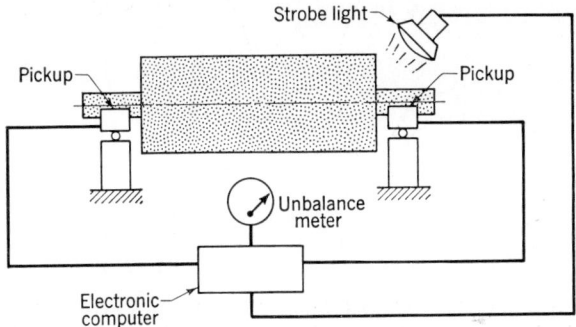

FIG. 19-21. Diagram of the electrical circuit in the Micro Dynamic Balancer.

or degree marking under the strobe light to locate the heavy spot. When the switch is shifted to the other correction plane, the meter again reads the amount and the strobe light illuminates the station. Sometimes as few as five station numbers distributed uniformly around the periphery are adequate for balancing.

The direction of the vibration is horizontal, and the phase angle is nearly 180°. Thus, rotation such that the top of the rotor moves away from the operator will cause the heavy spot to be in a horizontal plane and on the near side of the axis when illuminated by the strobe lamp. A pointer is usually placed here to indicate its location. If, during production balancing, it is found that the phase angle is less than 180°, then the pointer can be shifted slightly to indicate the proper position to observe.

19-8. Mechanical Compensation. An unbalanced rotating rotor located in a balancing machine develops a vibration. One can introduce, in the balancing machine, counterforces in each correction plane which exactly balance the forces causing the vibration. The result of introducing these is a smooth-running motor. Upon stopping, the location and amount of the counterforce are measured to give the exact correction required. This is called *mechanical compensation*.

When mechanical compensation is used, the speed of the rotor during balancing is not important because the equipment will be in calibration for all speeds. The rotor may be driven by a belt, from a universal joint, or it may be self-driven if, for example, it is a gasoline engine. The electronic equipment is simple, no built-in damping is necessary, and the machine is easy to operate because the unbalance in both correction planes is measured simultaneously and the magnitude and location read directly.

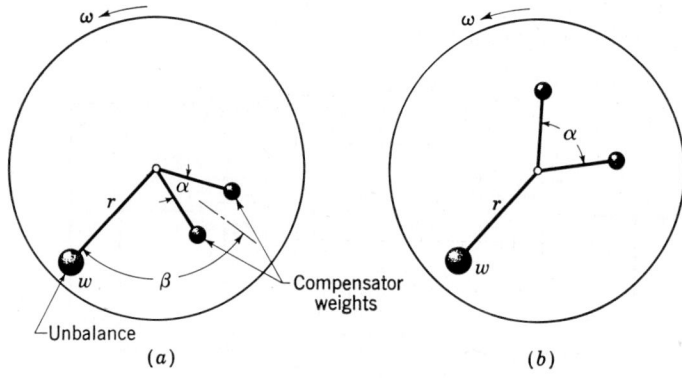

FIG. 19-22. Correction plane viewed from along the axis of rotation to show the unbalance and the compensator weights. (a) Position of compensator weights increases the vibration. (b) Compensated.

We can understand how mechanical compensation is applied by examining Fig. 19-22a. Looking at the end of the rotor, we see one of the correction planes with the unbalance to be corrected represented by wr. Two compensator weights are also shown in the figure. All three of these weights are to rotate with the same angular velocity ω, but by means to be explained shortly, the position of the compensator weights relative to one another and their position relative to the unbalanced weight can be varied by two controls. One of these controls changes the angle α, that is, the angle between the compensator weights. The other control changes the angular position of the compensator weights relative to the unbalance, that is, the angle β. The knob which changes the angle β is the *location* control, and when the rotor is compensated (balanced) in this plane, a pointer on the knob indicates the exact angular location of the unbalance. The knob which changes the angle α is the *amount* control, and it also gives a direct reading when the rotor unbalance is compensated. The magnitude of the vibration is measured electrically and displayed on a voltmeter. Thus compensation is secured when the controls are manipulated to make the voltmeter read zero.

The Tinius Olsen balancer shown in Fig. 19-23 permits vibration in a horizontal plane. The amplitude of this vibration is measured by

Fig. 19-23. View of a compensator-type engine balancer. (*Tinius Olsen Testing Machine Company, Willow Grove, Pa.*)

electrical pickups in each correction plane, amplified, and displayed on meters in the control console (Fig. 19-24). When an unbalance is found, controls in the console are operated to produce a second unbalance which compensates for the original unbalance. These controls are manipulated to cause the meter to read zero. The positions of the controls then indicate the amount and the location of the correction.

An a-c generator is attached to the shaft of the part to be balanced and used to drive selsyns located in the control cabinet and also selsyns to which the compensator weights are attached, as shown in Fig. 19-23. These are all in synchronism with the a-c generator, but the stators of the control selsyns are connected to the amount and location knobs. When these knobs are moved, the position of the compensating weights can be changed relative to the rotating unbalance. The block diagram of Fig. 19-25 will be helpful in understanding the operation.

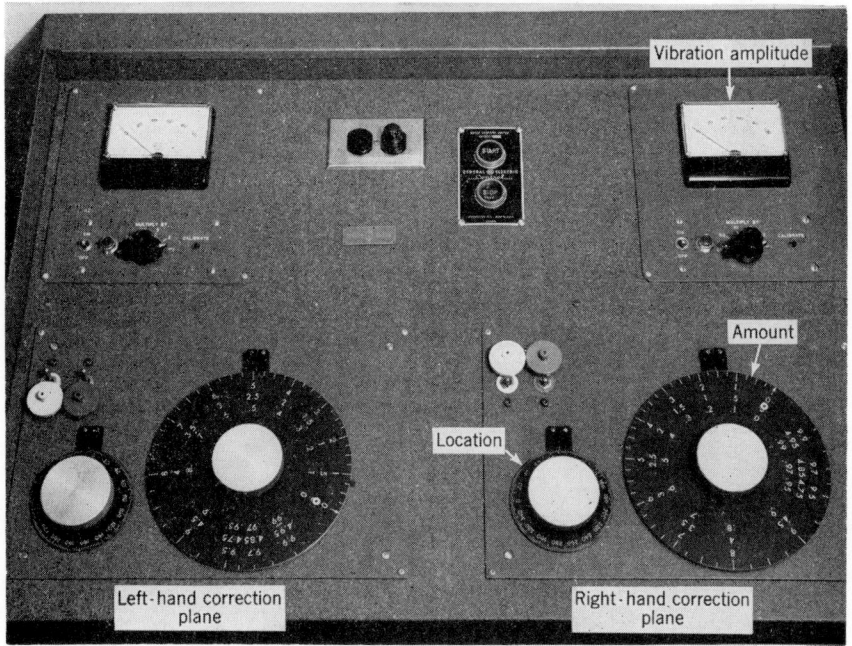

FIG. 19-24. The control console. (*Tinius Olsen Testing Machine Company, Willow Grove, Pa.*)

19-9. Balancing Flexible Rotors. The balancing procedures discussed in this chapter have assumed the existence of a rigid rotor, that is, that the shaft supporting the rotating mass or masses experiences no deformation at the operating speed. When the speed of the shaft is well below the critical speed of the system, say not over 50 per cent, this assumption of rigidity gives very good results.

When the shaft speed is near or is more than the critical speed of the system, then the shaft should be balanced in its own bearings at the operating speed. Otherwise, the addition of balancing corrections in the usual procedure will not correct the unbalance and may actually create more unbalance than was originally present (Fig. 19-26).

It is noted that in some instances the part may operate at speeds beyond the first and, sometimes, beyond even the second or third critical speed. In all these cases balancing can be achieved at the operating speed, but the part will usually be out of balance at any other speed.

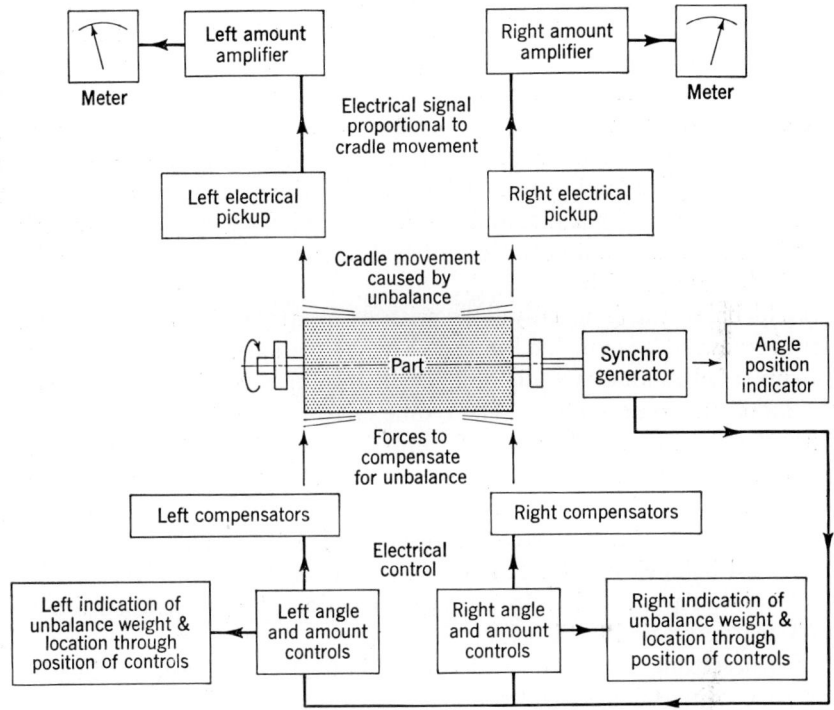

FIG. 19-25. Block diagram of Tinius Olsen compensator-type electronic balancing machine. (*Tinius Olsen Testing Machine Company, Willow Grove, Pa.*)

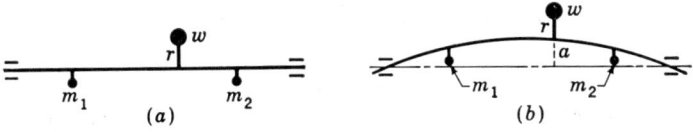

FIG. 19-26. (a) Original unbalance wr corrected by $m_1 m_2$ employing the usual procedures for a rigid shaft. (b) Shaft deflects at the critical speed, and the new unbalance is $w(r + a)$. Correction weights are no longer effective.

19-10. Field Balancing. While large rotors should be balanced in the shop before assembly into a machine, it is frequently necessary to balance them in the field too. The reasons for this are that the bearings may have some elasticity, that slight deformations may occur during assembly or in shipping, or that residual strains may be built in or relieved during the preliminary testing period. Changes also occur in machinery during long periods of operation, and vibration is a consequence of these changes. This is often due to temperature or wear but may also be caused by erosion in the case of centrifugal machinery.

Before rebalancing machinery it is always wise to check for other sources of the vibration. The frequency can quickly be found with a vibration analyzer. If there are a number of machines involved, shut them all down and run them individually until the trouble is located. Check the vibration of each machine as it coasts to a stop to see if there are large amplitude changes. Check all bearings for alignment and fit and the gears for backlash and wear. Sometimes a resonant condition develops from a remote exciting force, and in these cases the mountings and isolators should be checked.

Field balancing is usually done with a vibration pickup and a strobe light somewhat as explained in the previous sections. The output of the pickup is fed to an electronic unit which amplifies the signal and displays the amplitude on a meter. The strobe lamp is operated in synchronism with the shaft so that a piece of colored tape fastened to an end of the rotor appears as a stationary reference mark. Thus the strobe lamp illuminates the phase angle as indicated by the angular location of the tape, and the meter shows the corresponding amplitude.

In field balancing, a trial-and-error system may be employed in which trial weights are attached to the two correction planes and their amount and location varied until balance is obtained. The equipment described in the preceding paragraph will usually indicate the directions in which the succeeding trials should proceed.

Another method of field balancing involves making three test runs. The rotor is first tested at its operating speed by observing the amplitude of vibration in a preselected direction at each bearing. A reference mark is painted on the rotor, and the angle that this mark makes with the direction of the measured amplitude recorded. From this information two vector equations relating the amplitude and the unbalance can be written. In the second test a trial weight is fastened to a correction plane and the resulting amplitudes and phase angles relative to the same reference mark again recorded. This results in two more vector equations. In the third test the trial weight or another weight is attached to the other correction plane, the amplitude and phase angles recorded, and this yields two vector equations also. These six vector equations can then be solved for the amount and location of the corrections for the two correction planes. Since the solution of six vector equations is equivalent to solving twelve algebraic equations simultaneously, the work gets rather tedious. Space requirements forbid the inclusion of a demonstration in this book. The method, however, utilizes fundamentals already studied; the solution is obtained by writing the usual equations of equilibrium.[1]

PROBLEMS

19-1. If the system shown in the figure rotates at 600 rpm, what are the bearing reactions at A and B? The system is to be balanced by a weight located at a radius of 12 in. Determine the magnitude and angular location of this weight.

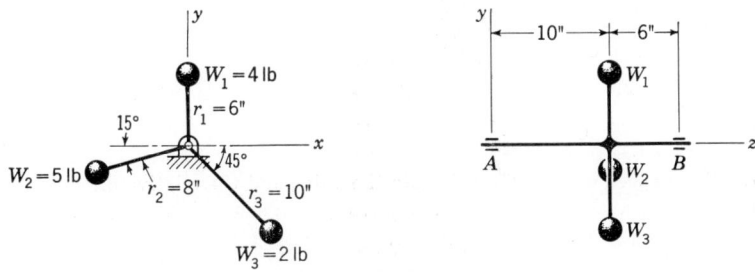

PROB. 19-1

19-2. Three weights are shown connected to a shaft which rotates in bearings at A and B. What is the magnitude of the bearing reactions if the shaft rotates at a speed of 180 rpm? The system is to be balanced by another weight 7 in. from the axis of rotation. Find the value of the weight and its angular location.

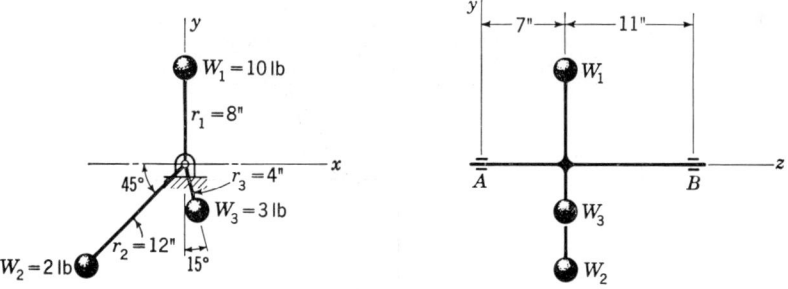

PROB. 19-2

[1] Myklestad solves this problem using complex notation. See N. O. Myklestad, "Fundamentals of Vibration Analysis," pp. 226–236, McGraw-Hill Book Company, Inc., New York, 1956. Graphical as well as analytical solutions can be found in John N. Macduff and John R. Curreri, "Vibration Control," pp. 233–240, McGraw-Hill Book Company, Inc., New York, 1958, and in James B. Hartman, "Dynamics of Machinery," pp. 150–156, McGraw-Hill Book Company, Inc., New York, 1956.

19-3. The figure shows two weights connected to a rotating shaft and mounted outboard of bearings at A and B. If the shaft rotates at a speed of 36 rpm, what is the magnitude of the bearing reactions produced by the rotating weights? The system is to be balanced by subtracting weight in one place at a radius of 5 in. Determine the amount and angular location of the weight to be removed.

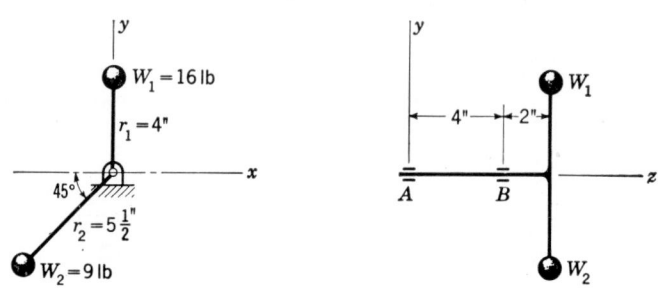

Prob. 19-3

19-4. For a speed of 180 rpm calculate the value and the relative direction of the bearing reactions at A and B of the two-mass system shown in the figure.

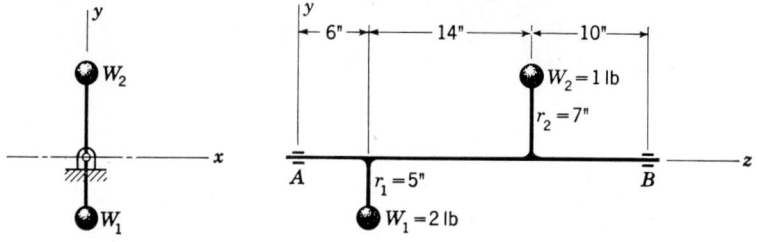

Prob. 19-4

19-5. Illustrated in the figure is a shaft rotating at a speed of 120 rpm and having three weights attached to it. The shaft rotates in radial bearings at A and B. Calculate the bearing reactions and their directions for the position shown.

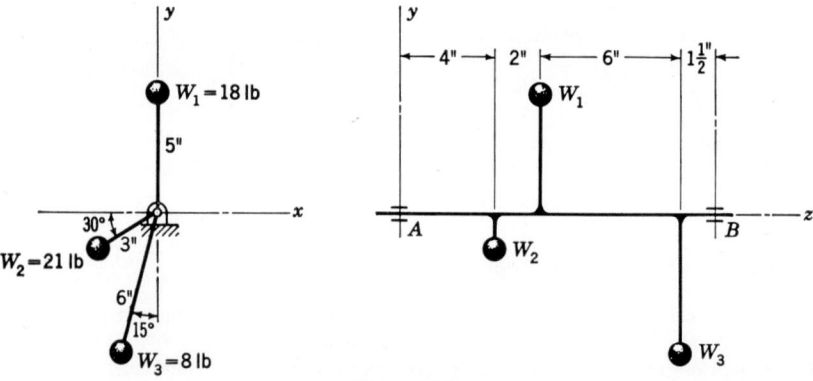

Prob. 19-5

19-6. The shaft shown in the figure is to be statically and dynamically balanced by placing correction weights in correction planes which are designated as L and R on the drawing. Calculate the magnitude of the two corrections in ounce-inches and their locations.

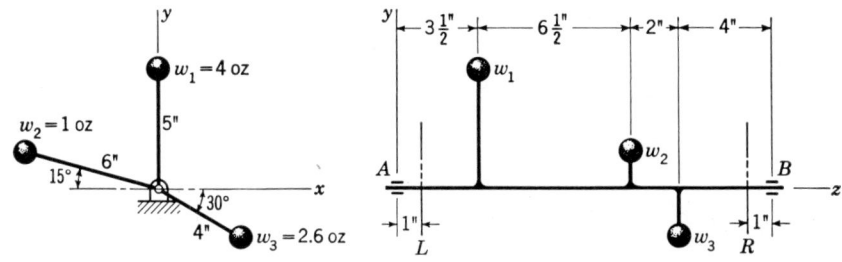

PROB. 19-6

19-7. Illustrated is a rotating system which is to be balanced by removing mass from the two correction planes which are designated as L and R. Calculate the magnitude in ounce-inches and the angular location of the two corrections.

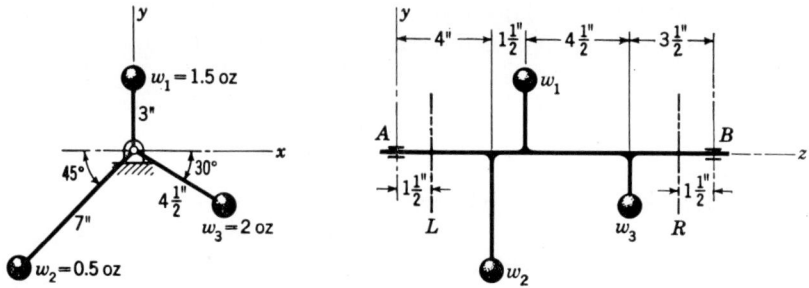

PROB. 19-7

19-8. The figure shows a rotating shaft connected to four unbalanced masses. The system is to be balanced by adding corrections to the two planes L and R. What are the magnitude and location of these two corrections?

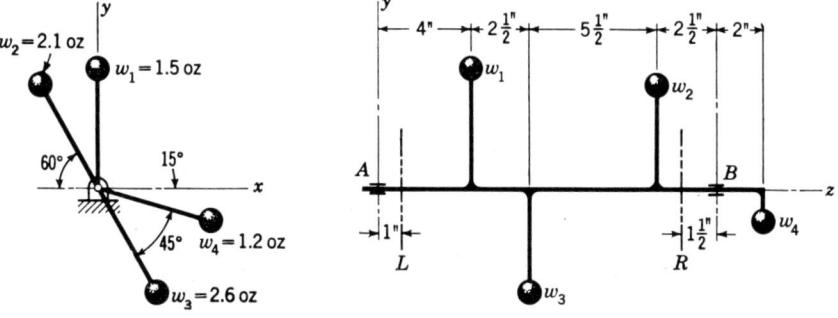

PROB. 19-8

CHAPTER 20

DYNAMICS OF THE RECIPROCATING ENGINE

The purpose of this chapter is to apply fundamentals—kinematic and dynamic analysis—in a complete investigation of a particular group of machines. The reciprocating engine has been selected for this purpose because it has reached such a high state of development and is of more general interest than other machines. For our purpose, however, any machine or group of machines involving interesting dynamical situations would serve just as well. The primary objective is to demonstrate methods of applying fundamentals to the analysis of any machine.

20-1. Engine Types. The description and characteristics of all the engines which have been conceived and constructed would fill many books. Here our purpose is to outline very briefly a few of the engine types which are currently popular and in general use. The exposition is not intended to be complete. Furthermore, the reader is expected to be mechanically inclined and generally familiar with internal-combustion engines; so the primary purpose of this section is merely to record things which the reader already knows and to furnish a nomenclature for the balance of the chapter.

We shall also include in this section, so as to have it all in one place, the descriptions and specifications of some of the more interesting engines. The material will then be easily available for use in problems and examples.

In this chapter we shall classify engines according to their intended use, the combustion cycle used, and the number and arrangement of the cylinders. Thus, we refer to aircraft engines, automotive engines, marine engines, and stationary engines, for example, all so named because of the purpose for which they were designed. Similarly, one might have in mind an engine designed on the basis of the *Otto cycle*, in which the fuel and air are mixed before compression, and in which combustion takes place with no excess air, or the *diesel engine*, in which the fuel is injected near the end of compression and combustion takes place with much excess air. The Otto-cycle engine utilizes quite volatile fuels and ignition is by spark, but the diesel-cycle engine operates on low-volatile fuels and ignition occurs because of compression.

The diesel- and Otto-cycle engines may be either *two-stroke-cycle* or *four-stroke-cycle*, depending upon the number of piston strokes required for the complete combustion cycle. Many outboard marine engines utilize the two-stroke-cycle, or simply the two-cycle, process. In these the piston uncovers exhaust ports in the cylinder wall near the end of the

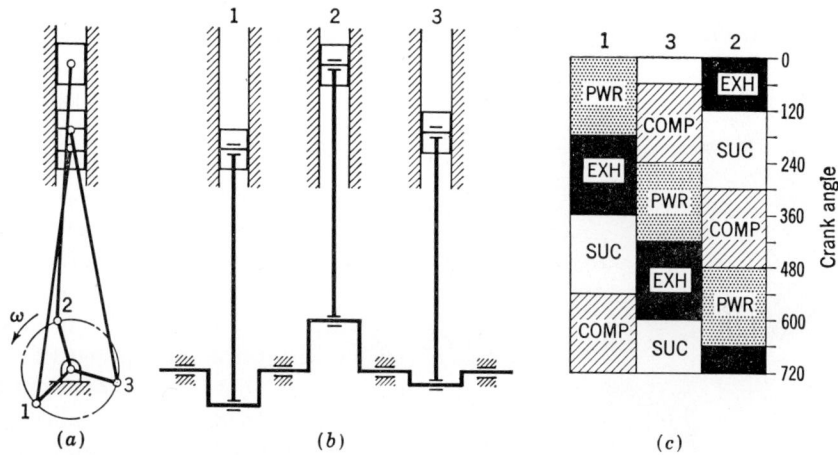

Fig. 20-1. A three-cylinder in-line engine. (a) Front view. (b) Side view. (c) Firing order.

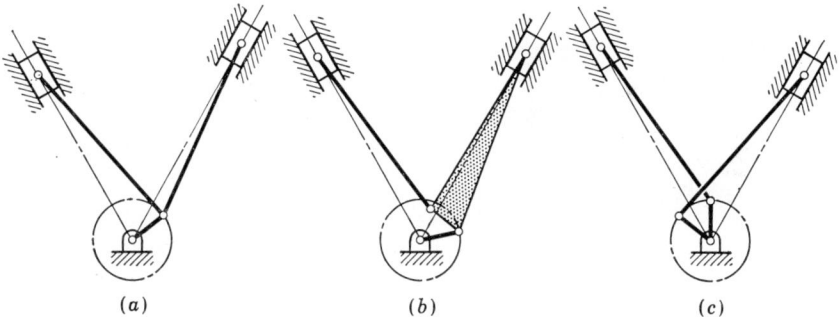

Fig. 20-2. Crank arrangements of V engines. (a) Single crank per pair of cylinders; connecting rods interlock with each other and are of *fork and blade* design. (b) Single crank per pair of cylinders; the *master connecting rod* carries a bearing for the *articulated rod*. (c) Separate cranks connect to staggered pistons.

expansion stroke and permits the exhaust gases to flow out. Soon after the exhaust ports are opened, the inlet ports open too and permit entry of a precompressed fuel-air mixture which also assists in expelling the remaining exhaust gases. The ports are then closed by the piston moving upward, and the fuel mixture again compressed. Then the cycle begins again. Note that the two-cycle engine has an expansion and a compression stroke and that these occur during one revolution of the crank.

The four-cycle engine has four piston strokes in a single combustion cycle corresponding to two revolutions of the crank. The events corresponding to the four strokes are (1) expansion or power stroke, (2) exhaust, (3) suction or intake stroke, (4) compression.

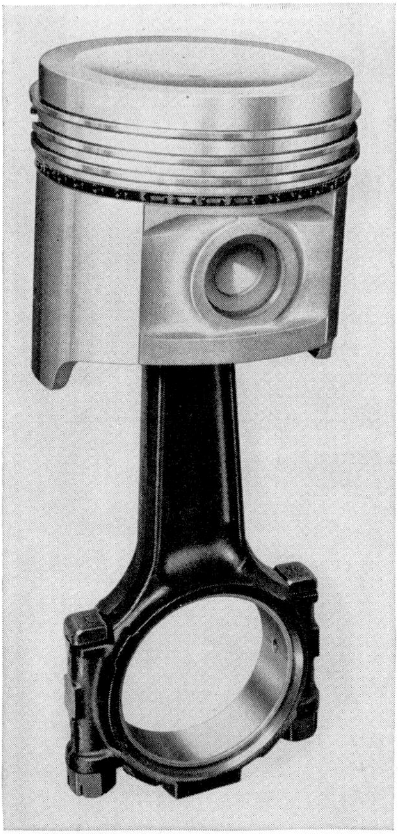

Multicylinder engines are broadly classified according to how the cylinders are arranged with respect to one another and to the crankshaft. Thus, an *in-line* engine is one in which the piston axes form a single plane coincident with the crankshaft and in which the pistons are all on the same side of the crankshaft. Figure 20-1 is a schematic drawing of a three-cylinder in-line engine with the cranks spaced 120°; a firing-order diagram for four-cycle operation is included for its interest.

A V-type engine is one utilizing two banks of one or more in-line cylinders each and a single crankshaft. Figure 20-2 illustrates several crank arrangements commonly employed. The pistons in the right and left banks of (*a*) and (*b*) are in the same plane, but those in (*c*) are in different planes.

If the V angle is increased to 180°, the result is called an *opposed-piston* engine. An opposed engine may have the two piston axes coincident or offset, and the rods may connect to the same crank or to separate cranks spaced 180° apart.

FIG. 20-3. A piston-connecting-rod assembly for a 351-in.[3] 1960 model V6 truck engine. (*GMC Truck and Coach Division, General Motors Corporation, Pontiac, Mich.*)

A *radial* engine is one having the pistons arranged in a circle about the crank center. Radial engines employ a master connecting rod for one cylinder, and the remaining pistons are connected to the master rod by articulated rods somewhat the same as for the V engine of Fig. 20-2*b*.[1]

Figures 20-3, 20-4, and 20-5 illustrate, respectively, the piston-connect-

---

[1] The book by A. R. Holowenko, "Dynamics of Machinery," John Wiley & Sons, Inc., New York, 1955, describes and illustrates a great many engine types and variations. It contains many firing orders and analyzes engine balance in excellent detail. See pp. 294–382.

Fig. 20-4. A cast crankshaft for a 305-in.[3] 1960 model V6 truck engine. (*GMC Truck and Coach Division, General Motors Corporation, Pontiac, Mich.*)

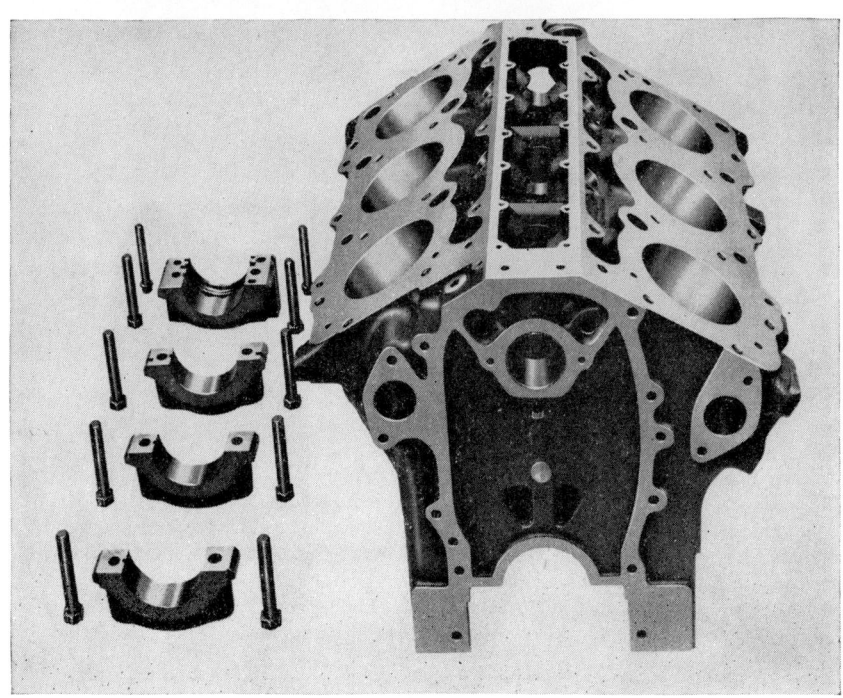

Fig. 20-5. Block for a 305-in.[3] 1960 model V6 truck engine. The same casting is used for a 351-in.[3] engine by boring for larger pistons. (*GMC Truck and Coach Division, General Motors Corporation, Pontiac, Mich.*)

530 DYNAMIC ANALYSIS OF MACHINES

ing-rod assembly, the crankshaft, and the block of a V6 truck engine. These are included as typical of modern design to show the form of important parts of an engine and for future reference.

The following engine specifications will provide data for the problems and examples in this chapter and will give a general idea of the performance of modern engines together with the sizes of the parts used in them.

FIG. 20-6. Model 803 single-cylinder engine. (*Briggs and Stratton Corporation, Milwaukee, Wis.*)

Briggs and Stratton Corporation, Milwaukee, Wis. Model 803 single-cylinder engine, illustrated in Fig. 20-6, has the following specifications: 1.73 hp at 2,200 rpm, 2.53 hp at 2,900 rpm, 3.0 hp at 3,600 rpm; rope or rewind starter; 25 lb net weight; 2⅜-in. bore; 1¾-in. stroke; 7.75-in.³ displacement; four-stroke cycle; air-cooled; counterclockwise rotation viewed from power take-off side; weight of piston assembly, 0.393 lb; weight of connecting-rod assembly, 0.214 lb; connecting-rod length, 3.125 in.; 0.40 in. from crankpin bearing to mass center of connecting rod; flywheel $wr^2 = 23.0$ lb-in.²

Chevrolet Motor Division, General Motors Corporation, Warren, Mich. The 1960 Corvair opposed-piston engine is shown in Fig. 20-7; Fig. 20-8 is a graph of the output characteristics. Specifications are as follows: 6 cylinders, horizontally opposed

FIG. 20-7. End view of 1960 Corvair engine. (*Chevrolet Motor Division, General Motors Corporation, Warren, Mich.*)

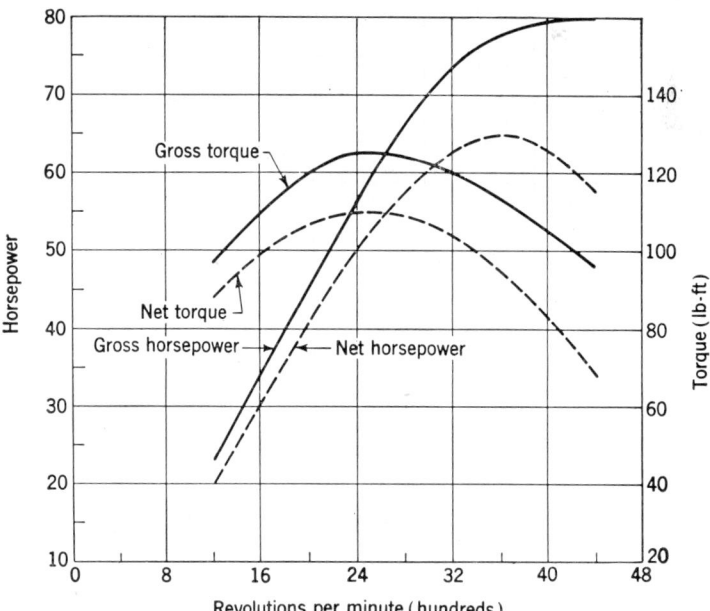

FIG. 20-8. Torque and horsepower relations for 1960 Corvair engine. (*Chevrolet Motor Division, General Motors Corporation, Warren, Mich.*)

with overhead valves; air-cooled; 3.375-in. bore; 2.60-in. stroke; 140-in.³ piston displacement; compression ratio, 8.0:1; 80 bhp (max) at 4,400 rpm; torque, 125 lb-ft (max) at 2,400 rpm; numbering system (front of vehicle to rear) is 6, 4, 2 for left bank, 5, 3, 1 for right bank; firing order 1, 4, 5, 2, 3, 6; intake valve timing, open 15° BTC,

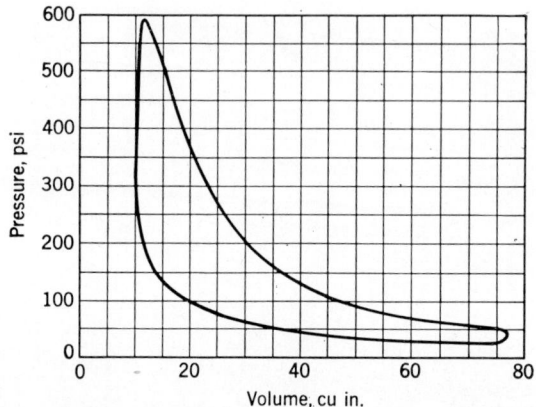

FIG. 20-9. Typical indicator diagram for 401-in.³ V6 truck engine; conditions unknown. (*GMC Truck and Coach Division, General Motors Corporation, Pontiac, Mich.*)

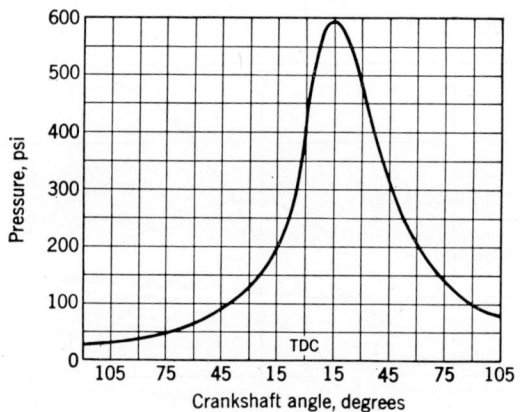

FIG. 20-10. A pressure-time curve for the 401-in.³ V6 truck engine. These data were taken from a running engine. (*GMC Truck and Coach Division, General Motors Corporation, Pontiac, Mich.*)

close 37° ABC; exhaust valve timing, open 59° BBC, close 13° ATC; weight of piston assembly, 1.266 lb; connecting-rod length, 7.20 in.; weight of reciprocating part of connecting rod, 0.196 lb; total weight of reciprocating parts per cylinder, 1.462 lb; weight of rotating part of connecting rod, 0.654 lb; rotating weight per crankpin, 0.732 lb; six-throw crankshaft with the throws arranged in pairs; members of each pair are 180° apart, and the pairs are 120° apart; rotation is clockwise when viewed from flywheel end (front of car).

GMC Truck and Coach Division, General Motors Corporation, Pontiac, Mich. One of the 1960 V6 truck engines is illustrated in Fig. 20-11. These engines are manufactured in four displacements, and they include one model, a V12 (702 in.³),

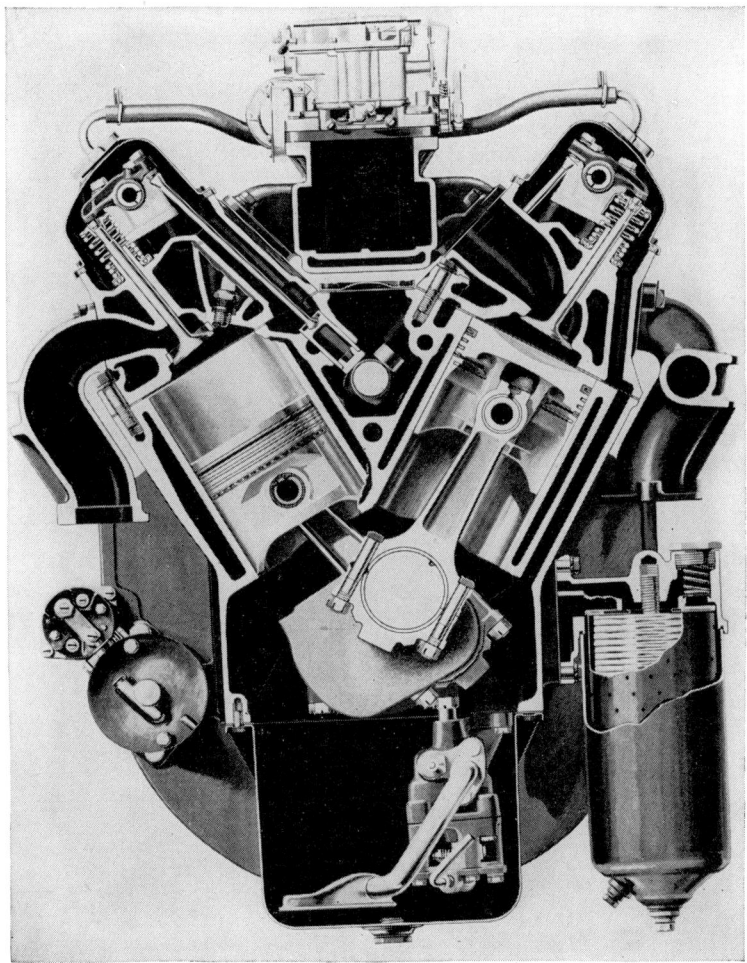

FIG. 20-11. Cross-sectional view of the 401-in.³ V6 truck engine. (*GMC Truck and Coach Division, General Motors Corporation, Pontiac, Mich.*)

which is described as a twin six because many of the V6 parts are interchangeable with it. Data to be included here are restricted to the 401-in.³ engine. Typical performance curves are exhibited in Figs. 20-9, 20-10, and 20-12. The specifications are as follows: bore, 4.875 in.; stroke, 3.56 in.; 60° vee design; 7.50:1 compression ratio; cylinders numbered from front to rear, 1, 3, 5 on the left bank, 2, 4, 6 on the

	Grams
Reciprocating weights:	
Weight of piston	1,560
Weight of piston pin	317.5
Weight of piston rings	127.0
Weight of retainers	0.34
Reciprocating weight of rod	360.0
Total weight	2,364.84
Reciprocating weight balanced	1,182.42
Rotating weights:	
Rotating weight of rod	926.00
Weight of bearings	101.28
Total balanced weight	2,209.70

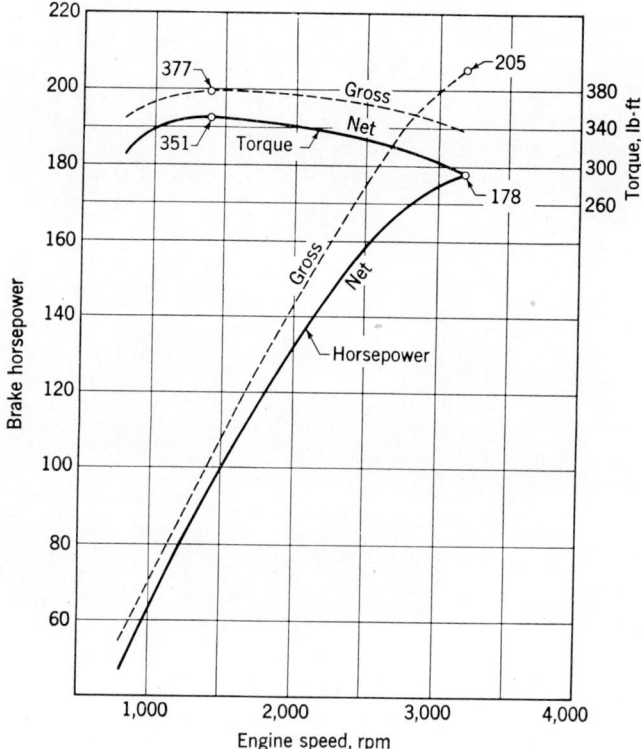

FIG. 20-12. Horsepower and torque characteristics of the 401-in.³ V6 truck engine. The solid curve is the net output as installed; dashed curve is the maximum output without accessories. Notice that the maximum torque occurs at a very low engine speed. (*GMC Truck and Coach Division, General Motors Corporation, Pontiac, Mich.*)

right bank; firing order 1, 6, 5, 4, 3, 2; crank arrangement, Fig. 20-13; connecting-rod length, 7.19 in.

20-2. Gas Forces.[1] In this section we shall assume that the moving parts are weightless so that the inertia forces and inertia torques are zero and also that there is no friction. These assumptions make it possible to trace the effect of the gas pressure from the piston to the crankshaft without the complicating effects of other forces.

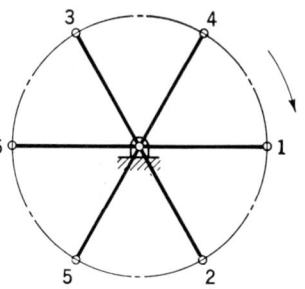

FIG. 20-13. Front view of V6 truck engine showing crank arrangement and direction of rotation.

A particular phase of the mechanism is chosen for investigation, as shown in Fig. 20-14a, and a gas force vector **P**, which is a function of time, designated as acting upon the piston 4. Starting with the force **P**, free-body diagrams of the piston, connecting rod, and crank are constructed successively as shown in (b), (c), and (d). In (d) note that the crank is in equilibrium because the couple formed by the connecting-rod force $\mathbf{F}'_{32}$ and the frame force $\mathbf{F}'_{12}$ is opposed by the crank torque T_2. Figure 20-14e shows the forces which act upon the frame. The gas force **P** exerted upon the cylinder head is completely balanced by $\mathbf{F}'^x_{21}$, which is a component of the force of the crankshaft journal against the main bearings. A force couple is formed by the lateral force of the piston against the cylinder wall $\mathbf{F}'_{41}$ and the other orthogonal component $\mathbf{F}'^y_{21}$ of the main bearing reaction. This couple is opposed to and balanced by the crankshaft torque.

The free-body diagram of the connecting rod in Fig. 20-14c gives directly the magnitude and direction of the forces acting upon the crank-pin bearing $\mathbf{F}'_{23}$ and the forces acting upon the wrist-pin bearing $\mathbf{F}'_{43}$. The force $\mathbf{F}'_{21}$ in Fig. 20-14e is the resultant force of the main bearings against the crankshaft journal.

This force analysis is, of course, valid only for the particular phase of the mechanism shown because the elements of the linkage will occupy another position when the crank rotates and also because the gas force **P** is a function of ωt. If the relationship between the gas force **P** and the crank angle ωt is known, then a graphical analysis like Fig. 20-14 can be made for all crank angles in a single cycle (360° for a two-cycle engine, 720° for a four-cycle engine) and the resulting forces and directions tabulated, or if preferred, the analytical methods presented in Chap. 15 can be used to make the analysis. Neither of these methods is difficult to apply, but when they must be applied a great many times to obtain a

[1] See also Appendix IV.

complete analysis of the forces, then the work gets rather tedious. For this reason it is preferable to develop algebraic expressions for these forces.

As shown in Fig. 20-15, the crank radius is given as r, the connecting-

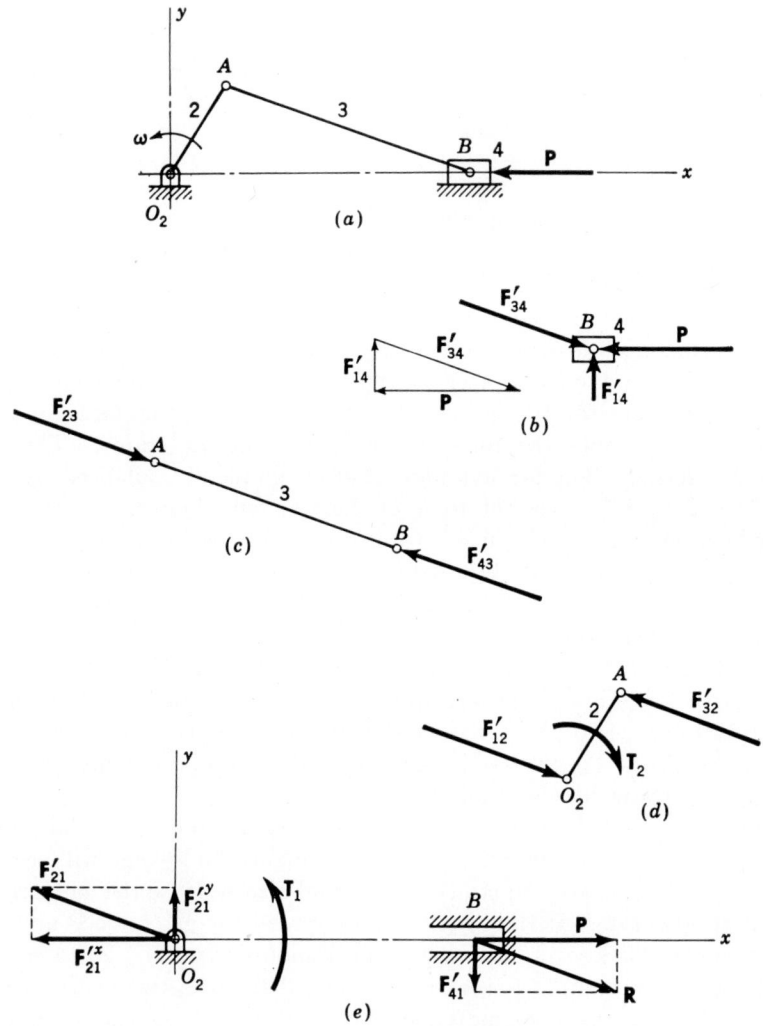

FIG. 20-14. Analysis of the forces in the engine mechanism due only to gas pressure.

rod length as l, and the location of the piston is given by the coordinate x for any value of the crank angle ωt. The reader should now review Sec. 10-8 and compare Fig. 20-15 with Fig. 10-17. Equations (10-2) to (10-4) are reproduced for convenience, making the substitution $\theta = \omega t$

DYNAMICS OF THE RECIPROCATING ENGINE

for the crank angle:

$$r \sin \omega t = l \sin \phi$$
$$x = r \cos \omega t + l \cos \phi$$
$$= r \cos \omega t + l \sqrt{1 - \left(\frac{r}{l} \sin \omega t\right)^2} \quad (20\text{-}1)$$

Equation (20-1) gives the exact position of the piston in terms of the crank angle ωt and the dimensions of the linkage. In the usual engine the ratio

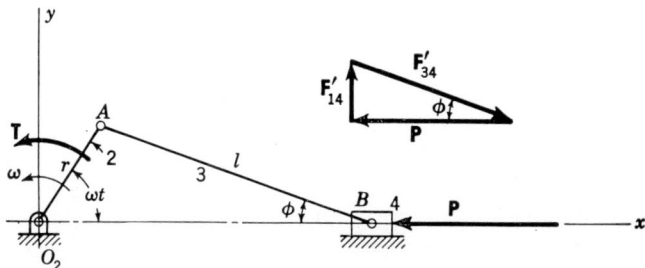

FIG. 20-15

r/l is about $\frac{1}{4}$, and so the maximum value of the second term under the radical will usually be about $\frac{1}{16}$ or less. If we expand the radical using the binomial theorem and neglect all but the first two terms, there results

$$\sqrt{1 - \left(\frac{r}{l} \sin \omega t\right)^2} = 1 - \frac{r^2}{2l^2} \sin^2 \omega t \quad (a)$$

Since
$$\sin^2 \omega t = \frac{1 - \cos 2\omega t}{2} \quad (b)$$

Eq. (20-1) becomes

$$x = l - \frac{r^2}{4l} + r\left(\cos \omega t + \frac{r}{4l} \cos 2\omega t\right) \quad (20\text{-}2)$$

Differentiating successively to obtain the velocity and acceleration gives

$$\dot{x} = -r\omega\left(\sin \omega t + \frac{r}{2l} \sin 2\omega t\right) \quad (20\text{-}3)$$

$$\ddot{x} = -r\alpha\left(\sin \omega t + \frac{r}{2l} \sin 2\omega t\right) - r\omega^2\left(\cos \omega t + \frac{r}{l} \cos 2\omega t\right) \quad (20\text{-}4)$$

The reader should compare these with Eqs. (10-6) and (10-8).

The piston-wall force can now be obtained directly from the trigonometry of the force polygon in Fig. 20-15:

$$\mathbf{F}'_{14} = P \tan \phi \, \mathbf{J} \quad (20\text{-}5)$$

DYNAMIC ANALYSIS OF MACHINES

The quantity $\tan \phi$ appears frequently in expressions throughout this chapter. It is therefore convenient to develop an expression in terms of the crank angle ωt. Thus

$$\tan \phi = \frac{(r/l) \sin \omega t}{\cos \phi} = \frac{(r/l) \sin \omega t}{\sqrt{1 - [(r/l) \sin \omega t]^2}} \quad (c)$$

Now, utilizing the binomial theorem again, we find that

$$\frac{1}{\sqrt{1 - [(r/l) \sin \omega t^2]}} = 1 + \frac{r^2}{2l^2} \sin^2 \omega t \quad (d)$$

where only the first two terms have been retained. Equation (c) now becomes

$$\tan \phi = \frac{r}{l} \sin \omega t \left(1 + \frac{r^2}{2l^2} \sin^2 \omega t\right) \quad (20\text{-}6)$$

The trigonometry of Fig. 20-14 shows that the wrist-pin (piston pin) bearing force has a magnitude of

$$F'_{34} = \frac{P}{\cos \phi} = \frac{P}{\sqrt{1 - [(r/l) \sin \omega t]^2}} = P\left(1 + \frac{r^2}{2l^2} \sin^2 \omega t\right) \quad (e)$$

or in vector notation,

$$\mathbf{F}'_{34} = P\mathbf{I} - F'_{14}\mathbf{J} = P\mathbf{I} - P \tan \phi \, \mathbf{J} \quad (20\text{-}7)$$

As illustrated in Fig. 20-14e, the torque can be obtained from a multiplication of the force F'_{14} and the piston coordinate x. Employing Eqs. (20-2), (20-5), and (20-6) yields

$$\mathbf{T} = F'_{14} x \mathbf{K} = P \frac{r}{l} \sin \omega t \left(1 + \frac{r^2}{2l^2} \sin^2 \omega t\right)$$
$$\left[l - \frac{r^2}{4l} + r\left(\cos \omega t + \frac{r}{4l} \cos 2\omega t\right)\right] \mathbf{K} \quad (f)$$

When the terms of Eq. (f) are multiplied, we can neglect those containing second or higher powers of r/l with only a very small error. Equation (f) then becomes

$$\mathbf{T} = Pr \sin \omega t \left(1 + \frac{r}{l} \cos \omega t\right) \mathbf{K} \quad (20\text{-}8)$$

This is the torque delivered to the crankshaft by the gas force and is positive when it has the counterclockwise direction as in Fig. 20-15. Merely by substituting various values of the crank angle ωt into Eqs. (20-5) to (20-8) one can calculate all the bearing reactions as well as the crankshaft torque for all phases of the linkage in a cycle of operation.

DYNAMICS OF THE RECIPROCATING ENGINE

20-3. Equivalent Masses. In analyzing the inertia forces due to the connecting rod of an engine it is convenient, in many cases, to concentrate a portion of the mass at the crankpin A and the remaining portion at the wrist pin B (Fig. 20-16). The reason for this is that the crankpin moves on a circle and the wrist pin on a straight line. Both of these motions are quite easy to analyze. However, the center of gravity G is somewhere between the crankpin and wrist pin, and its motion is more complicated and consequently more difficult to determine in algebraic form. We shall see that this simplification also introduces some errors into the analysis.

The problem is illustrated in Fig. 20-16. The mass of the connecting rod m_3 is assumed to be concentrated at the center of gravity G_3. We divide this mass into two parts; one, m_{3B}, is then concentrated at the

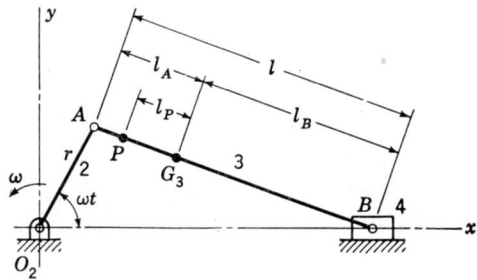

FIG. 20-16

wrist pin B. The other part, m_{3P}, is concentrated at the *center of percussion* P for oscillation of the rod about B. This disposition of the mass of the rod is dynamically equivalent to the original rod if the total mass is the same, if the position of the center of gravity G_3 is unchanged, and if the moment of inertia is the same. Writing these three conditions, respectively, in equation form produces

$$m_3 = m_{3B} + m_{3P} \qquad (a)$$
$$m_{3B} l_B = m_{3P} l_P \qquad (b)$$
$$I = m_{3B} l_B^2 + m_{3P} l_P^2 \qquad (c)$$

Solving Eqs. (a) and (b) simultaneously gives the portion of mass to be concentrated at each point:

$$m_{3B} = m_3 \frac{l_P}{l_B + l_P} \qquad m_{3P} = m_3 \frac{l_B}{l_B + l_P} \qquad (20\text{-}9)$$

Substituting Eqs. (20-9) into (c) gives

$$I = m_3 \frac{l_P}{l_B + l_P} l_B^2 + m_3 \frac{l_B}{l_B + l_P} l_P^2 = m_3 l_P l_B \qquad (d)$$

or
$$l_P l_B = \frac{I}{m_3} \qquad (20\text{-}10)$$

Equation (20-10) shows that the two distances l_P and l_B are dependent on each other. Thus, if l_B is specified in advance, then l_P is fixed in length by Eq. (20-10).

In the usual connecting rod the center of percussion is close to the crankpin and it is assumed that they are coincident. Thus, letting $l_A = l_P$, Eqs. (20-9) reduce to

$$m_{3B} = \frac{m_3 l_A}{l} \qquad m_{3A} = \frac{m_3 l_B}{l} \qquad (20\text{-}11)$$

We note again that the equivalent masses, obtained by Eqs. (20-11), are not exact because of the assumption made but are close enough for ordinary connecting rods. The approximation, for example, is not valid

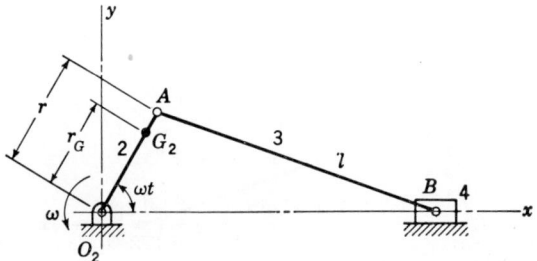

Fig. 20-17

for the master connecting rod of a radial engine because the crankpin end has bearings for all of the other connecting rods as well as its own bearing.

Church[1] states that, for estimating and checking purposes, about two-thirds of the mass should be concentrated at A and the remaining third at B.

Figure 20-17 illustrates an engine linkage in which the mass of the crank m_2 is not balanced, as evidenced by the fact that the center of gravity G_2 is displaced outward along the crank a distance r_G from the axis of rotation. In the inertia-force analysis, simplification is obtained by locating an equivalent mass m_{2A} at the crankpin. Thus, for equivalence

$$m_2 r_G = m_{2A} r \qquad \text{or} \qquad m_{2A} = m_2 \frac{r_G}{r} \qquad (20\text{-}12)$$

20-4. Inertia Forces in the Single-cylinder Engine. The inertia-force analysis of the slider-crank mechanism can, of course, be accomplished using either of the methods of Chap. 16. The simplifications of the

[1] Austin H. Church, "Mechanical Vibrations," p. 185, John Wiley & Sons, Inc., New York, 1957.

preceding section can be used if desired but are not at all necessary for these methods. Since a demonstration of these methods here would be repetitious, we shall, instead, demonstrate the algebraic approach, which has the additional advantage of permitting one to see the general force situation for an entire cycle of operation.

Using the methods of the preceding section we begin by locating equivalent masses at the crankpin and at the wrist pin. Thus

$$m_A = m_{2A} + m_{3A} \qquad (20\text{-}13)$$
$$m_B = m_{3B} + m_4 \qquad (20\text{-}14)$$

Equation (20-13) states that the mass m_A located at the crankpin is made up of the equivalent masses m_{2A} of the crank and m_{3A} of part of the connecting rod. Of course, if the crank is balanced, then all

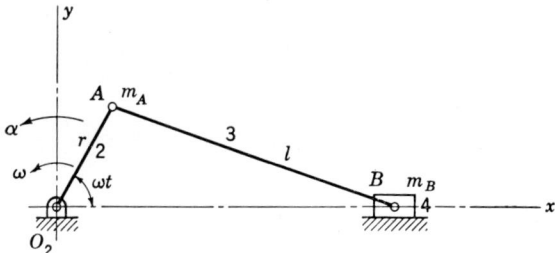

FIG. 20-18

its mass is assumed to be located at the axis of rotation and m_{2A} is then zero. Equation (20-14) indicates that the reciprocating mass m_B located at the wrist pin is composed of the equivalent mass m_{3B} of the other part of the connecting rod and the mass m_4 of the piston assembly.

In Fig. 20-18 is shown the slider-crank mechanism with masses m_A and m_B located at points A and B, respectively. Designating the angular velocity of the crank as ω and the angular acceleration as α, the position vector of the crankpin relative to the origin O_2 is

$$\mathbf{R}_A = r \cos \omega t \, \mathbf{I} + r \sin \omega t \, \mathbf{J} \qquad (a)$$

Differentiating twice to obtain the acceleration gives

$$\mathbf{A}_A = (-r\alpha \sin \omega t - r\omega^2 \cos \omega t)\mathbf{I} + (r\alpha \cos \omega t - r\omega^2 \sin \omega t)\mathbf{J} \qquad (20\text{-}15)$$

The inertia force of the rotating parts is then

$$-m_A\mathbf{A}_A = m_A r(\alpha \sin \omega t + \omega^2 \cos \omega t)\mathbf{I} + m_A r(-\alpha \cos \omega t + \omega^2 \sin \omega t)\mathbf{J} \qquad (20\text{-}16)$$

542   DYNAMIC ANALYSIS OF MACHINES

Since the analysis is usually made at constant angular velocity ($\alpha = 0$), Eq. (20-16) reduces to

$$-m_A \mathbf{A}_A = m_A r \omega^2 \cos \omega t \, \mathbf{I} + m_A r \omega^2 \sin \omega t \, \mathbf{J} \qquad (20\text{-}17)$$

The acceleration of the piston has already been determined [Eq. (20-4)] and is repeated here for convenience in a slightly different form:

$$\mathbf{A}_B = \left[-r\alpha \left(\sin \omega t + \frac{r}{2l} \sin 2\omega t \right) - r\omega^2 \left(\cos \omega t + \frac{r}{l} \cos 2\omega t \right) \right] \mathbf{I} \qquad (20\text{-}18)$$

The inertia force of the reciprocating parts is, therefore,

$$-m_B \mathbf{A}_B = \left[m_B r \alpha \left(\sin \omega t + \frac{r}{2l} \sin 2\omega t \right) \right.$$
$$\left. + \, m_B r \omega^2 \left(\cos \omega t + \frac{r}{l} \cos 2\omega t \right) \right] \mathbf{I} \qquad (20\text{-}19)$$

or, for constant angular velocity,

$$-m_B \mathbf{A}_B = m_B r \omega^2 \left(\cos \omega t + \frac{r}{l} \cos 2\omega t \right) \mathbf{I} \qquad (20\text{-}20)$$

Adding Eqs. (20-17) and (20-20) gives the total inertia force for all the moving parts. The components in the x and y directions are

$$F^x = (m_A + m_B) r \omega^2 \cos \omega t + \left(m_B \frac{r}{l} \right) r \omega^2 \cos 2\omega t \qquad (20\text{-}21)$$

$$F^y = m_A r \omega^2 \sin \omega t \qquad (20\text{-}22)$$

It is customary to refer to the portion of the force occurring at the circular frequency ω rad/sec as the *primary inertia force* and the portion occurring at 2ω rad/sec as the *secondary inertia force*. We note that the vertical component has only a primary part and that it therefore varies directly with the crankshaft speed. On the other hand the horizontal component, which is in the direction of the cylinder axis, has a primary part varying directly with the crankshaft speed and a secondary part which moves at twice the crankshaft speed.

We proceed now to a determination of the inertia torque. As shown in Fig. 20-19, the inertia force due to the mass at A has no moment arm about O_2 and therefore produces no torque. Consequently, we need consider only the inertia force given by Eq. (20-20) due to the reciprocating part of the mass. From the force polygon of Fig. 20-19 the inertia torque exerted by the engine on the crankshaft is

$$\mathbf{T} = -(-m_B \ddot{x} \tan \phi) x \mathbf{K} \qquad (b)$$

Expressions for x, $\ddot{x}$, and $\tan \phi$ appear in Sec. 20-2. Making appropriate substitutions for these yields the following for the torque:

$$\mathbf{T} = -m_B r\omega^2 \left(\cos \omega t + \frac{r}{l}\cos 2\omega t\right)\left[1 - \frac{r^2}{4l}\right.$$

$$\left. + r\left(\cos \omega t + \frac{r}{4l}\cos 2\omega t\right)\right]\frac{r}{l}\sin \omega t \left(1 + \frac{r^2}{2l^2}\sin^2 \omega t\right)\mathbf{K} \quad (c)$$

Terms which are proportional to the second or higher powers of r/l can be neglected in performing the indicated multiplication. Equation (c) then can be written

$$\mathbf{T} = -m_B r^2 \omega^2 \sin \omega t \left(\frac{r}{2l} + \cos \omega t + \frac{3r}{2l}\cos 2\omega t\right)\mathbf{K} \quad (d)$$

Then, using the identities

$$2 \sin \omega t \cos 2\omega t = \sin 3\omega t - \sin \omega t \quad (e)$$
and
$$2 \sin \omega t \cos \omega t = \sin 2\omega t \quad (f)$$

results in an equation having only sine terms, and Eq. (d) finally becomes

$$\mathbf{T} = \frac{m_B}{2} r^2 \omega^2 \left(\frac{r}{2l}\sin \omega t - \sin 2\omega t - \frac{3r}{2l}\sin 3\omega t\right)\mathbf{K} \quad (20\text{-}23)$$

This is the inertia torque exerted by the engine on the shaft in the positive direction. A clockwise or negative inertia torque of the same magnitude is, of course, exerted on the frame of the engine.

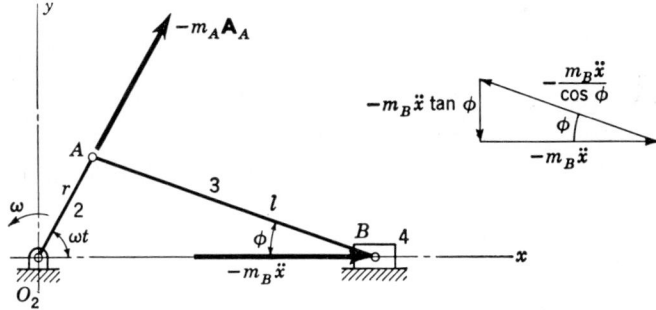

FIG. 20-19

The assumed distribution of the connecting-rod mass (Sec. 20-3) results in a moment of inertia which is greater than the true value. Consequently, the torque given by Eq. (20-23) is not the exact value. In addition, terms proportional to the second- or higher-order powers of r/l were dropped in simplifying Eq. (c). These two errors are about the same magnitude and are quite small for ordinary connecting rods having r/l ratios near $\frac{1}{4}$.

20-5. Bearing Loads in the Single-cylinder Engine.

The designer of a reciprocating engine must know the values of the forces acting upon the bearings and how these forces vary in a cycle of operation. This is necessary to proportion and select the bearings properly, and it is also needed for the design of other engine parts. This section comprises an investigation of the force exerted by the piston against the cylinder wall and the forces acting against the piston pin and against the crankpin. Main bearing forces will be investigated in a later section because they depend upon the action of all the cylinders of the engine.

The resultant bearing loads are made up of the following components:
1. The gas force components, designated by a single prime
2. Inertia force due to the weight of the piston assembly, designated by a double prime
3. Inertia force of that part of the connecting rod assigned to the piston-pin end, triple primed
4. Connecting-rod inertia force at the crankpin end, quadruple primed

Equations for the gas force components have been determined in Sec. 20-2, and reference shall be made to them in finding the total bearing loads.

Figure 20-20 is a graphical analysis of the forces in the engine mechanism with a zero gas force and subjected to an inertia force resulting only from the weight of the piston assembly. Figure 20-20a shows the position of the mechanism selected for analysis, and the inertia force $-m_4\mathbf{A}_B$ is shown acting upon the piston. In (b) the free-body diagram of the piston forces is shown together with the force polygon from which they were obtained. Figures 20-20c, d, and e illustrate, respectively, the free-body diagrams of forces acting upon the connecting rod, crank, and frame.

In Fig. 20-20e notice that the torque $\mathbf{T}_1''$ balances the force couple formed by the forces $\mathbf{F}_{41}''$ and $\mathbf{F}_{21}''^y$. But the force $\mathbf{F}_{21}''^x$ at the crank center remains unopposed by any other force. This is a very important observation which we shall reserve for discussion in a separate section.

The following forces are of interest to us:
1. The force $\mathbf{F}_{41}''$ of the piston against the cylinder wall
2. The force $\mathbf{F}_{34}''$ of the connecting rod against the piston pin
3. The force $\mathbf{F}_{32}''$ of the connecting rod against the crankpin

These forces can be determined graphically or analytically using the methods of Chap. 16. By methods similar to those employed previously in this chapter the analytical expressions are found to be

$$\mathbf{F}_{41}'' = -m_4\ddot{x} \tan \phi \, \mathbf{J} \qquad (20\text{-}24)$$
$$\mathbf{F}_{34}'' = m_4\ddot{x}\mathbf{I} - m_4\ddot{x} \tan \phi \, \mathbf{J} \qquad (20\text{-}25)$$
$$\mathbf{F}_{32}'' = -\mathbf{F}_{34}'' \qquad (20\text{-}26)$$

where $\ddot{x}$ is the acceleration of the piston as given by Eq. (20-4) and

DYNAMICS OF THE RECIPROCATING ENGINE 545

m_4 is the mass of the piston assembly. The quantity $\tan \phi$ can be evaluated in terms of the crank angle with the use of Eq. (20-6).

In Fig. 20-21 we neglect all forces except those which result because of the inertia of that part of the mass of the connecting rod which is

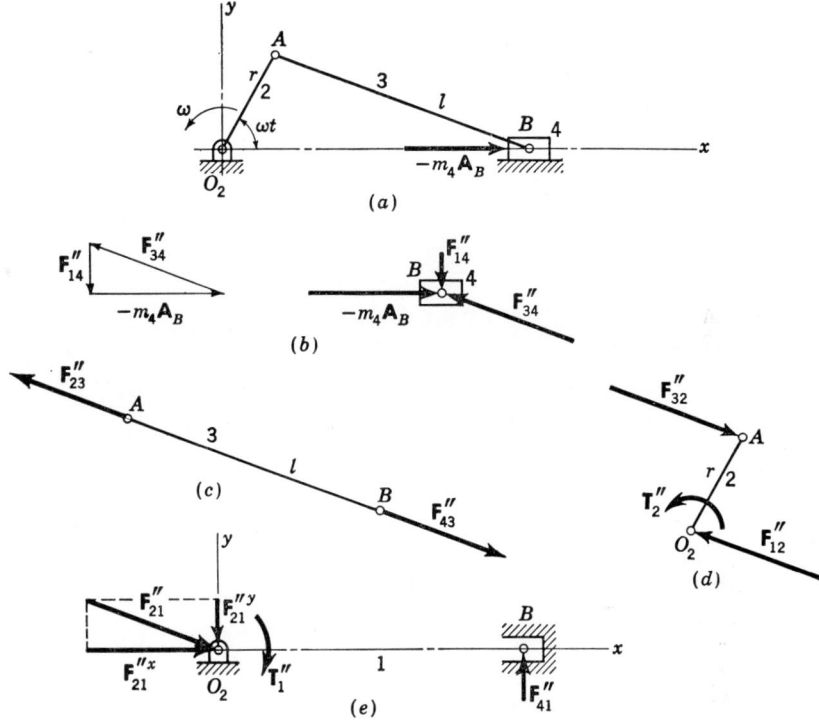

FIG. 20-20. Analysis of the forces in the engine mechanism when only the inertia force due to the weight of the piston assembly is considered.

assumed to exist at the piston-pin center. Thus (b) is a free-body diagram of the connecting rod showing the inertia force $-m_{3B}\mathbf{A}_B$ acting at the piston-pin end.[1] The forces on the piston pin, the crank, and the frame are illustrated in Fig. 20-21c, d, and e, respectively. The equations for these forces for a crank having uniform angular velocity are found

---

[1] It is incorrect to add m_{3B} and m_4 together and then to calculate a resultant inertia force in finding the bearing loads, although such a procedure would seem to be simpler. The reason is that m_4 is the mass of the piston assembly and the corresponding inertia force acts on the piston side of the wrist pin. But m_{3B} is part of the connecting-rod mass, and hence, its inertia force acts on the connecting-rod side of the wrist pin. Thus, adding the two will yield correct results for the crankpin load and the force of the piston against the cylinder wall but will give *incorrect* results for the piston-pin load.

to be

$$F'''_{41} = -m_{3B}\ddot{x} \tan \phi \, \mathbf{J} \quad (20\text{-}27)$$
$$F'''_{34} = F'''_{41} \quad (20\text{-}28)$$
$$F'''_{32} = -m_{3B}\ddot{x}\mathbf{I} + m_{3B}\ddot{x} \tan \phi \, \mathbf{J} \quad (20\text{-}29)$$

Figure 20-22 illustrates the forces which result because of that part of the connecting-rod mass which is concentrated at the crankpin end. While a counterweight attached to the crank balances the reaction at

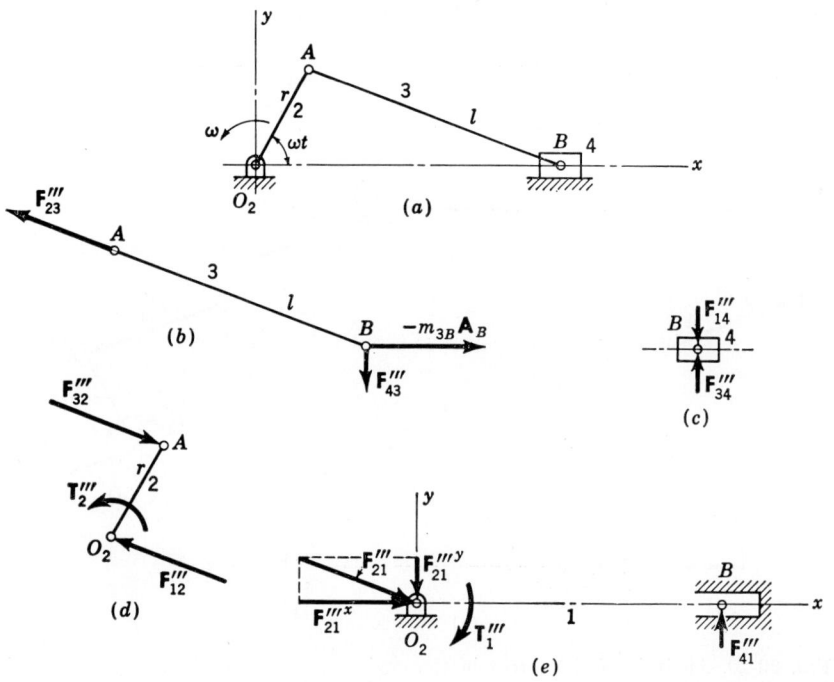

Fig. 20-21. Graphical analysis of forces resulting solely from the mass of the connecting rod assumed to be concentrated at the wrist-pin end.

O_2, it cannot make F''''_{32} zero. Thus the crankpin force exists no matter whether the rotating mass of the connecting rod is balanced or not. This force is

$$F''''_{32} = m_{3A}r\omega^2(\cos \omega t \, \mathbf{I} + \sin \omega t \, \mathbf{J}) \quad (20\text{-}30)$$

As previously indicated, the bearing loads exist due to the gas force and the inertia forces. The total force of the piston against the cylinder wall is found by summing Eqs. (20-5), (20-24), and (20-27). When simplified, the answer is

$$\begin{aligned} \mathbf{F}_{41} &= \mathbf{F}'_{41} + \mathbf{F}''_{41} + \mathbf{F}'''_{41} \\ &= -[(m_{3B} + m_4)\ddot{x} + P] \tan \phi \, \mathbf{J} \end{aligned} \quad (20\text{-}31)$$

DYNAMICS OF THE RECIPROCATING ENGINE 547

The forces on the piston pin and crankpin are found similarly and are

$$\mathbf{F}_{34} = (m_4\ddot{x} + P)\mathbf{I} - [(m_{3B} + m_4)\ddot{x} + P]\tan\phi\,\mathbf{J} \tag{20-32}$$
$$\mathbf{F}_{32} = [m_{3A}r\omega^2 \cos\omega t - (m_{3B} + m_4)\ddot{x} - P]\mathbf{I}$$
$$+ \{m_{3A}r\omega^2 \sin\omega t + [(m_{3B} + m_4)\ddot{x} + P]\tan\phi\}\mathbf{J} \tag{20-33}$$

20-6. Crankshaft Torque Delivered by the Single-cylinder Engine.
Inspection of Figs. (20-14), (20-20), and (20-21) shows that the crank-

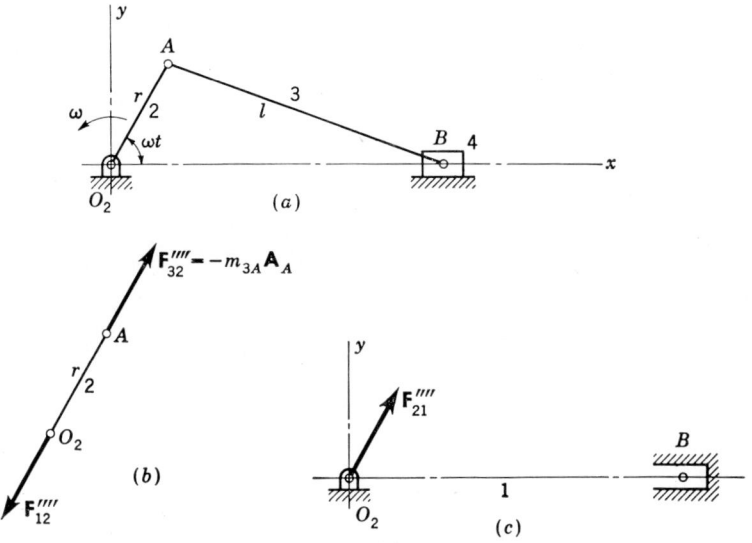

FIG. 20-22. Graphical analysis of forces resulting solely from the mass of the connecting rod assumed to be concentrated at the crankpin end.

shaft torque is obtained from the equation

$$\mathbf{T} = -F_{41}x\mathbf{K} \tag{a}$$

where **T** is the torque delivered by the crankshaft to the load. Substituting the magnitude of the force $\mathbf{F}_{41}$ from Eq. (20-31) yields

$$\mathbf{T} = [(m_{3B} + m_4)\ddot{x} + P]x \tan\phi\,\mathbf{K} \tag{20-34}$$

20-7. Balancing Single-cylinder Engines. The balancing of rotating masses was considered in detail in Chap. 19, and for this reason, it is sufficient here to assume that the crank and the rotating part of the connecting rod are or can be balanced. It turns out that, in a single-cylinder engine, the reciprocating masses cannot be balanced at all, and so our problem really is a study of unbalance in the single-cylinder engine rather than that of balance.

The inertia force due to the reciprocating masses is shown acting in the positive direction in Fig. 20-23a. In (b) the forces acting upon the engine block due to these inertia forces are shown. The resultant forces are $\mathbf{F}_{21}$, the force exerted by the crankshaft on the main bearings, and a positive couple formed by the forces $\mathbf{F}_{41}$ and $\mathbf{F}_{21}{}^y$. The force $\mathbf{F}_{21}{}^x = -m_B\mathbf{A}_B$ is frequently termed a *shaking force*, and the couple $T = xF_{41}$ a *shaking couple*. As indicated by Eqs. (20-20) and (20-23), the magnitude and direction of this force and couple change with ωt; consequently, the shaking force induces linear vibration of the block in the x direction, and the shaking couple a torsional vibration of the block about the crank

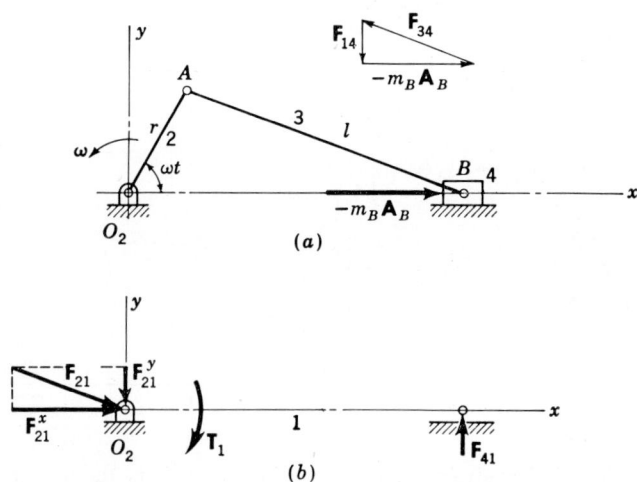

FIG. 20-23. Inertia forces due to the reciprocating masses. (The primes have been omitted for simplification.)

center. If the block is mounted on springs or flexible supports, the magnitude of the resulting vibrations can be predicted by the methods of Chap. 18. Finally, then, the inertia force and couple are resisted by the inertia of the block itself because of the vibrations which are induced in it as a result of these.

A graphical representation of the inertia force is possible if Eq. (20-20) is rearranged as follows:

$$F = m_B r\omega^2 \cos \omega t + m_B r\omega^2 \frac{r}{l} \cos 2\omega t \qquad (20\text{-}35)$$

where $F = F_{21}{}^x$ for simplicity of notation. The first term of Eq. (20-35) is represented by the x projection of a vector $m_B r\omega^2$ in length rotating at ω rad/sec. This is the primary part of the inertia force. The second term is, similarly, represented by the x projection of a vector $m_B r \omega^2 (r/l)$ in length rotating at 2ω rad/sec; this is the secondary part. Such a diagram is shown in Fig. 20-24 for $r/l = \frac{1}{4}$. The total inertia or shaking force is the algebraic sum of the horizontal projections of the two vectors.

While the reciprocating masses of a single-cylinder engine cannot be balanced using a simple counterweight,[1] it is possible to modify the shaking forces by *unbalancing* the rotating masses. As an example of this let us add a counterweight opposite the crankpin whose mass exceeds the rotating mass by one-half of the reciprocating mass (from one-half to two-thirds of the reciprocating mass is usually added to the counterweight to alter the balance characteristics in single-cylinder engines). We designate the mass of the counterweight by m_C, substitute this mass in Eq. (20-17), use a negative sign because the counterweight is opposite the crankpin, and then the inertia force due to this counterweight is

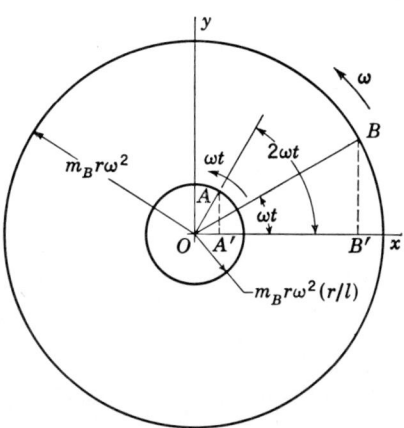

Fig. 20-24. Circle diagram for finding inertia force. Total inertia force is $OA' + OB'$.

$$\mathbf{F}_C = -m_C r \omega^2 \cos \omega t \, \mathbf{I} - m_C r \omega^2 \sin \omega t \, \mathbf{J} \qquad (a)$$

Note that the balancing mass and the crankpin both have the same radius. Since

$$m_C = m_A + \frac{m_B}{2} \qquad (b)$$

Eq. (a) can be written

$$\mathbf{F}_C = -\left(m_A + \frac{m_B}{2}\right) r \omega^2 \cos \omega t \, \mathbf{I} - \left(m_A + \frac{m_B}{2}\right) r \omega^2 \sin \omega t \, \mathbf{J} \qquad (c)$$

The inertia force due to the rotating and reciprocating masses is, from Eqs. (20-21) and (20-22),

$$\mathbf{F}_{A,B} = F^x \mathbf{I} + F^y \mathbf{J} = \left[(m_A + m_B) r \omega^2 \cos \omega t + m_B r \omega^2 \frac{r}{l} \cos 2\omega t\right] \mathbf{I} \\ + m_A r \omega^2 \sin \omega t \, \mathbf{J} \qquad (d)$$

[1] Several schemes for balancing single-cylinder engines are discussed by Holowenko, *op. cit.*, pp. 308–310.

The resulting force is now obtained by adding Eqs. (c) and (d). This gives

$$\mathbf{F} = \mathbf{F}_C + \mathbf{F}_{A,B} = \left(\frac{m_B}{2} r\omega^2 \cos \omega t + m_B r\omega^2 \frac{r}{l} \cos 2\omega t\right) \mathbf{I} - \frac{m_B}{2} r\omega^2 \sin \omega t \, \mathbf{J}$$
(20-36)

It is readily verified that the vector

$$\frac{m_B}{2} r\omega^2 (\cos \omega t \, \mathbf{I} - \sin \omega t \, \mathbf{J})$$

has a magnitude $m_B r\omega^2/2$ and rotates *backward* (clockwise) with an angular velocity ω rad/sec. The secondary component, as before, is the x projection of a vector of length $m_B r\omega^2(r/l)$ rotating *forward* at an angular velocity of 2ω rad/sec. The maximum inertia force occurs when $\omega t = 0$ and, from Eq. (20-36), is seen to be

$$F_{\max} = m_B r\omega^2 \left(\frac{r}{l} + \frac{1}{2}\right) \quad (e)$$

since $\cos \omega t = \cos 2\omega t = 1$. Before the extra counterweight was added, the maximum inertia force was

$$F_{\max} = m_B r\omega^2 \left(\frac{r}{l} + 1\right) \quad (f)$$

Thus the effect of the additional counterweight, in this instance, is to reduce the maximum shaking force by 50 per cent of the primary component and to add vertical inertia forces where formerly none existed. Equation (20-36) is plotted as a polar diagram in Fig. 20-25 for an r/l value of $\frac{1}{4}$. Here, the vector OA rotates counterclockwise at 2ω rad/sec. The horizontal projection of this vector OA' is the secondary component. The vector OB rotates clockwise at ω rad/sec. The total shaking force $\mathbf{F}$ is shown for the 30° position and is the sum of the vectors OB and $BB' = OA'$.

20-8. Balance of Multicylinder Engines. In order to obtain a clear understanding of the balancing problem in multicylinder engines, let us consider a two-cylinder in-line engine having cranks 180° apart and the rotating parts already balanced by counterweights. Equation (20-35) applies for cylinder 1, and we repeat it here for convenience:

$$F_1 = m_B r\omega^2 \cos \omega t + m_B r\omega^2 \frac{r}{l} \cos 2\omega t \quad (20\text{-}37)$$

For cylinder 2 we replace ωt by $\omega t + \pi$. Since $\cos(\omega t + \pi) = -\cos \omega t$ and $\cos 2(\omega t + \pi) = \cos 2\omega t$, Eq. (20-35) becomes

$$F_2 = -m_B r\omega^2 \cos \omega t + m_B r\omega^2 \frac{r}{l} \cos 2\omega t \quad (20\text{-}38)$$

We cannot add Eqs. (20-37) and (20-38) directly because the cylinder axes are not coincident. The equations do show, however, that we have obtained something akin to balance, for the primary forces are always opposed to each other. On the other hand, even though the cranks are 180° apart, the secondary forces are always in the same direction. These forces are illustrated in Fig. 20-26 for one phase of the linkage and in their proper directions according to the relations expressed above. The figure shows main bearing reactions R_A and R_B, which are due to the primary forces as well as the secondaries. The reader should have no difficulty in calculating the reactions R_A and R_B, using the methods already developed, when the distances a, b, and c are known.

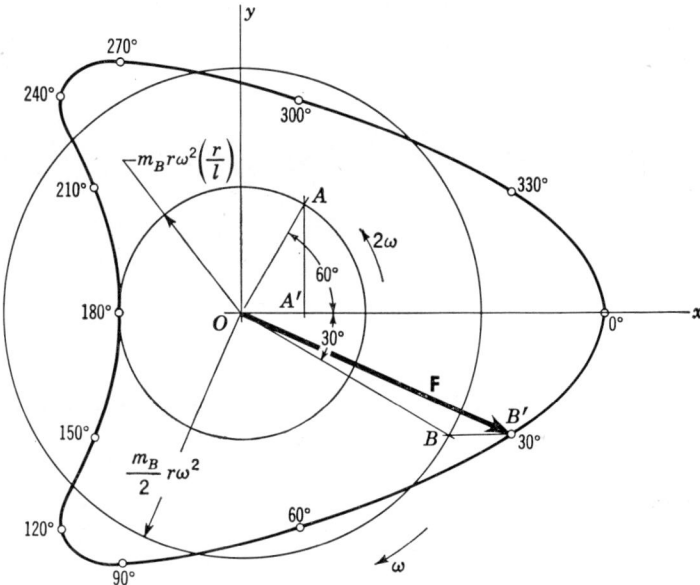

Fig. 20-25. Circle (polar) diagram of inertia forces in a single-cylinder engine for $r/l = \frac{1}{4}$. Counterweight includes one-half of reciprocating mass.

In-line Engines. In developing general relationships for the balance of multicylinder engines we shall assume that all the cranks have the same radius, that connecting-rod lengths are all the same, and that the masses of the parts of each cylinder are equal. In addition we assume that the rotating parts have already been balanced. We now consider an n-cylinder in-line engine and specify the crank angle of the first cylinder as ωt, of the second cylinder as $\omega t + \phi_2$, of the third cylinder as $\omega t + \phi_3$, etc. Thus, the angle ϕ_n is the angle from crank 1 to crank n measured in the direction of rotation. Then the primary inertia force due to cylinder n is

$$F_{n,pri} = m_B r \omega^2 \cos(\omega t + \phi_n) \qquad (a)$$

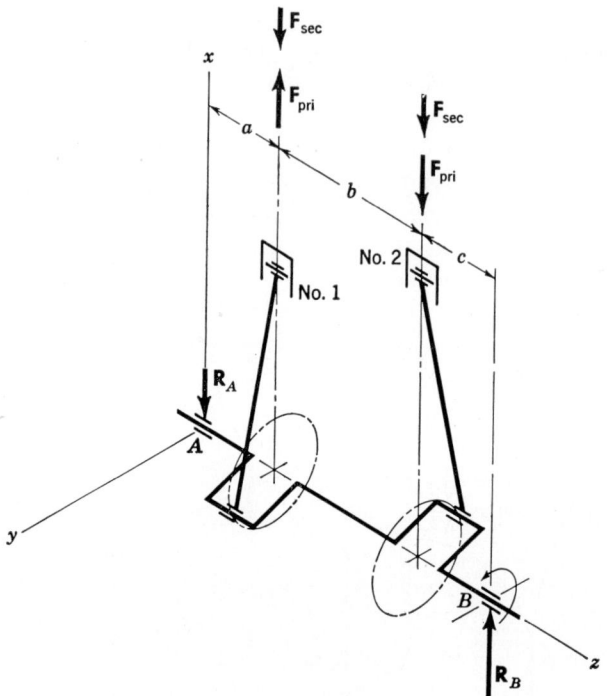

Fig. 20-26. A two-cylinder engine with 180° cranks and main bearings at A and B. The primary and secondary inertia forces are shown on the extensions of the cylinder axes.

One of the conditions for balance is that the crank angles of all the cylinders must be arranged so that the sum of these inertia forces is zero. Thus, for balance

$$m_B r \omega^2 \sum_{1}^{n} \cos(\omega t + \phi_n) = 0 \qquad (b)$$

The factor $m_B r \omega^2$ is not zero. Therefore, expanding the remaining portion of Eq. (b) and equating it to zero give

$$\sum_{1}^{n} (\cos \omega t \cos \phi_n - \sin \omega t \sin \phi_n) = 0 \qquad (c)$$

Neither $\cos \omega t$ nor $\sin \omega t$ is, in general, zero. Therefore, Eq. (c) requires that

$$\sum_{1}^{n} \cos \phi_n = 0 \qquad \sum_{1}^{n} \sin \phi_n = 0 \qquad (20\text{-}39)$$

Next, designating a_n as the distance from the axis of the nth cylinder to any preselected transverse reference plane, the moment of the primary inertia force of the nth cylinder about this plane is

$$T_{n,pri} = m_B r \omega^2 a_n \cos(\omega t + \phi_n) \tag{d}$$

Similarly, then, balance of the primary moments requires that

$$\sum_1^n a_n \cos \phi_n = 0 \qquad \sum_1^n a_n \sin \phi_n = 0 \tag{20-40}$$

Equations (20-39) and (20-40) must be satisfied to obtain balance of the primary inertia forces.

The secondary inertia force due to the nth cylinder is

$$F_{n,sec} = m_B r \omega^2 \frac{r}{l} \cos 2(\omega t + \phi_n) \tag{e}$$

Balance of the secondary forces is obtained if the sum of the forces of each cylinder is zero. Thus

$$m_B r \omega^2 \frac{r}{l} \sum_1^n \cos 2(\omega t + \phi_n) = 0 \tag{f}$$

The factor $m_B r \omega^2 (r/l)$ is not zero and can be dropped. The remaining portion of the equation can be written

$$\sum_1^n (\cos 2\omega t \cos 2\phi_n - \sin 2\omega t \sin 2\phi_n) = 0 \tag{g}$$

Again we note that $\cos 2\omega t$ and $\sin 2\omega t$ are not, in general, zero. Therefore, using the identities

$$\cos 2\phi_n = \cos^2 \phi_n - \sin^2 \phi_n$$
$$\sin 2\phi_n = 2 \sin \phi_n \cos \phi_n$$

and

results in

$$\sum_1^n (\cos^2 \phi_n - \sin^2 \phi_n) = 0$$
$$\sum_1^n \sin \phi_n \cos \phi_n = 0 \tag{20-41}$$

as one set of conditions for secondary balance. The secondary moments must also be balanced, and so

$$\sum_1^n a_n(\cos^2 \phi_n - \sin^2 \phi_n) = 0$$
$$\sum_1^n a_n \sin \phi_n \cos \phi_n = 0 \qquad (20\text{-}42)$$

If the engine is not balanced, then one or more of Eqs. (20-39) to (20-42) will not be zero. The resulting forces or moments can then be calculated from the equations

$$F_{pri} = m_B r \omega^2 \left(\cos \omega t \sum_1^n \cos \phi_n - \sin \omega t \sum_1^n \sin \phi_n \right) \qquad (20\text{-}43)$$

$$T_{pri} = m_B r \omega^2 \left(\cos \omega t \sum_1^n a_n \cos \phi_n - \sin \omega t \sum_1^n a_n \sin \phi_n \right) \qquad (20\text{-}44)$$

$$F_{sec} = m_B r \omega^2 \frac{r}{l} \left[\cos 2\omega t \sum_1^n (\cos^2 \phi_n - \sin^2 \phi_n) \right.$$
$$\left. - 2 \sin 2\omega t \sum_1^n \sin \phi_n \cos \phi_n \right] \qquad (20\text{-}45)$$

$$T_{sec} = m_B r \omega^2 \frac{r}{l} \left[\cos 2\omega t \sum_1^n a_n(\cos^2 \phi_n - \sin^2 \phi_n) \right.$$
$$\left. - 2 \sin 2\omega t \sum_1^n a_n \sin \phi_n \cos \phi_n \right] \qquad (20\text{-}46)$$

EXAMPLE 20-1. Investigate the balance of a four-cylinder engine having the crank spacing 0, 180, 180, 0°. Use a unit distance between the cylinder axes.

Solution. It is customary to solve balancing problems using a tabulation method; then the summations are taken simply by adding the appropriate rows in a table. In this example the solution is shown in Table 20-1.

The table shows that the primary forces and moments are balanced but that the secondary ones are not. Taking $m_B r \omega^2 = 1$ and $r/l = \frac{1}{4}$, Eqs. (20-45) and (20-46) are

$$F_{sec} = \cos 2\omega t \qquad T_{sec} = 1.5 \cos 2\omega t$$

or

$$z = \frac{T_{sec}}{F_{sec}} = 1.5 \text{ units}$$

Therefore the resultant inertia force occurs at the center of the crankshaft.

V Engines. In analyzing the balance of V engines each bank should be treated as an in-line engine and the results added vectorially to

78 TANGENT FUNCTIONS

Example 5.8 Find tan 3° = .0524

Operation   Opposite 3° on T (black) of Sec T SRT scale read 0.0524 on C.

TANGENT 5.74° TO 45° (Using T (black) of TT Scale)

To find the tangent of an angle from 5.74° to 45°, the TT scale is used with the C scale. The desired angle is set on the T (black) scale and the tangent is read directly opposite on the C scale. The decimal range can be found at the right hand scale instruction symbol, (0.1 ← 1.0).

Example 5.9 Find tan 16.9° = 0.304

Operation   Opposite 16.9° on T (black) of TT scale read 0.304 on C.

TANGENT 45° TO 84.27° (Using T (red) of TT Scale)

To find the tangent of an angle from 45° to 84.27°, the TT scale is used with the CI scale. The desired angle is set on the T (red) scale and the tangent is read directly opposite on the CI scale. The decimal range can be found at the right hand scale instruction symbol, (10.0 → 1.0).

Example 5.10 Find tan 75° = 3.73

Operation   Opposite 75° on T (red) of TT scale read 3.73 on CI.

TANGENT 84.27° TO 89.43° (Using T (red) of Sec T SRT)

To find the tangent of an angle from 84.27° to 89.43°, the Sec T SRT scale is used with the CI scale. The desired angle is set on T (red) scale and the tangent is read directly opposite on the CI scale. The decimal range can be found at the right hand scale instruction symbol, (100.0 → 10.0).

Example 5.11 Find tan 89° = 57.3

Operation   Opposite 89° on T (red) of Sec T SRT scale read 57.3 on CI.

TANGENT 89.43° TO 90° (Using R of Sec T SRT)

To find the tangent of an angle from 89.43° to 90°, the Sec T SRT scale is used with the CI scale. For this range of angles use the formula $\tan A = \dfrac{1}{\sin (90° - A)}$. The expression $\sin (90° - A)$ is

TANGENT FUNCTIONS 77

Exercise 5.4
The Cosine Function of Angles from 0° to 90°

Find the cosine values of the following angles:

1. Cos 34.5°
2. Cos 0.4°
3. Cos 1.4°
4. Cos 13.5°
5. Cos 2° 48′
6. Cos 6.4°
7. Cos 8.4°
8. Cos 12° 42′
9. Cos 20° 30′
10. Cos 24° 36′
11. Cos 46.5°
12. Cos 60°
13. Cos 80°
14. Cos 87.5°
15. Cos 80° 18′

Find the angle values of the following cosine functions:

16. Cos ___ = 0.052	20. Cos ___ = 0.809	24. Cos ___ = 0.00200
17. Cos ___ = 0.201	21. Cos ___ = 0.913	25. Cos ___ = 0.00350
18. Cos ___ = 0.400	22. Cos ___ = 0.200	26. Cos ___ = 0.800
19. Cos ___ = 0.588	23. Cos ___ = 0.020	27. Cos ___ = 0.900

5.5 THE TANGENT FUNCTION OF ANGLES FROM 0° TO 90°

The tangent function of angles from 0° to 90° are found by the use of five different scales. The radian scale is used to find tangent functions of extremely small angles and angles extremely close to 90°. The red and black T values of the Sec T SRT scale are used to find tangent functions of angles between 0.574° and 5.74° as well as those between 84.27° and 89.43°. All remaining tangent functions of angles between 5.74° and 84.27° are found by using the TT scale. Each scale and technique is described in detail below.

TANGENT 0° TO 0.574° (Using R of Sec T SRT Scale)

The technique for finding tangent of angles between 0° and 0.574° is the same as described in Section 5.3, since the tangent and sine of extremely small angles are considered equal. The tangent is approximately equal to the size of the angle expressed in radians.

Example 5.7 Find tan 0.5° = 0.00872

Operation Opposite 0.5° on R of Sec T SRT scale read 0.00872 on C.

TANGENT 0.574° TO 5.74° (Using T (black) of Sec T SRT)

To find the tangent of an angle from 0.574° to 5.74°, the Sec T SRT scale is used with the C scale. The desired angle is set on the T (black) scale and the tangent is read directly opposite on the C scale. The decimal range can be found at the right hand scale instruction symbol, (0.01 → 0.1).

DYNAMICS OF THE RECIPROCATING ENGINE

TABLE 20-1

Term	Crank				Sum
	1	2	3	4	
ϕ	0	180	180	0	
a_n	0	1	2	3	
$\cos \phi_n$	1.000	−1.000	−1.000	1.000	0
$\sin \phi_n$	0	0	0	0	0
$a_n \cos \phi_n$	0	−1.000	−2.000	3.000	0
$a_n \sin \phi_n$	0	0	0	0	0
$\cos^2 \phi_n$	1.000	1.000	1.000	1.000	
$\sin^2 \phi_n$	0	0	0	0	
$\cos^2 \phi_n - \sin^2 \phi_n$	1.000	1.000	1.000	1.000	4.000
$\sin \phi_n \cos \phi_n$	0	0	0	0	0
$a_n (\cos^2 \phi_n - \sin^2 \phi_n)$	0	1.000	2.000	3.000	6.000
$a_n \sin \phi_n \cos \phi_n$	0	0	0	0	0

obtain the resultant primary and secondary quantities. If the crank angle is referred to one of the banks of cylinders, then, when the other bank is treated, the angle ωt must be adjusted for the V angle.

20-9. Flywheels. The reciprocating engine serves as a convenient vehicle to demonstrate the principles of flywheel analysis, but our investigation of this subject is to be quite general.

A flywheel is a mechanical filtering element in a circuit through which power is flowing. It acts as a smoothing or equalizing element in any mechanical power-transmission circuit which has a back-and-forth flow of energy. In particular, a flywheel is an energy-storage device and is analogous to the capacitor in electrical circuits. Energy is stored in a flywheel by speeding it up and delivered by a flywheel by slowing it down.

The kinetic energy of a flywheel rotating at ω rad/sec is given by the equation

$$U = \frac{I}{2} \omega^2 \qquad (a)$$

Since a flywheel must change its velocity in order to store or deliver energy, there must exist a relation between the energy change and the velocity change. Designating initial and final conditions, respectively, by the subscripts 1 and 2, then, from Eq. (a) the energy change must be

$$U_2 - U_1 = \frac{I}{2} (\omega_2^2 - \omega_1^2) \qquad (20\text{-}47)$$

Next, designating
$$\omega_{av} = \frac{\omega_2 + \omega_1}{2} \tag{20-48}$$
we can define
$$C_s = \frac{\omega_2 - \omega_1}{2} \tag{20-49}$$
as the *coefficient of speed fluctuation*. Equation (20-47) can now be factored to give
$$U_2 - U_1 = \frac{I}{2}(\omega_2 - \omega_1)(\omega_2 + \omega_1) \tag{b}$$

Since $\omega_2 - \omega_1 = C_s\omega_{av}$ and $\omega_2 + \omega_1 = 2\omega_{av}$, we have
$$U_2 - U_1 = C_s I \omega_{av}^2 \tag{20-50}$$

Equation (20-50) permits us to select an appropriate value of C_s and then to solve for the mass moment of inertia I when the speed and energy variation are known.

Proceeding next to a study of torque variation we add Eq. (20-8) for gas torque to Eq. (20-23) for the inertia torque. This produces

$$T = Pr \sin \omega t \left(1 + \frac{r}{l}\cos \omega t\right) + \frac{m_B}{2}r^2\omega^2\left(\frac{r}{2l}\sin \omega t - \sin 2\omega t - \frac{3r}{2l}\sin 3\omega t\right) \tag{20-51}$$

which is valid for a single cylinder. This equation should be solved by tabulation employing crank-angle intervals of 5 to 10°. In multicylinder engines the quantities repeat themselves, and so the equation need be solved only for a single cylinder. The firing order will then indicate how the quantities are to be combined. As an example, an eight-cylinder engine will fire four times for each revolution of the crankshaft if it is a four-cycle engine. Consequently, the curve of torque vs. crank angle will repeat itself every 90°, as shown in Fig. 20-27. The line of output torque is obtained, in this figure, using the condition that the area enclosed by the curve should be the same above the line as it is below the line.

Since the loops above the output torque line of Fig. 20-27 contain the same area as do those below the line, the energy represented by the area of any one loop is the deviation from the mean. Consequently,

$$U_2 - U_1 = \int_{\theta_1}^{\theta_2} T\, d\theta \tag{20-52}$$

where θ_1 and θ_2 are the crank angles, in radians, at the beginning and end of one of the loops. Equation (20-52) cannot be solved analytically because no expression is available relating the gas forces to the crank angle. After the torque curve is plotted, the area can be obtained quickly using a planimeter or, if desired, the methods of graphical integration discussed in Sec. 4-12 can be employed.

The value of the coefficient of speed fluctuation depends upon the purpose for which the engine was designed. Usual values range between $\frac{1}{40}$ and $\frac{1}{200}$.

20-10. Torsional Models. The torque fluctuations of Fig. 20-27 suggest that high-amplitude torsional oscillations may occur in the crankshaft unless it is carefully designed. The procedures discussed in Sec. 18-14 make it possible to find an equivalent torsional model of the engine which can then be analyzed for oscillations using the Holzer method.

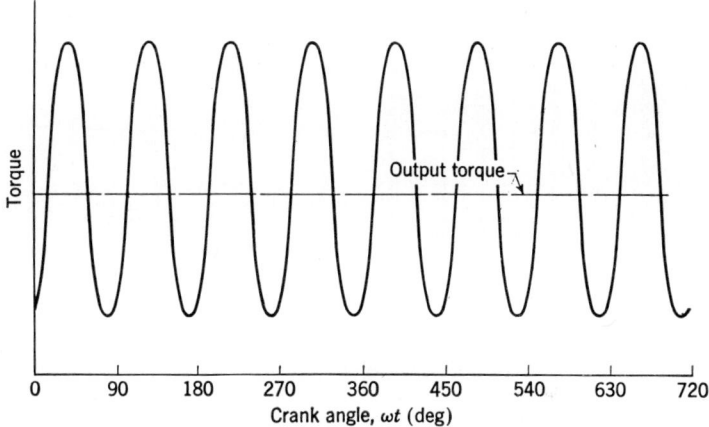

FIG. 20-27. Graph of torque variation in an eight-cylinder engine.

The Holzer method yields the natural frequencies of oscillation of the crankshaft. The terms $\sin \omega t$, $\sin 2\omega t$, and $\sin 3\omega t$ in Eq. (20-51) indicate that resonance will occur when one of these frequencies is equal to one of the natural frequencies. In this book our purpose is only to analyze the engine to discover the natural frequencies. Once they are known, the engineer can redesign to avoid them or select dampers or vibration absorbers to minimize their effects.

The masses associated with the rotating crankshaft can be calculated by considering it as a combination of cylinders, prisms, and other simple shapes. The moments of inertia of these parts are then easily calculated, as will be shown in the example to follow.

The mass moment of inertia due to the reciprocating elements of the engine does make a contribution, but the full effect of this inertia is apparent only near midstroke. Consequently, it is customary to consider that only half of the reciprocating mass acts at the crankpin during the cycle. Thus

$$I = \frac{m_B}{2} r^2 \tag{20-53}$$

gives the equivalent moment of inertia of the reciprocating masses.

The torsional spring constant of a crankshaft is usually expressed in terms of a solid round shaft having an equivalent length l_e and an equivalent diameter d_e. The equations for making this conversion are empirical, are based on experiment, and several of them are available.[1] The Wilson[2] equation is representative of these and is

$$l_e = d_e^4 \left[\frac{a + 0.4 D_1}{D_1^4 - d_1^4} + \frac{b + 0.4 D_2}{D_2^4 - d_2^4} + \frac{r - 0.2(D_1 + D_2)}{tw^3} \right] \tag{20-54}$$

Figure 20-28 is a section of a hollow crankshaft and illustrates the notation for Eq. (20-54); the quantities are all in inches. The equation is

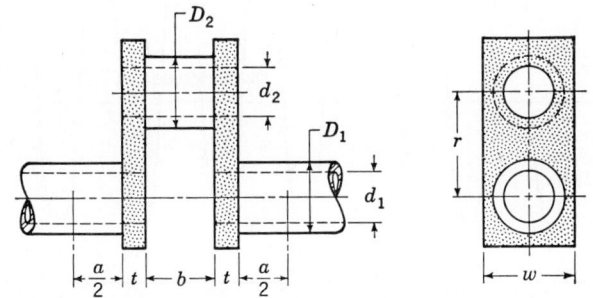

FIG. 20-28. Section of crankshaft showing notation for finding equivalent length and diameter. a = length of journal; b = length of crankpin; t = thickness of web; w = width of web; r = crank radius; D_1 = journal OD; d_1 = journal ID; D_2 = crankpin OD; d_2 = crankpin ID; all dimensions in inches.

also valid for a solid crankshaft when the appropriate terms are made zero.

EXAMPLE 20-2. Figure 20-29 shows a four-throw crankshaft with a flywheel and pinion connected to one end. The pinion meshes with a gear on another shaft which drives a load having a mass moment of inertia of 0.92 lb-in.-sec². The weight of the reciprocating parts is 1.70 lb, and the rotating portion of the connecting rod has a weight of 0.68 lb. Find the equivalent torsional system assuming that the flywheel and two gears can be lumped as a single mass. All materials are steel having a unit weight of 0.285 lb/in.³, a modulus of elasticity of 30,000,000 psi, and a modulus of rigidity of 12,000,000 psi.

Solution. We shall begin by calculating the mass moment of inertia and equivalent

---

[1] See John N. Macduff and John R. Curreri, "Vibration Control," pp. 305–308, McGraw-Hill Book Company, Inc., New York, 1958.

[2] W. K. Wilson, "Practical Solution of Torsional Vibration Problems," p. 597, John Wiley & Sons, Inc., New York, 1956.

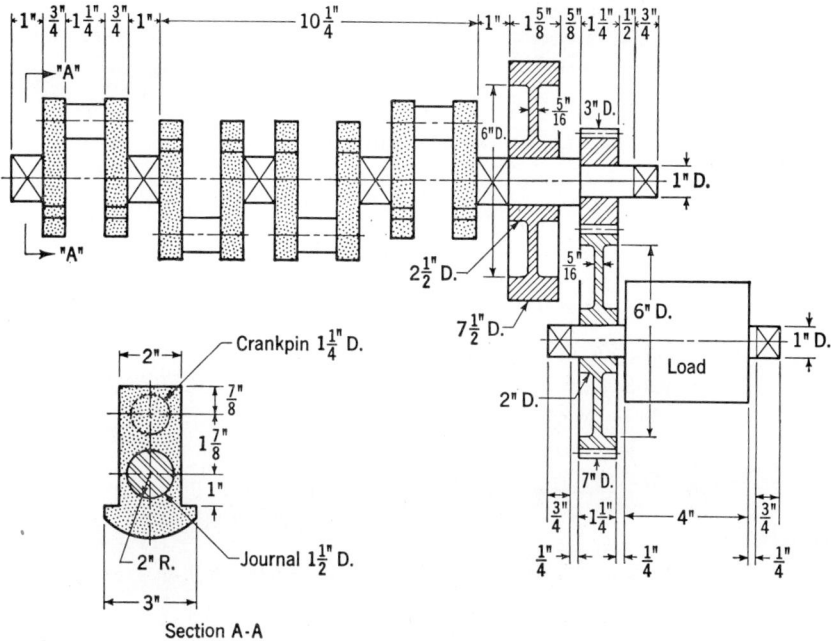

Fig. 20-29

shaft length for one of the cylinders. The journal has a diameter of 1½ in. and a length of 1 in. Its weight is

$$w = 0.285 \frac{\pi d^2}{4} l = (0.285) \frac{\pi (1.5)^2}{4} (1) = 0.504 \text{ lb}$$

From Appendix III its moment of inertia is

$$I = \frac{mr^2}{2} = \frac{0.504}{386} \frac{(1.5/2)^2}{2} = 0.000366 \text{ lb-in.-sec}^2$$

The crankpin has a 1¼-in. diameter and is 1¼ in. long. Its weight is

$$w = 0.285 \frac{\pi d^2}{4} l = (0.285) \frac{\pi (1.25)^2}{4} (1.25) = 0.437 \text{ lb}$$

The moment of inertia about its own mass center is

$$I_G = \frac{mr^2}{2} = \frac{0.437}{386} \frac{(1.25/2)^2}{2} = 0.000221 \text{ lb-in.-sec}^2$$

The moment of inertia about the crankshaft center is

$$I = I_G + mr^2 = 0.000221 + \frac{0.437}{386} (1.875)^2 = 0.00410 \text{ lb-in.-sec}^2$$

where the moment of inertia is transferred by the amount of the crank radius $r = 1.875$ in.

The web can be divided into a rectangular portion and an irregular or curved section. Taking the rectangular section first, we find that its dimensions are $3\frac{3}{4}$ by 2 by $\frac{3}{4}$ in. and that its mass center is $\frac{7}{8}$ in. from the crankshaft center. The weight of this portion of the web is

$$w = (0.285)(3.75)(2)(0.75) = 1.60 \text{ lb}$$

and so the moment of inertia is, from Appendix III,

$$I_G = \frac{m(a^2 + c^2)}{12} = \frac{1.60}{386} \frac{(3.75)^2 + (2)^2}{12} = 0.00624 \text{ lb-in.-sec}^2$$

Transferring through $\frac{7}{8}$ in. to the crankshaft axis gives

$$I = I_G + mr^2 = 0.00624 + \frac{1.60}{386}(0.875)^2 = 0.00941 \text{ lb-in.-sec}^2$$

Here we shall, rather loosely,[1] assume that the irregular-shaped portion is rectangular, having the dimensions 3 by $\frac{3}{4}$ by $\frac{3}{4}$ in. with the mass center $1\frac{1}{2}$ in. from the crankshaft center. Then the weight and moment of inertia of this section is

$$w = (0.285)(3)(0.75)(0.75) = 0.480 \text{ lb}$$
$$I_G = \frac{m(a^2 + c^2)}{12} = \frac{0.480}{386} \frac{(3)^2 + (0.75)^2}{12} = 0.000990 \text{ lb-in.-sec}^2$$

Transferring,

$$I = I_G + mr^2 = 0.000990 + \frac{0.480}{386}(1.5)^2 = 0.00379 \text{ lb-in.-sec}^2$$

so the moment of inertia of a pair of webs is

$$I = 2(0.00941 + 0.00379) = 0.0264 \text{ lb-in.-sec}^2$$

Next, the moment of inertia of the rotating part of the connecting rod is

$$I = \frac{w}{g} r^2 = \frac{0.68}{386}(1.875)^2 = 0.00620 \text{ lb-in.-sec}^2$$

Also, as previously noted, we consider that one-half of the reciprocating mass is effective at the crankpin. Thus, its moment of inertia about the crankshaft is

$$I = \frac{w/2}{g} r^2 = \frac{1.70/2}{386}(1.875)^2 = 0.00775 \text{ lb-in.-sec}^2$$

Thus the total moment of inertia associated with a single cylinder is

$$I = 0.000366 + 0.00410 + 0.0264 + 0.00620 + 0.00775$$
$$= 0.0448 \text{ lb-in.-sec}^2$$

when rounded off.

We use an equivalent crankshaft diameter equal to the diameter of the journal; then the equivalent length associated with a single cylinder is calculated from Eq. (20-54). When the substitutions are made in order, the equation reads

$$l_e = (1.5)^4 \left[\frac{1 + (0.4)(1.5)}{(1.5)^4} + \frac{1.25 + (0.4)(1.25)}{(1.25)^4} + \frac{1.875 - (0.2)(1.5 + 1.25)}{(0.75)(2)^3} \right]$$
$$= 6.35 \text{ in.}$$

[1] This is a poor assumption and would not be tolerated in a good design office. If a drawing of the part is constructed at a scale of two to four times size and divided into small easily measurable sections, then the moments of inertia of each of these sections can be calculated and the transfer distances measured.

The flywheel rim is considered as a hollow section 7½ in. OD by 6 in. ID by 1⅝ in. long. Its weight is

$$w = 0.285 \frac{\pi}{4}(D^2 - d^2)l = (0.285)\frac{\pi}{4}[(7.5)^2 - (6)^2](1.625)$$
$$= 7.35 \text{ lb}$$

From Appendix III, its moment of inertia is

$$I = \frac{m(a^2 + b^2)}{2} = \frac{7.35}{386}\frac{(3)^2 + (3.75)^2}{2} = 0.220 \text{ lb-in.-sec}^2$$

The web of the flywheel is another hollow section 6 in. OD by 2½ in. ID by 5/16 in. thick, and its weight is

$$w = 0.285 \frac{\pi}{4}(D^2 - d^2)l = (0.285)\frac{\pi}{4}[(6)^2 - (2.5)^2](0.3125)$$
$$= 2.08 \text{ lb}$$

and the moment of inertia is

$$I = \frac{m(a^2 + b^2)}{2} = \frac{2.08}{386}\frac{(3)^2 + (1.25)^2}{2} = 0.0284 \text{ lb-in.-sec}^2$$

The hub, together with the portion of the shaft it encloses, can be considered the final portion of the flywheel. Thus it is a cylinder 2½ in. in diameter by 1⅝ in. long and has a weight of

$$w = 0.285 \frac{\pi d^2}{4} l = (0.285)\frac{\pi (2.5)^2}{4}(1.625) = 2.275 \text{ lb}$$

The moment of inertia is

$$I = \frac{mr^2}{2} = \frac{2.275}{386}\frac{(2.5/2)^2}{2} = 0.00460 \text{ lb-in.-sec}^2$$

so the total moment of inertia of the flywheel is

$$I = 0.220 + 0.0284 + 0.00460 = 0.253 \text{ lb-in.-sec}^2$$

Segments of the shaft on each side of the flywheel total 1⅛ in. in length. The weight and inertia are

$$w = 0.285 \frac{\pi d^2}{4} l = (0.285)\frac{\pi (1.5)^2}{4}(1.125) = 0.567 \text{ lb}$$

$$I = \frac{mr^2}{2} = \frac{0.567}{386}\frac{(1.5/2)^2}{2} = 0.000413 \text{ lb-in.-sec}^2$$

The pinion is considered as a cylinder having a diameter equal to the pitch diameter and a length of 1¼ in.; so its weight and inertia are

$$w = 0.285 \frac{\pi d^2}{4} l = (0.285)\frac{\pi (3)^2}{4}(1.25) = 2.52 \text{ lb}$$

$$I = \frac{mr^2}{2} = \frac{2.52}{386}\frac{(3/2)^2}{2} = 0.00735 \text{ lb-in.-sec}^2$$

562   DYNAMIC ANALYSIS OF MACHINES

To the right of the pinion is a shaft segment 1 in. in diameter by $1\frac{1}{4}$ in. long.

$$w = 0.285 \frac{\pi d^2}{4} l = (0.285) \frac{\pi (1)^2}{4} (1.25) = 0.279 \text{ lb}$$

$$I = \frac{mr^2}{2} = \frac{0.279}{386} \frac{(1/2)^2}{2} = 0.000090 \text{ lb-in.-sec}^2$$

The rim of the gear is 7 in. OD by 6 in. ID by $1\frac{1}{4}$ in. long, and so its weight is

$$w = 0.285 \frac{\pi}{4} (D^2 - d^2) l = (0.285) \frac{\pi}{4} [(7)^2 - (6)^2] (1.25)$$
$$= 3.63 \text{ lb}$$

The moment of inertia of this segment is

$$I = \frac{m(a^2 + b^2)}{2} = \frac{3.63}{386} \frac{(3.5)^2 + (3)^2}{2} = 0.0996 \text{ lb-in.-sec}^2$$

The web of the gear is 6 in. OD by 2 in. ID by $\frac{5}{16}$ in. thick. Therefore

$$w = 0.285 \frac{\pi}{4} (D^2 - d^2) l = (0.285) \frac{\pi}{4} [(6)^2 - (2)^2] (0.3125)$$
$$= 2.24 \text{ lb}$$

$$I = \frac{m(a^2 + b^2)}{2} = \frac{2.24}{386} \frac{(3)^2 + (1)^2}{2} = 0.029 \text{ lb-in.-sec}^2$$

The hub and shaft segment it encloses is 2 in. in diameter by $1\frac{1}{4}$ in. long; so its weight and inertia are

$$w = 0.285 \frac{\pi d^2}{4} l = (0.285) \frac{\pi (2)^2}{4} (1.25) = 1.12 \text{ lb}$$

$$I = \frac{mr^2}{2} = \frac{1.12}{386} \frac{(1)^2}{2} = 0.00145 \text{ lb-in.-sec}^2$$

The total moment of inertia associated with the gear is the sum of its parts. Therefore

$$I = 0.0996 + 0.029 + 0.00145 = 0.130 \text{ lb-in.-sec}^2$$

The speed of the gear is less than that of the pinion, and our equivalent system is to be based on the crankshaft speed. The equivalent moment of inertia of the gear operating at the crankshaft speed is found from Eqs. (18-52) and is

$$I_e = \left(\frac{\omega}{\omega_e}\right)^2 I = \left(\frac{3}{7}\right)^2 (0.130) = 0.0239 \text{ lb-in.-sec}^2$$

Here ω_e is the speed of the crankshaft and ω the speed of the countershaft; the ratio of the gear diameters is such that $\omega = (\frac{3}{7})\omega_e$. The total inertia for the flywheel and gears, when rounded off, is

$$I = 0.253 + 0.000413 + 0.00735 + 0.000090 + 0.0239$$
$$= 0.285 \text{ lb-in.-sec}^2$$

The load has a moment of inertia of 0.92 lb-in.-sec². The equivalent moment of inertia corresponding to the crankshaft speed is

$$I_e = \left(\frac{\omega}{\omega_e}\right)^2 I = \left(\frac{3}{7}\right)^2 (0.92) = 0.169 \text{ lb-in.-sec}^2$$

The length of the 1-in.-diameter shaft between the gear center and load center is

$4\frac{7}{8}$ in. The torsional spring constant is found from the formula

$$\theta = \frac{Tl}{GJ} \quad \text{where } J = \frac{\pi d^4}{32}$$

which can be found in any text on strength of materials. These equations give the angle of twist θ in radians of a solid round shaft d in. in diameter and l in. long when subjected to a torque T. The quantity G is the modulus of rigidity. When these equations are solved for the torsional spring constant, there results

$$k_t = \frac{T}{\theta} = \frac{GJ}{l} = \frac{\pi d^4 G}{32l}$$

Therefore the spring constant for the section of shaft between the gear and the load is

$$k_t = \frac{\pi d^4 G}{32l} = \frac{\pi(1)^4(12,000,000)}{(32)(4.875)} = 242,000 \text{ lb-in./rad}$$

When this is transferred to the crankshaft speed, the equivalent spring constant is, from Eq. (18-53),

$$k_{te} = \left(\frac{\omega}{\omega_e}\right)^2 k_t = \left(\frac{3}{7}\right)^2 (242,000) = 44,500 \text{ lb-in./rad}$$

We shall assume a length of $2\frac{3}{4}$ in. between the center of the masses associated with the flywheel and the center of the main bearing adjacent to the last crank. Then the length of shaft from the last cylinder axis to the assumed center of the flywheel masses is

$$l_e = \frac{6.35}{2} + 2.75 = 5.925 \text{ in.}$$

and the spring constant is

$$k_t = \frac{\pi d^4 G}{32l} = \frac{\pi(1.5)^4(12,000,000)}{(32)(5.925)} = 1,008,000 \text{ lb-in./rad}$$

The spring constant corresponding to the 6.35-in. equivalent length between crank centers is

$$k_t = \frac{\pi d^4 G}{32l} = \frac{\pi(1.5)^4(12,000,000)}{(32)(6.35)} = 940,000 \text{ lb-in./rad}$$

The results of all these calculations are summarized in Fig. 20-30, which is the equivalent system. This is all the information necessary to determine the natural torsional frequencies using the Holzer method.

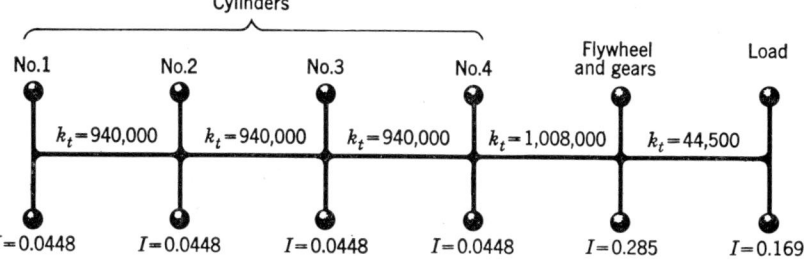

Fig. 20-30. The equivalent torsional system of Example 20-2. The speed is the same as the crankshaft speed. Units of k_t are pound-inches per radian, and those of I are lb-in.-sec².

PROBLEMS

20-1. Choose one of the engines described in this chapter and develop an indicator diagram for a typical set of operating conditions.

20-2. Using the indicator diagram obtained as a solution to Prob. 20-1, make a force analysis of the engine at crank intervals of 15°. Plot polar diagrams showing the piston-pin and crankpin forces for a complete cycle. Plot a chart of torque vs. crank angle for a single cylinder and another for all cylinders. Note that the data are incomplete for some of the engines listed and assumptions will be necessary.

20-3. Construct an indicator diagram for a four-cylinder four-cycle gasoline engine having a 3⅜-in. bore, a 3½-in. stroke, and a compression ratio of 6.25. The operating conditions to be used are 30 hp at 1,900 rpm. Use a mechanical efficiency of 72 per cent and a card factor of 0.90.

20-4. The phase of the engine mechanism shown in the figure is near the end of the compression stroke. For this phase, make a graphical analysis to determine the crankshaft torque, all bearing reactions, and the piston-wall force. The rotating weights are balanced.

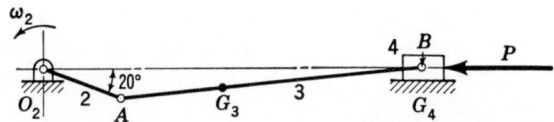

PROB. 20-4. $n = 1{,}900$ rpm; $W_3 = 1.80$ lb; $W_4 = 1.562$ lb; $O_2A = 1.75$ in.; $AG_3 = 2.17$ in.; $AB = 6.50$ in.; $I_3 = 0.0438$ lb-sec²-in.; $P = 1{,}320$ lb.

20-5. The same as Prob. 20-4 but use the algebraic method.

20-6. Use the data in Prob. 20-4 except that the phase of the mechanism is 45° after top dead center and occurs during the expansion stroke. Employing a force $P = 620$ lb, find, graphically, the crankshaft torque, all bearing reactions, and the piston-wall force for this phase.

20-7. The phase of the engine linkage shown in the figure is nearing the end of the expansion stroke. Using graphical procedures determine the crankshaft torque, all bearing reactions, and the piston-wall force. The rotating weights are balanced.

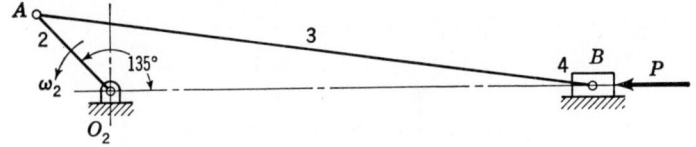

PROB. 20-7. $n = 2{,}000$ rpm; $W_{3A} = 3.30$ lb; $W_{3B} = 1.70$ lb; $W_4 = 4.50$ lb; $O_2A = 2.25$ in.; $AB = 12$ in.; $P = 355$ lb.

20-8. Work Prob. 20-7 using an analytical method.

20-9. The same as Prob. 20-7 except that the crank is in the 330° position and the exhaust valve is open.

DYNAMICS OF THE RECIPROCATING ENGINE

20-10. Illustrated is an engine mechanism which is just beginning the suction stroke, and the gas forces may be assumed to be zero. Using graphical methods find the crankshaft torque, the bearing reactions, and the piston-wall force. The crank has a counterweight which balances the rotating weights and also includes 50 per cent of the reciprocating weight to alter the balance characteristics.

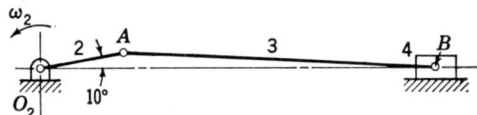

PROB. 20-10. $n_2 = 2{,}500$ rpm; $W_{3A} = 0.641$ lb; $W_{3B} = 0.32$ lb; $W_4 = 1.25$ lb; $O_2A = 1.375$ in.; $AB = 5$ in.; $P = 0$ lb.

20-11. The same as Prob. 20-10 except that the engine is in the expansion stroke and the gas force is 2,730 lb.

20-12. Illustrated is an engine linkage in the compression stroke. Calculate the crankshaft torque and bearing reactions using the algebraic method. The rotating weights are balanced.

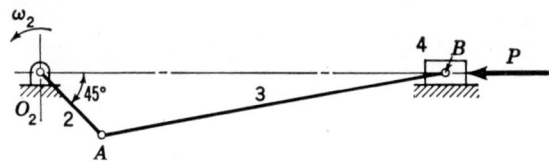

PROB. 20-12. $n = 1{,}350$ rpm; $W_{3A} = 13.2$ lb; $W_{3B} = 4.41$ lb; $W_4 = 16.5$ lb; $O_2A = 3.75$ in.; $AB = 15$ in.; $P = 2{,}640$ lb.

20-13. The same as Prob. 20-12 except that the mechanism is on top dead center and the gas force is 32,600 lb.

20-14. The same as Prob. 20-12 except that the piston has advanced 20 per cent of its stroke in the expansion phase and the gas force is 8,700 lb.

20-15. Plot a circle diagram of the inertia forces for the engine of Prob. 20-12.

20-16. Plot a circle diagram of the inertia forces for the engine of Prob. 20-12 after a counterweight equal to 60 per cent of the reciprocating weight has been added to the crank.

20-17. The single-cylinder engine of Prob. 20-7 has a counterweight added equal to two-thirds of the weight of the reciprocating weights. Plot a circle diagram of the inertia forces.

20-18. The same as Prob. 20-17 except that the counterweight is equal to all the reciprocating weight.

20-19. Investigate the balance of a three-cylinder in-line engine with equally spaced cranks.

20-20. Investigate the balance of a six-cylinder in-line engine. The crank spacing is 0, 240, 120, 120, 240, 0°.

20-21. The figure is a schematic drawing of a crankshaft for a four-cylinder engine. The cranks are spaced 0, 180, 180, 0°. Also shown in the figure are the locations of the three main bearings relative to the cylinder axes. Each piston weighs 25 oz, and each rod 29 oz. The engine has a $3\frac{1}{8}$-in. bore and a $3\frac{1}{2}$-in. stroke and develops 30 hp at 1,900 rpm. The connecting rod is $6\frac{3}{4}$ in. long; two-thirds of its weight may be assigned to the crankpin. (a) Calculate the reaction at each main bearing due to the inertia forces when $\theta_1 = 45°$. (b) What are the maximum values of the main bearing reactions to the inertia forces?

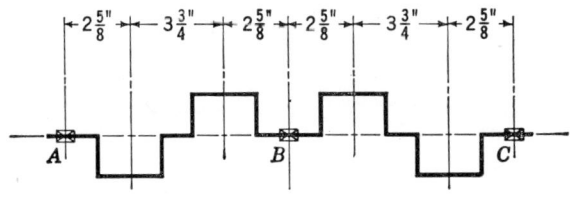

PROB. 20-21

20-22. An eight-cylinder in-line engine has the cranks arranged 0, 270, 90, 180, 135, 225, 45, 315°. The reciprocating weight is 29.2 lb per cylinder. The engine has a 9-in. stroke and a connecting rod 18 in. long. Distance between cranks is 8 in. The operating speed is 1,200 rpm. (a) Investigate the balance. (b) Calculate the magnitude of the unbalanced couple for $\theta_1 = 0°$.

20-23. Using the data of Prob. 20-22 calculate the unbalanced couple for all θ_1's from 0 to 360° in 15° intervals and plot the results.

20-24. Investigate the balance of the V6 truck engine described in Sec. 20-1.

20-25. Investigate the balance of the Corvair engine.

20-26. The figure shows the crank arrangement in a 90° V8 automotive engine. It is assumed that the engine has five equally spaced main bearings. Typical data for a $4\frac{1}{8}$-in.-bore by $3\frac{1}{4}$-in.-stroke four-cycle engine are 230 hp at 4,400 rpm, 35.4 oz reciprocating weight per cylinder, $6\frac{5}{64}$-in.-long connecting rod. (a) Investigate the balance of this engine. (b) Can the unbalance be corrected by application of the methods of Chap. 19? If so, choose suitable correction planes and specify the magnitude and location of the corrective weights.

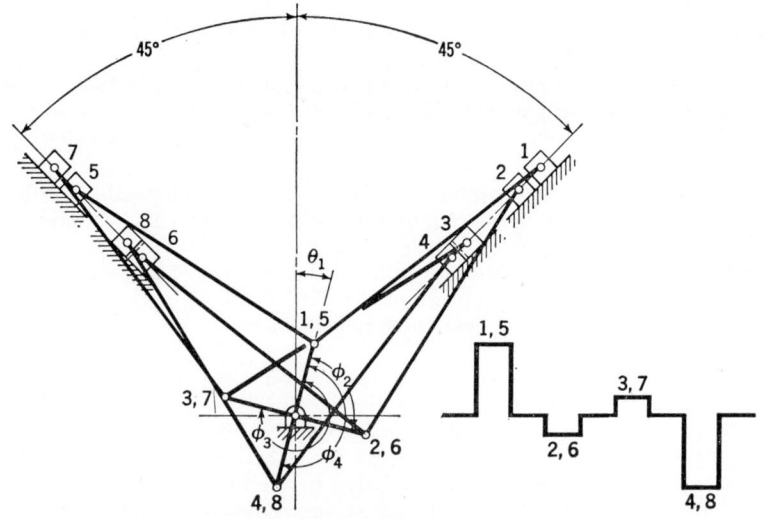

PROB. 20-26

20-27. The indicator card shown in the figure was approximated from published data for the Hercules GO-169-H three-cylinder, four-stroke-cycle, in-line gasoline engine.[1] Additional data are 4-in. bore, 4½-in. stroke, 52 hp at 2,600 rpm, 6.50 compression ratio, piston weight 43.4 oz, connecting rod 8 in. long and 50.9 oz. Using

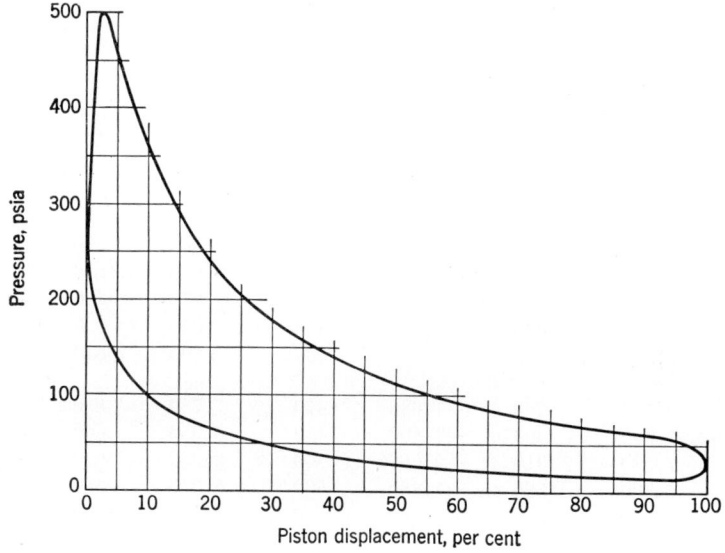

PROB. 20-27

whatever approximations or assumptions are necessary, make a torque analysis of a single cylinder for all crank angles in a cycle. Plot a curve showing the torque for the entire engine. Using a coefficient of speed fluctuation of $\frac{1}{200}$, design a cylindrical steel flywheel for the engine not over 16 in. in diameter. Steel weighs 0.285 lb/in.³

[1] From the statistical issue of *Automotive Ind.*, vol. 122, no. 6, p. 184, Mar. 15, 1960.

20-28. Table 20-2 is the output torque for a single-cylinder engine running at 4,600 rpm. Using $C_s = \frac{1}{40}$ design a cylindrical or disk-type flywheel for the engine.

TABLE 20-2

Crank angle, deg	Torque, lb-in.	Crank angle, deg	Torque, lb-in.	Crank angle, deg	Torque, lb-in.	Crank angle, deg	Torque, lb-in.
0	0	190	−3,040	370	−1,280	550	−3,040
10	1,550	200	−4,780	380	−1,330	560	−4,780
20	7,190	210	−5,100	390	−60	570	−5,110
30	8,520	220	−5,040	400	1,450	580	−5,060
40	8,990	230	−5,650	410	2,080	590	−5,690
50	8,290	240	−6,950	420	1,800	600	−7,020
60	6,850	250	−7,780	430	1,680	610	−7,900
70	5,670	260	−7,200	440	2,870	620	−7,400
80	6,170	270	−5,050	450	5,050	630	−5,350
90	7,510	280	−2,870	460	7,200	640	−3,350
100	9,120	290	−1,680	470	7,780	650	−2,330
110	9,090	300	−1,800	480	6,950	660	−2,650
120	7,980	310	−2,080	490	5,650	670	−3,260
130	6,300	320	−1,450	500	5,040	680	−2,960
140	5,370	330	60	510	5,100	690	−1,750
150	5,260	340	1,330	520	4,780	700	−500
160	4,810	350	1,280	530	3,040	710	20
170	3,050	360	0	540	0	720	0
180	0						

20-29. Using the data of Table 20-2 design a flywheel for a two-cylinder engine having a 180° spacing between cranks. Use $C_s = \frac{1}{80}$, and design a steel disk-type flywheel. The speed is 4,600 rpm.

CHAPTER 21

CAM DYNAMICS

In this chapter we shall apply the material learned earlier to an investigation of the forces present in cam-and-follower systems. We shall find, too, that the motion machined into the cam surface is not always faithfully reproduced by the follower because of the speed and the elasticity, or flexibility, of the members of the follower train. Thus, vibration theory will be applied in order to determine the manner in which the follower responds to the motion prescribed by the cam. Space limitations permit only an introduction to the study of cam dynamics; the interested reader should consult the references listed throughout this chapter.[1]

21-1. Forces in Rigid Systems. The typical cam-and-follower system illustrated in Fig. 21-1 is analyzed using the methods discussed in previous chapters. The free-body diagram of the follower shown in (b) should be constructed first, and the unknown forces calculated by writing the two conditions for dynamic equilibrium. Frictional forces μN_B and μN_D are shown acting at the bearings of the translating follower. Other forces shown are the force P due to the load, the spring force F_S, and the inertia force $-m_4 \mathbf{A}_A$, which is the force resulting from the combined mass of all parts connected to the follower. In this example the effect of the mass of the roller m_3 has been neglected.

Note in Fig. 21-1b and c that the inertia force is upward on the follower. This occurs near the end of the upward follower motion. The inertia force must be balanced by the spring force and other external forces acting; otherwise the follower will jump out of contact with the cam.

A simple relationship exists between the torque and the velocity in the case of a radial cam with translating follower. In Fig. 21-2 the common instantaneous center P of the cam and follower is located on the line of centers joining O_2 and 14 at infinity. The pressure angle ϕ is the angle between lines AO_2 and AP. The follower velocity $\dot{y}$ is, therefore,

$$\dot{y} = \omega(a \tan \phi) \qquad (a)$$

---

[1] See Harold A. Rothbart, "Cams," John Wiley & Sons, Inc., New York, 1956. This is the principal treatise on cams in the United States. It contains an extensive bibliography.

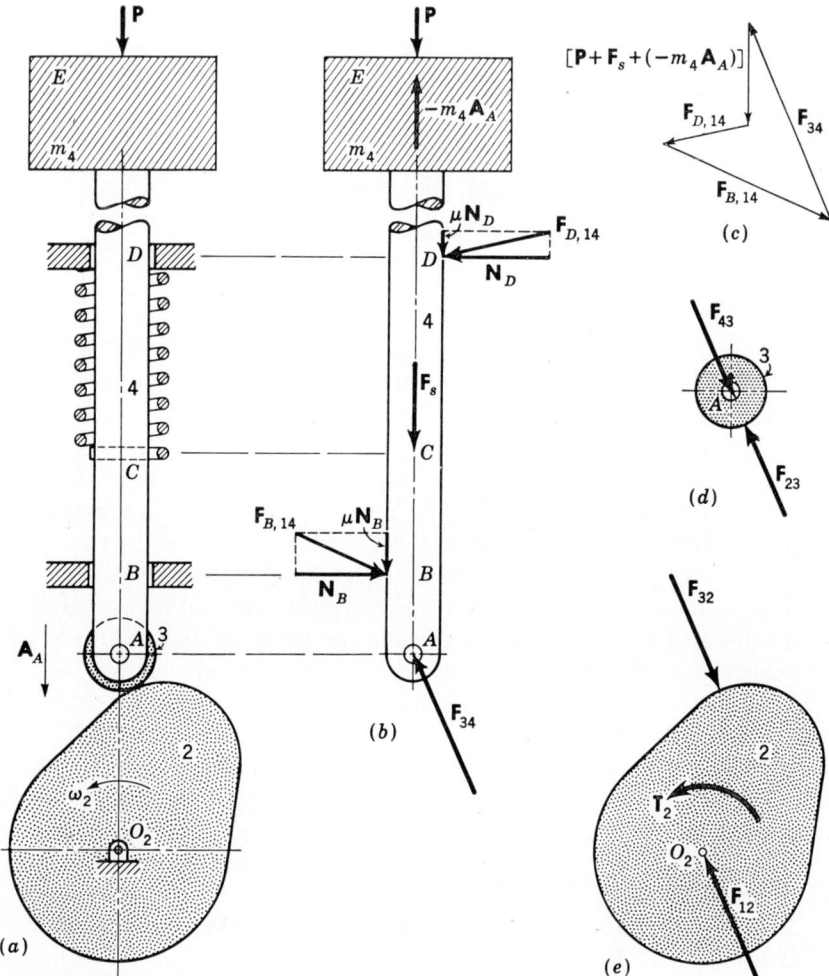

Fig. 21-1. Force analysis of a cam-and-follower system. The mass m_4 is that of the follower and load combined.

Next, designating the vertical component of the resultant follower forces acting on the cam as $F_{32}{}^y$, the torque is

$$T_2 = F_{32}{}^y(a \tan \phi) \qquad (b)$$

Then solving Eq. (a) for $a \tan \phi$ and substituting the results in (b) give for the torque

$$T_2 = \frac{\dot{y}}{\omega} F_{32}{}^y \qquad (21\text{-}1)$$

CAM DYNAMICS

EXAMPLE 21-1. A plate cam drives a radial translating follower with cycloidal motion at 1,200 rpm. The rise and return strokes are each 120° with two equal dwells. The follower is retained against the cam by a compression spring with a scale of 150 lb/in. The spring is compressed ¼ in. in assembly to provide an initial load. For a rise of 1 in. and a follower mass of 1.8 lb compute the radial component of the cam force during rise and the camshaft torque.

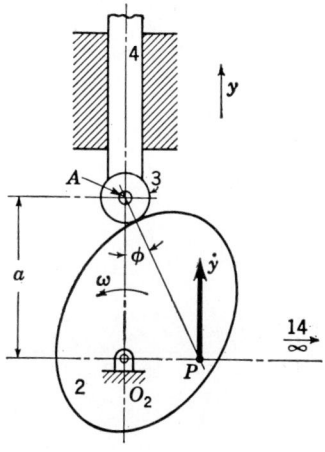

FIG. 21-2

Solution. The results of the analysis are tabulated in Table 21-1 and illustrated in Figs. 21-3 and 21-4. The following sample calculations are for the 100° position:

The displacement, velocity, and acceleration are calculated using Eqs. (11-20), (11-21), and (11-22). The values for use in these equations are

$$d = 1 \text{ in.} \qquad \theta = 100° \text{ or } 1.74 \text{ rad} \qquad \beta = 120° \text{ or } 2.10 \text{ rad}$$

$$\omega = \frac{2\pi n}{60} = \frac{(2\pi)(1,200)}{60} = 125.7 \text{ rad/sec}$$

Then

$$y = d\left(\frac{\theta}{\beta} - \frac{1}{2\pi}\sin\frac{2\pi\theta}{\beta}\right) = 1\left[\frac{1.74}{2.10} - \frac{1}{2\pi}\sin\frac{(2\pi)(1.74)}{2.10}\right]$$
$$= 0.663 \text{ in.}$$

$$\dot{y} = \frac{d\omega}{\beta}\left(1 - \cos\frac{2\pi\theta}{\beta}\right) = \frac{(1)(125.7)}{2.10}\left[1 - \cos\frac{(2\pi)(1.74)}{2.10}\right]$$
$$= 111.5 \text{ in./sec}$$

$$\ddot{y} = 2\pi d\left(\frac{\omega}{\beta}\right)^2 \sin\frac{2\pi\theta}{\beta} = (2\pi)(1)\left(\frac{125.7}{2.10}\right)^2 \sin\frac{(2\pi)(1.74)}{2.10}$$
$$= -11,200 \text{ in./sec}^2$$

The inertia force is

$$-m_4\ddot{y} = -\left(\frac{1.80}{386}\right)(-11,200) = 52.2 \text{ lb}$$

The initial spring compression is ¼ in. Therefore

$$F_S = k(y_0 + y) = -150(0.25 + 0.663) = -137.0 \text{ lb}$$

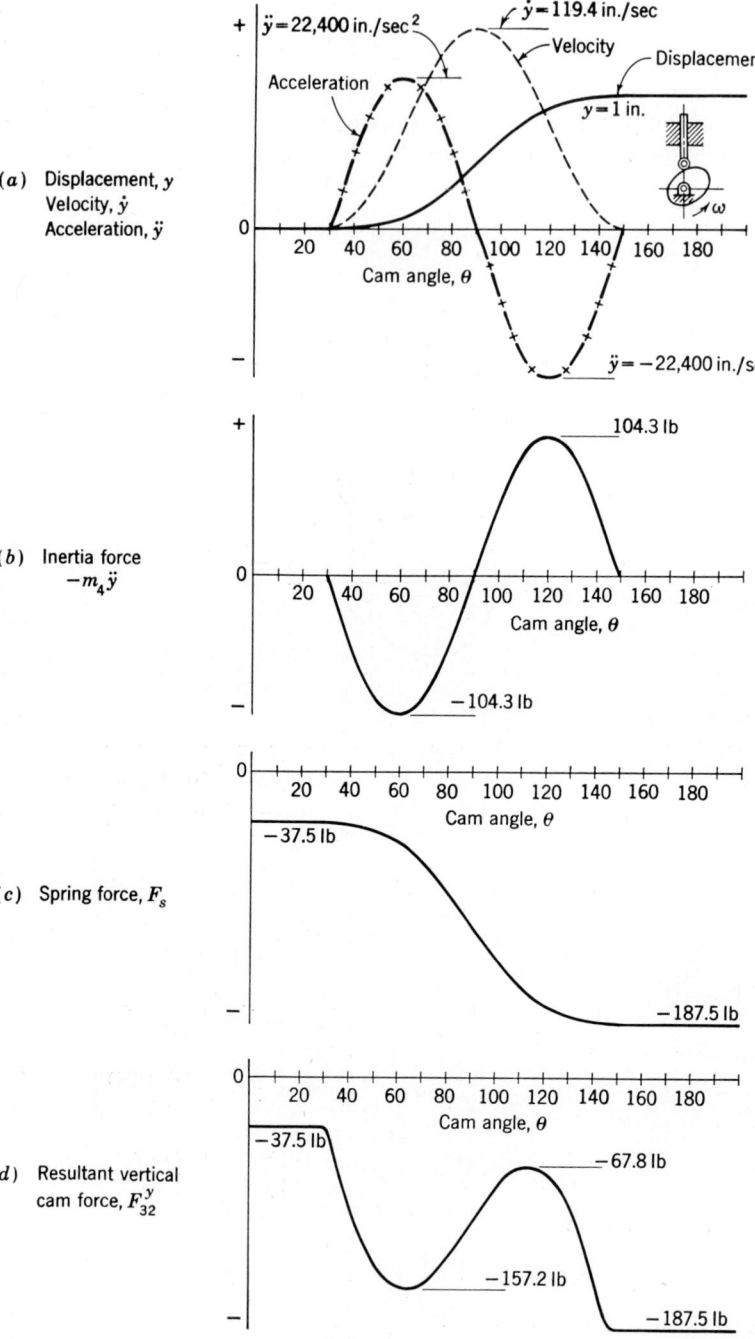

(a) Displacement, y
Velocity, $\dot{y}$
Acceleration, $\ddot{y}$

(b) Inertia force $-m_4\ddot{y}$

(c) Spring force, F_s

(d) Resultant vertical cam force, F_{32}^y

Fig. 21-3

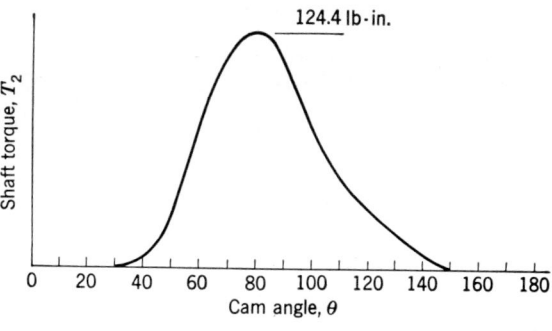

Fig. 21-4

The negative sign is used because the spring acts downward. The total radial (vertical) force acting on the cam is the sum of the spring and inertia forces because no external load was specified.

$$F_{32}{}^y = F_S - m_4\ddot{y} = -137.0 + 52.2 = -84.8 \text{ lb}$$

The torque is calculated from Eq. (21-1) and is

$$T_2 = \frac{\dot{y}}{\omega} F_{32}{}^y = \frac{111.5}{125.7} (84.8) = 75.2 \text{ lb-in.}$$

TABLE 21-1

Cam angle θ, deg	Displacement y, in.	Velocity $\dot{y}$, in./sec	Acceleration $\ddot{y}$, in./sec²	Inertia force $-m_4\ddot{y}$, lb	Spring force F_S, lb	Resultant force $F_{32}{}^y$, lb	Shaft torque T_2, lb-in.
0	0	0	0	0	−37.5	−37.5	0
10	0	0	0	0	−37.5	−37.5	0
20	0	0	0	0	−37.5	−37.5	0
30	0	0	0	0	−37.5	−37.5	0
40	0.003	8.0	11,200	−52.2	−38.0	−90.2	5.8
50	0.029	29.8	19,400	−90.5	−41.9	−132.4	31.4
60	0.091	59.7	22,400	−104.3	−51.1	−155.4	74.0
70	0.195	89.6	19,400	−90.5	−66.7	−157.2	112.2
80	0.336	111.5	11,200	−52.2	−87.9	−140.1	124.4
90	0.500	119.4	0	0	−112.5	−112.5	107.0
100	0.663	111.5	−11,200	52.2	−137.0	−84.8	75.2
110	0.805	89.6	−19,400	90.5	−158.3	−67.8	48.4
120	0.909	59.7	−22,400	104.3	−173.8	−69.5	33.1
130	0.971	29.8	−19,400	90.5	−183.3	−92.8	22.0
140	0.996	8.0	−11,200	52.2	−187.0	−134.8	8.6
150	1.000	0	0	0	−187.5	−187.5	0
160	1.000	0	0	0	−187.5	−187.5	0

21-2. Mathematical Models. In order that one can analyze the vibrations of an elastic system, such as a cam and follower, it is first necessary to reduce the mechanical system to a mathematical model. It follows that a simple model can be analyzed simply and quickly, but the simplification may have been carried to such an extent that there is no agreement between calculated results and actual measurements of the quantities at all. On the other hand it may be possible to create a very good model of the mechanical system, but the result may be so complex as to require expensive time on digital computers or a setup on an analog computer to obtain a solution. If this type of solution is not justified by the budget for engineering, then the designer must employ a simpler model and alter his results by judgment and experience factors to obtain a more rapid solution.

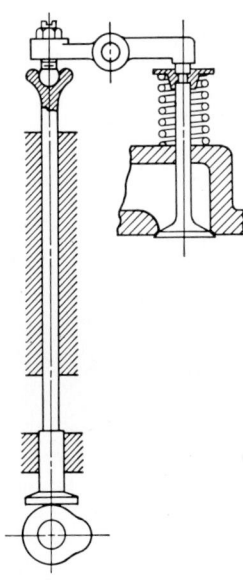

Fig. 21-5. Cross section showing the overhead valves in an automotive engine.

A follower train may consist of one or more rods, levers, gears, and springs, all of which are elastic. These members all have mass which is distributed or arranged in accordance with their physical dimensions. Various kinds of friction or damping will exist because of relative sliding, air resistance, and internal or molecular effects. Figure 21-5 is a schematic drawing of the overhead valves in an automotive engine and is an example of such a follower train; thus it is quite typical of what one may expect to find in analyzing any cam mechanism.

A mathematical model of a cam mechanism is shown in Fig. 21-6. This is a vibrating system having three degrees of freedom. It is far too complicated to analyze in a first study of cam dynamics, and so it is presented as a goal, that is, a problem we should like to solve.

The figure suggests that m_3 is the mass of the cam and part of the camshaft and k_4 represents the springiness or stiffness (which is the same thing) of the shaft. The coordinate y is the instantaneous resultant of the motion machined into the cam and the vibrational motion of the cam and shaft.

The mass m_1 can be used to represent the load to be moved, and it is to be retained by the spring k_1. The coordinate x_1 is the motion of the load; this is the desired motion and is the one the designer hopes to obtain when he specifies the shape of the cam curve.

The follower train mass is m_2, and the springs, k_2 and k_3, represent the elasticity of members of the follower train.

Dashpots are inserted between masses to account for friction and damping.

The growing use of digital and analog computers to solve mechanical engineering problems suggests that many solutions to the problem of Fig. 21-6 should soon be available for use.

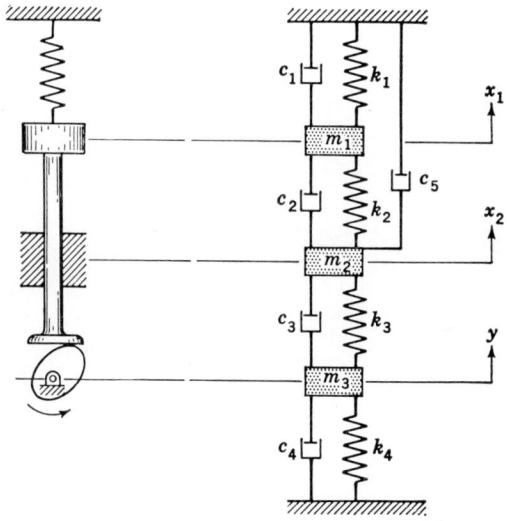

FIG. 21-6

The system described above can be reduced to two degrees of freedom by assuming that the camshaft is absolutely rigid. This produces the vibrating system of Fig. 21-7a. With this configuration the motion y is the motion machined into the cam. Johnson, of Yale University, has solved this problem for zero damping using the method of finite differences.[1] It is still too complicated for consideration here, though, and so we proceed to reduce to one of the single-degree-of-freedom systems of Fig. 21-7b or c. Here the assumption is made that the load and the entire follower train can be reduced to a single mass. The damped system gives quite good results in comparison with experiment and can be used for a great many cam mechanisms. The undamped system, how-

---

[1] R. C. Johnson, The Dynamic Analysis and Design of Relatively Flexible Cam Mechanisms Having More than One Degree of Freedom, *Trans. ASME*, vol. 81, ser. B, no. 4, pp. 323–331, 1959. The method of analysis should not be difficult to program for digital computation. See also Sec. 21-7.

ever, is not realistic and should be used only for estimating purposes unless it is definitely known that the damping factor is small.

For most systems the damping will range from about 5 to nearly 25 per cent of critical damping. About the only satisfactory method of determining the damping factor is to disturb the physical system itself and observe the decay on an oscilloscope. The damping factor is then calculated according to the methods of Chap. 18. If these observations

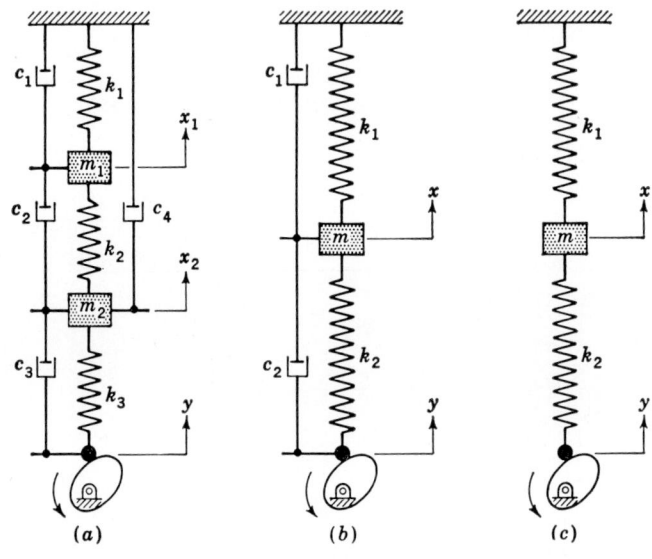

Fig. 21-7

cannot be made, then a good average value to use in analysis is about 10 per cent of critical damping.

Computation of equivalent masses and spring constants is accomplished using the methods explained in Sec. 18-13.

21-3. Response of a Uniform-motion Undamped Cam Mechanism—Analytical Method. Let us begin our investigation of follower response after the values of the lumped masses and spring constants have been obtained. At this point in the analysis we can obtain a good estimate of the motion with the assumption of no damping. These assumptions result in the single-degree-of-freedom vibrating system of Fig. 21-8a. Here the angular velocity of the cam is ω and the motion machined into the cam curve is y. The lumped follower mass is m, and the spring k_1 serves to hold the follower against the cam curve. This spring is always assembled with initial force or preload acting against the follower mass. The spring k_2 represents the elasticity of the follower train and so will be much stiffer than k_1.

CAM DYNAMICS

In order to write the conditions of dynamic equilibrium, we assume at some instant in time that the displacement x is greater than y. This gives the free-body diagram of Fig. 21-8b with the two spring forces acting downward on the mass; the conditions are then written

$$m\ddot{x} = -k_1 x - k_2(x - y)$$

or

$$\ddot{x} + \frac{k_1 + k_2}{m} x = \frac{k_2 y}{m} \qquad (21\text{-}2)$$

where $y = f(t)$ and is the motion imparted by the cam. This is the general differential equation for motion of the follower. We note particularly that $y = f(t)$ is not usually a continuous function for 360° of

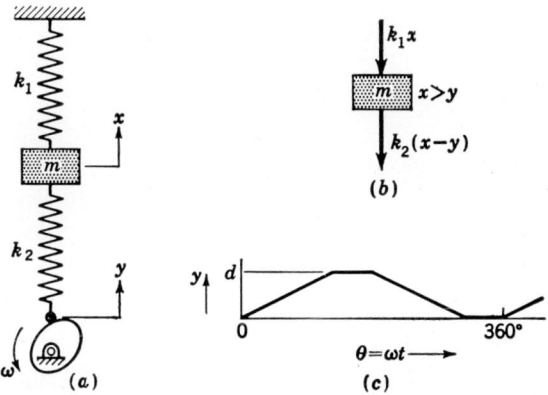

FIG. 21-8. (a) Undamped model of a cam mechanism. (b) Free-body diagram of the mass. (c) Displacement diagram.

cam rotation. Thus a dwell-rise-dwell cam with cycloidal motion will have a discontinuity at the beginning and end of each event. This makes it necessary to solve the equation separately for each region in which y is continuous. In doing so we must be careful to use the ending conditions of one era as the starting conditions for the next succeeding era, etc.

In Chap. 18 it was found that the solution of a differential equation of the form of Eq. (21-2) consisted of two parts, a complementary function obtained by making the right side zero and a particular solution of the equation. We can write the equation in the form

$$\ddot{x} + \omega_n^2 x = \frac{k_2 y}{m} \qquad (21\text{-}3)$$

where

$$\omega_n = \sqrt{\frac{k_1 + k_2}{m}} \qquad (21\text{-}4)$$

578   DYNAMIC ANALYSIS OF MACHINES

and is the natural circular frequency of the vibration. We have already solved equations for the complementary function, and so this solution can be written directly. It is

$$x' = A \cos \omega_n t + B \sin \omega_n t \tag{21-5}$$

where A and B are the constants of integration and depend upon the starting conditions. The particular solution depends upon the function $y = f(t)$ and will have to be found separately for each cam motion.

A rise with uniform motion is, by Eq. (11-2),

$$y = \frac{d}{\beta} \theta = \frac{d\omega t}{\beta} \tag{21-6}$$

where d is the rise in inches and β is the angle in radians required for the rise to take place. With this motion a particular solution of Eq. (21-3) is

$$x'' = \frac{k_2 y}{m\omega_n^2} \tag{21-7}$$

because $\ddot{y} = 0$. The validity of Eq. (21-7) can be readily demonstrated by substituting it and its second time derivative into Eq. (21-3). The complete solution for uniform motion is now obtained by adding Eqs. (21-5) and (21-7). Thus

$$x = A \cos \omega_n t + B \sin \omega_n t + \frac{k_2 y}{m\omega_n^2} \tag{21-8}$$

In the discussions of Chap. 18 on steady-state vibration the transient terms, represented by the first two terms of Eq. (21-8), were ignored because the damping would cause them to die out quickly. But in this example the force-time relationship itself must be treated as a transient disturbance; so even if the damping were not zero, it is necessary to investigate the effect of all the terms.

To study the meaning of Eq. (21-8) let us initiate cam rotation at the beginning of the rise when the follower displacement and velocity are both zero. Substituting the conditions $t = 0$, $x = 0$, and $\dot{x} = 0$ into Eq. (21-8) gives for the constants

$$A = 0 \qquad B = -\frac{k_2 \dot{y}}{m\omega_n^3}$$

Equation (21-8) then is

$$x = \frac{k_2}{m\omega_n^2}\left(y - \frac{\dot{y}}{\omega_n} \sin \omega_n t\right) \tag{21-9}$$

The velocity imparted by the cam during rise is, from Eq. (21-6),

$$\dot{y} = \frac{d\omega}{\beta} = \text{const} \tag{a}$$

CAM DYNAMICS

In most cam mechanisms the natural frequency ω_n of vibration will be large compared with the angular velocity ω of the cam. Therefore we define the ratio of these two quantities by the constant

$$K = \frac{\omega_n}{\omega} \qquad (21\text{-}10)$$

Then
$$\omega_n t = K\omega t = K\theta \qquad (b)$$

Substituting Eqs. (21-6), (a), and (b) into Eq. (21-9) gives the equation of motion for the first rise:

$$x = \frac{k_2 d}{Km\omega_n^2 \beta}(K\theta - \sin K\theta) \qquad (21\text{-}11)$$

This equation is valid only for $0 < \theta < \beta$.

In analyzing follower motion the conditions at the end of one event are used as the initial conditions for the next event. At the end of the rise described by Eq. (21-11), $\theta = \beta$, and the displacement is

$$x_1 = \frac{k_2 d}{Km\omega_n^2 \beta}(K\beta - \sin K\beta) \qquad (21\text{-}12)$$

The velocity is obtained by differentiating Eq. (21-9):

$$\dot{x} = \frac{k_2}{m\omega_n^2}(\dot{y} - \dot{y}\cos \omega_n t)$$

Then, substituting $K\beta$ for $\omega_n t$ and $\dot{y}$ from Eq. (a) produces the velocity at the end of the rise:

$$\dot{x}_1 = \frac{k_2 d}{Km\omega_n \beta}(1 - \cos K\beta) \qquad (21\text{-}13)$$

The next step is to evaluate the constants in Eq. (21-8) for the new initial conditions given by Eqs. (21-12) and (21-13). We shall choose a dwell at the end of the first rise. Thus y is constant and equal to d. When Eq. (21-8) is solved for its two constants, it is found that

$$A = x_1 - \frac{k_2 d}{m\omega_n^2} \qquad B = \frac{\dot{x}_1}{\omega_n}$$

Substituting these constants into Eq. (21-8) gives the equation of motion for the dwell period:

$$x = \left(x_1 - \frac{k_2 d}{m\omega_n^2}\right)\cos \omega_n t + \frac{\dot{x}_1}{\omega_n}\sin \omega_n t + \frac{k_2 d}{m\omega_n^2} \qquad (21\text{-}14)$$

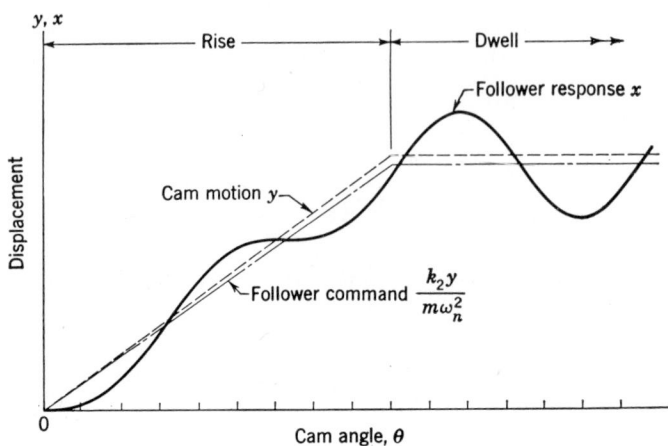

Fig. 21-9. Displacement diagram of a uniform-motion cam mechanism showing the follower response.

Transforming to a single trigonometric term and a phase angle gives

$$x = \sqrt{\left(x_1 - \frac{k_2 d}{m\omega_n{}^2}\right)^2 + \left(\frac{\dot{x}_1}{\omega_n}\right)^2} \cos(\omega_n t - \phi) + \frac{k_2 d}{m\omega_n{}^2} \quad (21\text{-}15)$$

$$\tan \phi = \frac{\dot{x}_1}{\omega_n \left(x_1 - \dfrac{k_2 d}{m\omega_n{}^2}\right)} \quad (21\text{-}16)$$

Finally, replacing $\omega_n t$ by $K\theta$ gives the equation of motion during the dwell:

$$x = \sqrt{\left(x_1 - \frac{k_2 d}{m\omega_n{}^2}\right)^2 + \left(\frac{\dot{x}_1}{\omega_n}\right)^2} \cos(K\theta - \phi) + \frac{k_2 d}{m\omega_n{}^2} \quad (21\text{-}17)$$

In using Eq. (21-17), θ is to begin at zero again; its maximum value is the angle through which the cam turns during the dwell period.

Equations (21-5), (21-11), and (21-17) are plotted in Fig. 21-9 so that the cam motion y can be compared with the follower response x. As shown, the follower has a steady-state free vibration of circular frequency ω_n during the dwell period. Note also that the follower command line is below the cam-motion line. This occurs because of the compression of the springs and exists regardless of any vibration. The effect of an external load or a preloaded retaining spring is to depress the follower command line still more if the original equilibrium position for the coordinate x is retained.

It is seldom necessary to analyze follower response beyond the first dwell because no new information is obtained. Equation (21-17) can be used to obtain the amplitude, and it can be differentiated twice to obtain the acceleration. The dynamic forces which exist are then readily calculated.

The method of analysis developed in this section can be applied to any cam motion for which an algebraic expression is available, but we shall not do so here.

The assumption is implicit, in the preceding discussion, that the follower always remains in contact with the cam. But we shall shortly discover that this is not always the case. Especially with high-speed cam mechanisms which include rather flexible follower systems, a phenomenon known as *jump* may occur in which the follower actually moves out of contact for short intervals of time.

21-4. Position Error. The vertical distance between the cam motion and the follower command in Fig. 21-9 is called the *position error*. Such an error arises because of the use of two springs in the mathematical model of the cam mechanism.

The spring system of Fig. 21-10a consists of a massless follower at A driven by a cam at B. The coordinates x and y are chosen such that the spring forces are zero when both x and y are zero. Now we permit the cam to rotate and lift the end B through the distance y (Fig. 21-10b). As shown in (c) the position error is $y - x$ and its value is

$$\varepsilon = y\left(1 - \frac{k_2}{k_1 + k_2}\right) \qquad (21\text{-}18)$$

Equation (21-18) is obtained by expressing algebraically that the spring forces are the same in each spring and that the sum of the spring deflections must equal y. Note that the error is proportional to y and is zero if y is zero.

21-5. Follower Response by Phase-plane Method. The analytical method of determining follower response gets rather complicated for most cam motions. If damping is introduced too, and it should be, then

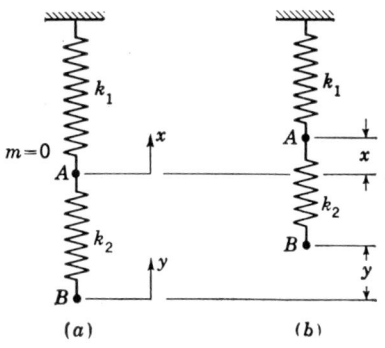
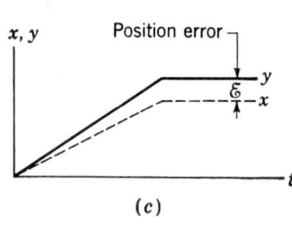

Fig. 21-10

the effort required to evaluate the initial conditions after each event is greatly multiplied. The solution is easily obtained, however, using the Laplace transformation of the differential equation, but we shall not introduce this method here. Instead, the phase-plane method shall be employed, because it provides a solution which is easy to apply and to understand. It also has the advantage that the cam motion need not be expressed in equation form.

The determination of the follower response of a cam mechanism by the phase-plane method is accomplished in the following steps:

Step 1. *Construct the Displacement Diagram.* This diagram should be the follower command rather than the cam motion; thus it accounts for position error. Expressed algebraically, the diagram is plotted according to the relation

$$y' = y(\theta) - \mathcal{E}(\theta) \qquad (21\text{-}19)$$

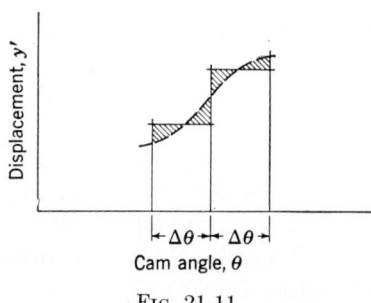

Fig. 21-11

In this equation y' is the follower command. The term $y(\theta)$ is the motion machined into the cam as a function of the cam angle θ. With the phase-plane solution $y(\theta)$ need not be expressible in algebraic form. The quantity $\mathcal{E}(\theta)$ is the position error for each value θ, determined according to the methods already investigated.

Step 2. *Calculate the Natural Frequency.* For a two-spring system without damping, the natural frequency is obtained from the equation

$$\omega_n = \sqrt{\frac{k_1 + k_2}{m}} \qquad (21\text{-}20)$$

as previously noted. For damping less than 20 per cent of critical, the same natural frequency can be used with only a very small error. Since this value falls within the range of most cam mechanisms, Eq. (21-20) can usually be employed for both damped and undamped systems.

Step 3. *Determine the Size of Steps to Be Used.* The quantity $K = \omega_n/\omega$ is the number of degrees of rotation of the vibration vector for each degree of cam rotation. Thus if a step $\Delta\theta°$ in width on the displacement diagram is chosen, then the vibration vector rotates through the angle $K(\Delta\theta)°$ from the beginning to the end of this step. In general the angle $K(\Delta\theta)$ should not exceed 90°, and more accurate results will be obtained if it is somewhat smaller.

Step 4. *Construct the Steps on the Displacement Diagram.* An appropriate method of doing this is shown in Fig. 21-11. The height of step is chosen so that the pairs of shaded triangles associated with each step

CAM DYNAMICS 583

have approximately the same area. It is also noted that these steps are necessary only during a rise or return motion.

Step 5. Construct the Phase-plane Diagram. Upon completion of this step the response curve is projected from the phase-plane diagram. It is often desirable to divide the angles $K(\Delta\theta)$ into several equal parts so as to obtain more points on the response curve.

The following example illustrates the application of this procedure to undamped motion. We shall defer an application of the method to damped motion long enough to consider jump phenomena.

EXAMPLE 21-2. A dwell-rise-dwell cam mechanism is to have a rise in 120° of cam rotation with parabolic motion. The cam is to be machined for a rise of 1.80 in. and is to have a speed of 2,580 rpm. The retaining spring is to be designed for a scale of 333 lb/in. and compressed 0.50 in. in assembly. A value of 4,200 lb/in. is estimated as the spring scale of the follower train. The equivalent weight of the follower mechanism to be used is 1.5 lb. Determine the response if the damping is assumed to be negligible.

Solution. The cam motion is given by Eqs. (11-6) and (11-14):

$$y = 2d\left(\frac{\theta}{\beta}\right)^2 \qquad 0 < \theta < \frac{\beta}{2} \qquad (1)$$

$$y = d\left[1 - 2\left(1 - \frac{\theta}{\beta}\right)^2\right] \qquad \frac{\beta}{2} < \theta < \beta \qquad (2)$$

The follower command function is obtained by combining Eqs. (21-18) and (21-19). Thus

$$y' = \frac{k_2 y}{k_1 + k_2} \qquad (3)$$

Equations (1), (2), and (3) can now be solved. In this case we choose to solve them for 7.5° intervals. The results are shown in Table 21-2 and plotted in Fig. 21-12.

TABLE 21-2

θ, deg	y, in.	y', in.	θ, deg	y, in.	y', in.
0	0	0	75.0	1.296	1.201
7.5	0.014	0.013	82.5	1.451	1.348
15.0	0.056	0.052	90.0	1.575	1.462
22.5	0.126	0.117	97.5	1.674	1.553
30.0	0.225	0.209	105.0	1.744	1.620
37.5	0.349	0.324	112.5	1.786	1.657
45.0	0.504	0.467	120.0	1.800	1.670
52.5	0.688	0.638			
60.0	0.900	0.835			
67.5	1.112	1.032			

The natural frequency is

$$\omega_n = \sqrt{\frac{k_1 + k_2}{m}} = \sqrt{\frac{(333 + 4{,}200)(386)}{1.5}} = 1{,}080 \text{ rad/sec}$$

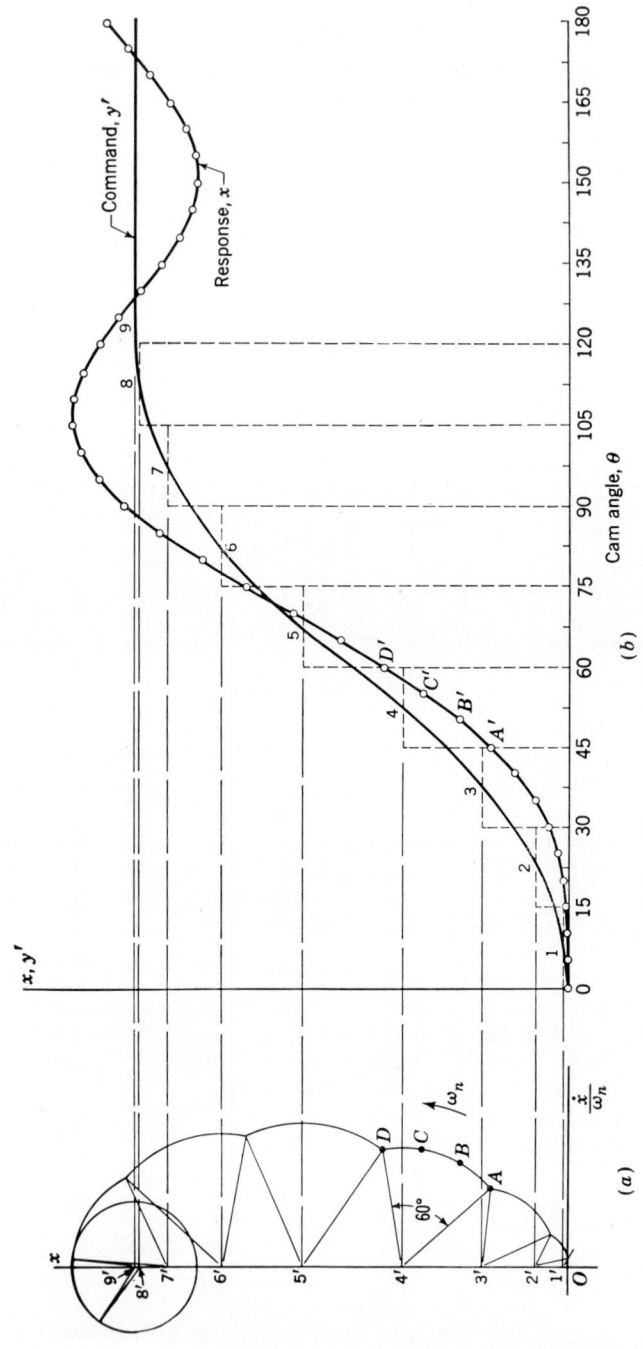

Fig. 21-12. Response of an undamped system with parabolic cam motion. (a) Phase-plane diagram. (b) Displacement diagram.

The angular velocity of the cam is

$$\omega = \frac{2\pi n}{60} = \frac{2\pi(2{,}580)}{60} = 270 \text{ rad/sec}$$

This gives the frequency ratio as

$$K = \frac{\omega_n}{\omega} = \frac{1{,}080}{270} = 4.0$$

We now select a convenient value for the angle $K(\Delta\theta)$, say 60°. With this choice the step width is

$$\Delta\theta = \frac{60°}{4.0} = 15.0°$$

This is all the information needed to determine the response.

The phase-plane diagram and response curve (Fig. 21-12) are constructed according to the procedure developed in Chap. 18. Referring to step 4 in the figure, an arc is constructed with center at 4' on the phase-plane diagram. The arc begins at A, corresponding to the end of step 3 and therefore the beginning of step 4, and is extended through an angle $K(\Delta\theta)$ of 60° to point D. Point D thus corresponds to the end of step 4 and the beginning of step 5. The arc AD can be divided into any number of equal parts as shown. The response curve is obtained by projecting points on the phase-plane arcs to their corresponding positions on the displacement diagram.

21-6. Jump and Crossover Shock. A cam follower retained against the cam with a compression retaining spring will, under certain conditions, *jump* or bounce out of contact with the cam. This condition is most likely to occur with low values of damping and with high-speed cams or quite flexible follower trains.

Crossover shock occurs in a positive-drive cam mechanism when contact moves from one side of the cam to the other. Clearance and backlash are taken up during the crossover, and impact occurs. Crossover takes place on the rise or return motion when the acceleration changes sign and when the velocity is at its peak. The effects can be reduced by preloading the system to remove backlash, by designing for a low peak velocity, and by using a rigid follower train.

Rothbart[1] states that jump will not occur in high-speed systems if at least two full cycles of vibration occur during the positive-acceleration time interval of the motion. If a smaller number of cycles exist during this period, then, he states, the system should be investigated mathematically to determine if jump exists. This condition can be expressed by the equation

$$\frac{\beta_1 K}{360} \geq 2 \qquad (21\text{-}21)$$

where β_1 is the angle through which the cam rotates during the positive-

[1] *Op. cit.*, p. 251.

acceleration period. This figure can probably be reduced slightly for appreciable amounts of damping.

As shown in Fig. 21-13, spring k_2 loses compression when jump begins and is carried in motion with the mass. The resulting motion now gets rather complicated because the mass, too, must be redistributed. Probably a good first approximation could be obtained by concentrating a portion of the mass at the bottom of spring k_2 and treating the motion as a system of two degrees of freedom. This, however, is beyond the

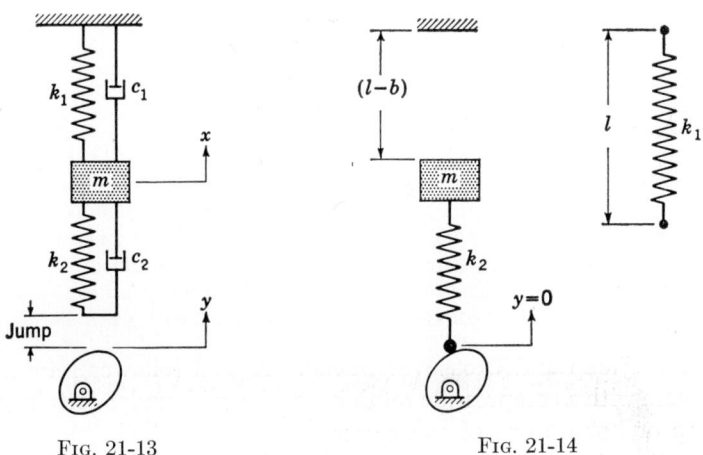

Fig. 21-13   Fig. 21-14

purposes of this book, and so we shall be interested only in the conditions which exist when jump begins. It must be noted, though, that the system will vibrate at a *new* frequency after jump begins and that an analysis of the motion using the old frequency is *not* a true description of the motion.

Spring k_2 loses its compression whenever x exceeds y by the amount that k_2 was initially compressed during assembly. Thus, to set up a criteria for jump, it is necessary to calculate the precompression of k_2. The problem is illustrated in Fig. 21-14, where we wish to assemble spring k_1 of length l into a space of length $l - b$. After assembly, the combined deflection of both springs is the distance b. Since the springs are in series, the force, from Eq. (18-58), is

$$F = k_{eq}b = \frac{k_1 k_2 b}{k_1 + k_2} \qquad (a)$$

Since the same force exists in each spring, the deflection of k_2 is

$$b_2 = \frac{F}{k_2} = \frac{k_1 b}{k_1 + k_2} \qquad (b)$$

Thus, the criterion for jump is that

$$x > y + \frac{k_1 b}{k_1 + k_2} \qquad (21\text{-}22)$$

EXAMPLE 21-3. A dwell-rise-dwell cam has a 1-in. rise with harmonic motion in 136° of cam rotation. The retaining spring has a scale of 250 lb/in. and is compressed 1.5 in. when assembled. The follower train is estimated to have a spring scale of 6,000 lb/in. and an equivalent mass of 2.20 lb. Determine the response of the follower for a cam speed of 3,340 rpm with the damping estimated at 15 per cent of critical.[1]

Solution. For simple harmonic motion Eq. (11-17) applies:

$$y = \frac{d}{2}\left(1 - \cos\frac{\pi\theta}{\beta}\right) \qquad (1)$$

Using $d = 1$ in. and substituting values for θ from 0 to 136° produce the values shown for y in Table 21-3. Using Eq. (3) of the preceding example for the follower command function gives

$$y' = \frac{k_2 y}{k_1 + k_2} = \frac{6{,}000y}{250 + 6{,}000} = 0.96y \qquad (2)$$

This quantity is also included in Table 21-3 and is plotted as the displacement diagram in Fig. 21-15.

TABLE 21-3

θ, deg	y, in.	y', in.	$y + 0.60$, in.
0	0	0	0.060
8	0.008	0.008	0.068
16	0.033	0.031	0.093
24	0.075	0.072	0.135
32	0.130	0.125	0.190
40	0.198	0.190	0.258
48	0.291	0.279	0.351
56	0.363	0.348	0.423
64	0.454	0.435	0.514
72	0.546	0.525	0.606
80	0.637	0.612	0.697
88	0.709	0.680	0.769
96	0.802	0.770	0.862
104	0.870	0.835	0.930
112	0.925	0.888	0.985
120	0.966	0.927	1.026
128	0.991	0.951	1.051
136	1.000	0.960	1.060

[1] This speed is too high for such an elastic system. It was chosen to produce a large phase-plane diagram, to illustrate jump, and to produce a clearly visible vibration.

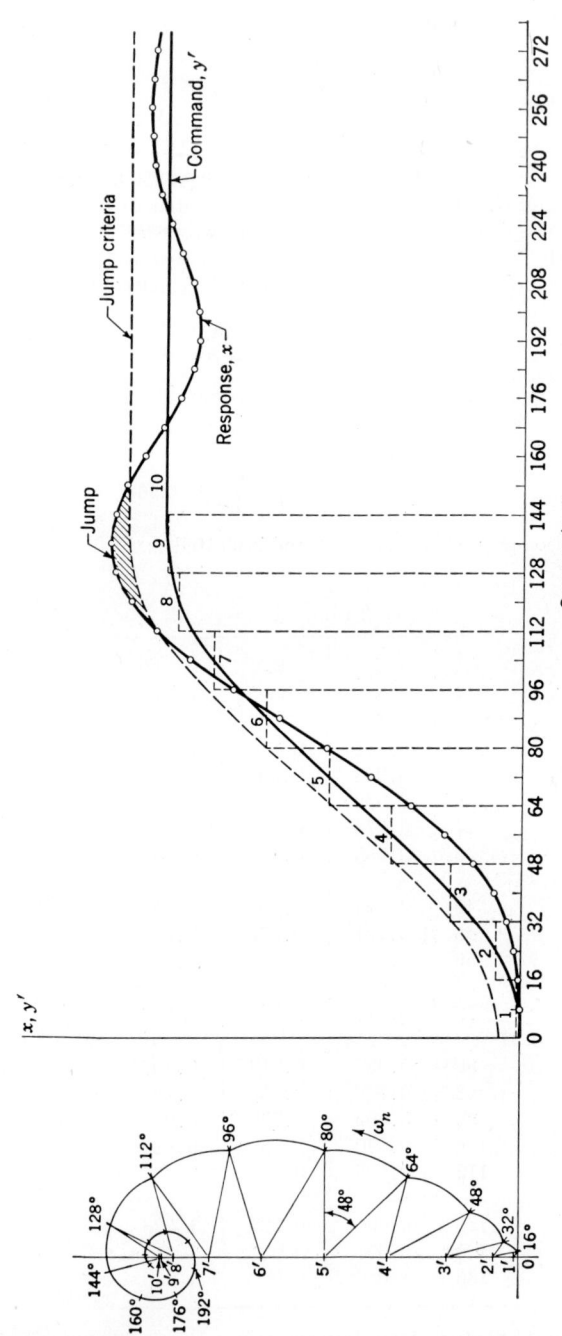

FIG. 21-15. A damped cam mechanism with jump; driving function is harmonic motion.

From Eq. (21-22), jump begins whenever x exceeds the quantity

$$y + \frac{k_1 b}{k_1 + k_2} = y + \frac{(250)(1.5)}{250 + 6{,}000} = y + 0.60$$

We plot this quantity in Fig. 21-15 too.

Assuming the same frequency for damped as for undamped motion,

$$\omega_n = \sqrt{\frac{k_1 + k_2}{m}} = \sqrt{\frac{(6{,}000 + 250)(386)}{2.2}} = 1{,}047 \text{ rad/sec}$$

The cam has an angular velocity of

$$\omega = \frac{2\pi n}{60} = \frac{2\pi(3{,}340)}{60} = 349 \text{ rad/sec}$$

and the frequency ratio is

$$K = \frac{\omega_n}{\omega} = \frac{1{,}047}{349} = 3.00$$

Choosing a step width of $\Delta\theta = 16°$ will cause the vibration vector to rotate through the angle

$$K(\Delta\theta) = (3.00)(16) = 48°$$

With this information and a spiral template constructed according to the methods of Sec. 18-7, the phase-plane diagram is drawn and points projected back to the displacement diagram to obtain the response curve.

We note, in Fig. 21-15, that jump begins when $\theta = 112°$. The response after this point is plotted as if the vibration vector had continued to rotate at the same velocity ω_n. It was pointed out earlier that this is not the case because the frequency of the vibration changes. Therefore, the response curve after 112° is not a true description of the motion. Nevertheless, it does demonstrate the effect of damping on the motion. The designer, of course, is responsible for creating a satisfactory cam mechanism. If he finds that jump may occur under certain circumstances, he should make some change in the design parameters.

21-7. Johnson's Numerical Analysis.[1] Here we shall explore a numerical procedure, due to R. C. Johnson of Yale University, for determining follower response. The method has the advantage that it can also be used for design purposes. The calculations should be easy to program for digital computation and are typical of the class of problems which are particularly suited to this procedure.

Three points on a portion of the displacement-time diagram of Fig. 21-16 are separated by the equal time intervals Δt. The rate of change of y', the follower command, with respect to time over the double interval is

$$\dot{y}'_0 = \frac{y'_2 - y'_1}{2 \Delta t} \tag{a}$$

---

[1] Ray C. Johnson, Flexible Cam Mechanisms, *Machine Design*, vol. 31, no. 18, pp. 140–145, September, 1959.

Since $\Delta\theta = \omega(\Delta t)$, where $\Delta\theta$ is in radians, Eq. (a) can also be written

$$\dot{y}'_0 = \frac{\omega}{2\,\Delta\theta}(y'_2 - y'_1) \tag{21-23}$$

where $\dot{y}'_0$ is the velocity in inches per second at the center of the double time interval.

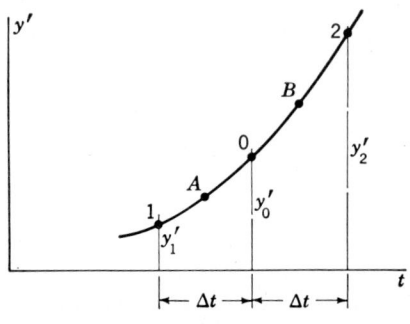

Fig. 21-16

Equation (a) can also be used to find the velocity at the center of each of the time intervals shown in Fig. 21-16. Thus, at points A and B we have

$$\dot{y}'_A = \frac{y'_0 - y'_1}{\Delta t} \qquad \dot{y}'_B = \frac{y'_2 - y'_0}{\Delta t} \tag{b}$$

so the acceleration at the center of the double time interval is

$$\ddot{y}'_0 = \frac{\dot{y}'_B - \dot{y}'_A}{\Delta t} = \frac{(y'_2 - y'_0)/\Delta t - (y'_0 - y'_1)/\Delta t}{\Delta t}$$

$$= \frac{1}{(\Delta t)^2}(y'_1 + y'_2 - 2y'_0) \tag{c}$$

Again substituting $\Delta\theta/\omega$ for Δt gives

$$\ddot{y}'_0 = \left(\frac{\omega}{\Delta\theta}\right)^2 (y'_1 + y'_2 - 2y'_0) \tag{21-24}$$

The relationships expressed by Eqs. (21-23) and (21-24) are also valid for the motion x of the follower mass. Thus, replacing the y' terms by x's gives

$$\dot{x}_0 = \frac{\omega}{\Delta\theta}\frac{x_2 - x_1}{2} \tag{21-25}$$

$$\ddot{x}_0 = \left(\frac{\omega}{\Delta\theta}\right)^2 (x_1 + x_2 - 2x_0) \tag{21-26}$$

for the velocity and acceleration of the follower.

CAM DYNAMICS 591

The set of equations just developed are termed *finite-difference equations*, and we shall solve them by selecting small values of $\Delta\theta$ and proceeding from the beginning of the rise.

Assuming that the displacement y' is greater than the displacement x, the equation for dynamic equilibrium is written

$$m\ddot{x}_0 = k_2(y'_0 - x_0) \qquad (d)$$

Substituting the acceleration from Eq. (21-26) we find that

$$m\left(\frac{\omega}{\Delta\theta}\right)^2 (x_1 + x_2 - 2x_0) = k_2(y'_0 - x_0) \qquad (e)$$

Equation (e) is now solved for x_2. This yields

$$x_2 = \frac{k_2(y_0 - x_0)}{m(\omega/\Delta\theta)^2} + 2x_0 - x_1 \qquad (21\text{-}27)$$

Equation (21-27) is solved sequentially beginning at any point, say the beginning of a rise, where the follower displacements x_0 and x_1 are known. The value of x_2 obtained is then used as x_0 for the next calculation, and so on, until the analysis is completed. Johnson recommends that the response be determined using a desk calculator with the results arranged in tabular form. The results should not be carried for more than four decimal places because of the limitation of cutting errors in producing the cams.

EXAMPLE 21-4. A dwell-rise-dwell cam has a 1.056-in. rise with cycloidal motion in 136° of cam rotation. The follower is assembled with a 200 lb/in. retaining spring with no preload. The follower train has an equivalent mass of 0.60 lb and an equivalent spring constant of 3,600 lb/in. Determine the response of the system for a cam speed of 3,500 rpm using Johnson's method.

Solution. The circular frequency of the vibrating system is

$$\omega_n = \sqrt{\frac{k_1 + k_2}{m}} = \sqrt{\frac{(200 + 3,600)(386)}{0.60}} = 1,560 \text{ rad/sec}$$

The cam velocity is

$$\omega = \frac{2\pi n}{60} = \frac{2\pi(3,500)}{60} = 367 \text{ rad/sec}$$

Thus

$$K = \frac{\omega_n}{\omega} = \frac{1,560}{367} = 4.25$$

Therefore, for one cycle of vibration, the cam rotates through the angle

$$\theta_n = \frac{360°}{4.25} = 84.8°$$

Johnson states that the interval $\Delta\theta$ should be selected according to the relation

$$\Delta\theta < \frac{\theta_n}{2\pi}$$

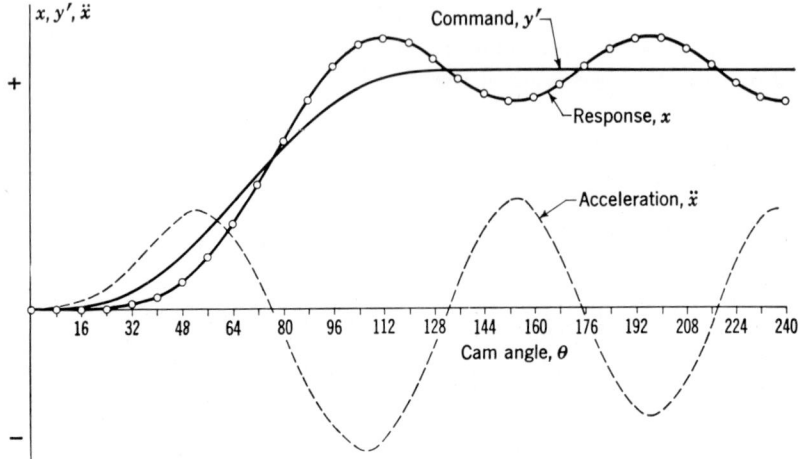

FIG. 21-17. Follower response determined by numerical procedures; command function is cycloidal.

In this example we shall choose a value $\Delta\theta = 8°$ which is well under the criterion. Substituting these values in Eq. (21-23) gives

$$x_2 = \frac{k_2(y_0' - x_0)}{m(\omega/\Delta\theta)^2} + 2x_0 - x_1$$

$$= \frac{3{,}600(y_0' - x_0)}{\dfrac{0.60}{386}\left(\dfrac{367}{8\pi/180}\right)^2} + 2x_0 - x_1$$

$$= 0.335(y_0' - x_0) + (2x_0 - x_1)$$

The follower command is

$$y' = \frac{k_2 y}{k_1 + k_2} = \frac{3{,}600 y}{200 + 3{,}600} = 0.947 y$$

With these equations Table 21-4 is prepared. As an example, the calculations for the 32° position are

$$x_2 = 0.335(0.0319 - 0.0042) + (2)(0.0042) - 0.0004$$
$$= 0.0173$$

Thus, subscript 1 is the 16° position, 0 the 24° position, and 2 the 32° position.

We note from Eq. (21-26) that the acceleration is proportional to the quantity $x_1 + x_2 - 2x_0$; so this quantity, too, is included in Table 21-4. The acceleration $\ddot{x}_0$ is then calculated by multiplying this quantity by the constant $(\omega/\Delta\theta)^2$. Note, however, that the acceleration factors correspond to the preceding positions. Thus a factor appearing in the 48° row is used to obtain the acceleration corresponding to the 40° position.

The results of these calculations are illustrated in Fig. 21-17.

Cam Design. Johnson shows, by working backward, how it is possible to design a cam to produce any follower motion. While the results are

TABLE 21-4

θ, deg	y', in.	x, in.	$y'_0 - x_0$, in.	$0.335(y'_0 - x_0)$	$2x_0 - x_1$	$x_1 + x_2 - 2x_0$
0	0	0	0	0	0	0
8	0.0012	0	0.0012	0.0004	0	0
16	0.0106	0.0004	0.0102	0.0034	0.0008	0.0004
24	0.0319	0.0042	0.0277	0.0093	0.0080	0.0034
32	0.0769	0.0173	0.0596	0.0200	0.0304	0.0093
40	0.1358	0.0504	0.0854	0.0286	0.0835	0.0200
48	0.2261	0.1121	0.1140	0.0382	0.1738	0.0286
56	0.3279	0.2120	0.1159	0.0388	0.3119	0.0382
64	0.4413	0.3507	0.0906	0.0304	0.4894	0.0388
72	0.5586	0.5198	0.0388	0.0130	0.6889	0.0304
80	0.6720	0.7019	−0.0299	−0.0100	0.8840	0.0130
88	0.7738	0.8740	−0.1012	−0.0340	1.0461	−0.0100
96	0.8597	1.0121	−0.1524	−0.0495	1.1502	−0.0340
104	0.9231	1.1007	−0.1776	−0.0595	1.1893	−0.0495
112	0.9680	1.1298	−0.1618	−0.0542	1.1589	−0.0595
120	0.9893	1.1047	−0.1154	−0.0387	1.0796	−0.0542
128	0.9983	1.0409	−0.0426	−0.0143	0.9771	−0.0387
136	1.0000	0.9628	0.0372	0.0125	0.8847	−0.0143
144	1.0000	0.8972	0.1028	0.0345	0.8316	0.0125
152	1.0000	0.8661	0.1339	0.0448	0.8350	0.0345
160	1.0000	0.8798	0.1202	0.0403	0.8935	0.0448
168	1.0000	0.9338	0.0662	0.0222	0.9878	0.0403
176	1.0000	1.0100	−0.0100	−0.0033	1.0862	0.0222
184	1.0000	1.0829	−0.0829	−0.0278	1.1558	−0.0033
192	1.0000	1.1280	−0.1280	−0.0429	1.1731	−0.0278
200	1.0000	1.1302	−0.1302	−0.0436	1.1324	−0.0429
208	1.0000	1.0888	−0.0888	−0.0298	1.0474	−0.0436
216	1.0000	1.0176	−0.0176	−0.0059	0.9464	−0.0298
224	1.0000	0.9405	0.0595	0.0200	0.8634	−0.0059
232	1.0000	0.8834	0.1166	0.0391	0.8263	0.0210
240	1.0000	0.8654				0.0391

valid only for a particular cam velocity, for flexible follower systems it is effective in eliminating vibration at this particular velocity. We begin by solving Eq. (e) for the follower command y'_0:

$$y'_0 = x_0 + \frac{m}{k_2}\left(\frac{\omega}{\Delta\theta}\right)^2 (x_1 + x_2 - 2x_0) \qquad (f)$$

Then, since

$$y = \frac{k_1 + k_2}{k_2} y' \qquad (g)$$

we have

$$y = \frac{k_1 + k_2}{k_2}\left[x_0 + \frac{m}{k_2}\left(\frac{\omega}{\Delta\theta}\right)^2 (x_1 + x_2 - 2x_0)\right] \qquad (21\text{-}28)$$

When the follower motion is specified, Eq. (21-28) can be solved directly, using small values of $\Delta\theta$, to yield the motion y to be machined into the cam.

21-8. Unbalance, Spring Surge, and Windup. As shown in Fig. 21-18a a disk cam produces unbalance because its mass is not symmetrical with the axis of rotation. This means that two sets of vibratory forces exist, one due to the eccentric cam mass and the other due to the reaction of the follower against the cam. By keeping these effects in mind

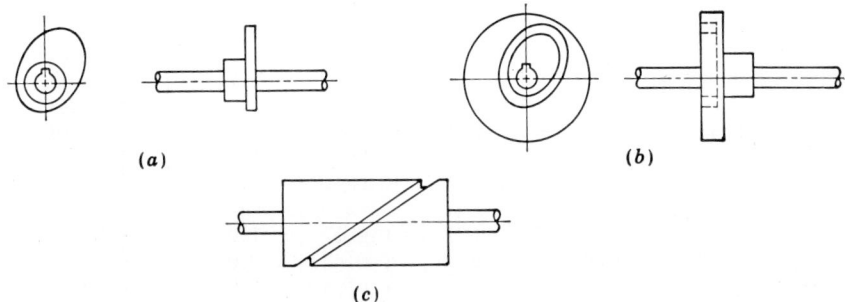

FIG. 21-18. (a) Disk cam is inherently unbalanced. (b) Face cam is usually well balanced. (c) Cylindrical cam has good balance.

during design, the engineer can do much to guard against difficulties during operation.

Figure 21-18b and c shows that the face and cylindrical cams have good balance characteristics. For this reason these are good choices when high-speed operation is involved.

Spring Surge. It is shown in texts on spring design that helical springs may themselves vibrate when subjected to rapidly varying forces. Poorly designed automotive valve springs, for example, when operating near the critical frequency range, permit the valve to open for short intervals during the period the valve is supposed to be closed. Such conditions result in very poor operation of the engine and rapid fatigue failure of the springs themselves.[1] This vibration of the retaining spring, called *spring surge*, has been photographed with high-speed motion-picture cameras, and the results exhibited in slow motion. When serious vibrations exist, a clear wave motion can be seen traveling up and down the valve spring.

[1] See Joseph E. Shigley, "Machine Design," p. 238, McGraw-Hill Book Company, Inc., New York, 1956.

CAM DYNAMICS 595

Windup. Camshaft windup, a very damaging effect, is discussed in some detail by Rothbart.[1] The camshaft is delivering a varying torque to the cam which causes the shaft to twist, or "wind up," as the torque increases during follower rise. Also, during this period, the angular cam velocity is slowed and so is the follower velocity. Near the end of rise the energy stored in the shaft is released, causing the follower velocity and acceleration to increase above the normal values. If the retaining spring is not strong enough to resist the forces, the follower may jump off the cam, or if the cam is a positive-drive mechanism, then impact will occur.

Rothbart states that shaft windup may cause trouble when:
1. Heavy loads or masses are being moved by the follower
2. The follower moves at a high speed and is driven at large pressure angles
3. The shaft is flexible

The suggested remedies are to use short stiff drive shafts and to include the torque effect in designing the cam profile.

PROBLEMS

21-1. A radial translating roller follower is driven by a plate cam at a speed of 600 rpm with harmonic motion during the rise stroke and parabolic motion during the return stroke. The follower events are rise in 150° of cam rotation, dwell in 30°, return in 180°. The follower is retained against the cam by a 80 lb/in. compression spring with an initial compression of ½ in. in assembly. The follower is to move an external load whose magnitude is related to the rise of ¾ in. as shown in the figure. The total follower mass is 3.6 lb. Neglecting friction, calculate and plot the radial component of the cam force during rise and the camshaft torque. Has the spring been compressed enough to hold the follower in contact with the cam?

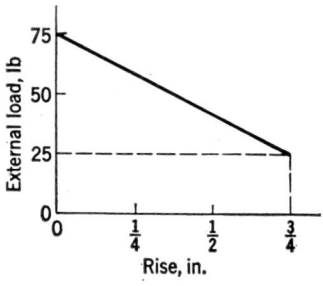

PROB. 21-1

21-2. Using the data of Prob. 21-1 calculate and plot the radial component of the cam force during the dwell and the return stroke. Also, plot the camshaft torque.

[1] Harold A. Rothbart, Cam Torque Curve, *Machine Design*, vol. 31, no. 15, pp. 127–129, July, 1959.

596   DYNAMIC ANALYSIS OF MACHINES

21-3. Using the data of Prob. 21-1 calculate the radial component of the cam force and the camshaft torque corresponding to a 120° and to a 270° position of the cam. *Ans.:* −114.5 lb, −30.3 lb-in. at 120°; −146.3 lb, 70.0 lb-in. at 270°.

21-4. A plate cam rotates at 870 rpm and drives a radial translating roller follower with a rise of $1\frac{3}{4}$ in. with parabolic motion in 120° of cam rotation, return with parabolic motion in 120°, and a dwell for the remaining cam angle. The total follower mass weighs 5.34 lb. Friction is to be neglected, and there is no external load. (*a*) What must be the initial compression of a 100 lb/in. retaining spring if the force of the roller on the cam is never to be less than 25 lb? (*b*) What is the largest value of the radial component of the follower force on the cam? (*c*) Calculate the peak value of the camshaft torque during the rise. *Ans.:* 0.695 in., 339 lb, 565 lb-in.

21-5. Using the data of Prob. 21-4 plot the radial cam force and the camshaft torque for all events as a function of the cam angle. Use an initial spring compression of $\frac{1}{2}$ in. and an external load which is related to rise according to the graph shown in the figure.

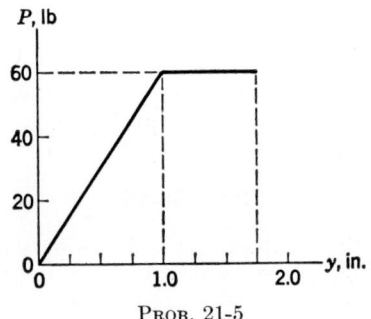

Prob. 21-5

21-6. The following data apply to a mathematical model of a simple undamped single-degree-of-freedom cam-and-follower system: $\beta = 150°$; $d = 1$ in.; $n = 1{,}200$ rpm; $k_1 = 80$ lb/in.; $k_2 = 1{,}600$ lb/in. The weight of the follower mass is 1.028 lb. The

TABLE 21-5. FOLLOWER DISPLACEMENTS FOR ELEMENTARY CAM MOTIONS
Based on $d = 1$ in. and $\beta = 1$

Cam angle θ/β	Parabolic y	Simple harmonic y	Cycloidal y	Cam angle θ/β	Parabolic y	Simple harmonic y	Cycloidal y
0	0	0	0	0.50	0.500	0.500	0.500
0.05	0.005	0.006	0.001	0.55	0.595	0.578	0.599
0.10	0.020	0.025	0.006	0.60	0.680	0.654	0.694
0.15	0.045	0.055	0.021	0.65	0.756	0.727	0.779
0.20	0.080	0.096	0.049	0.70	0.820	0.794	0.851
0.25	0.125	0.148	0.091	0.75	0.875	0.853	0.909
0.30	0.180	0.206	0.149	0.80	0.920	0.904	0.951
0.35	0.244	0.273	0.221	0.85	0.955	0.946	0.979
0.40	0.320	0.346	0.306	0.90	0.980	0.975	0.994
0.45	0.405	0.422	0.401	0.95	0.995	0.994	0.999
				1.00	1.000	1.000	1.000

80 lb/in. spring is assembled into a space $\frac{1}{4}$ in. shorter than its free length. Using the phase-plane method plot the response of a dwell-rise-dwell cam when it is cut for (a) parabolic motion, (b) simple harmonic motion, and (c) cycloidal motion. It is suggested that Table 21-5 be used to plot the displacement-time diagrams and that 15° intervals be employed in plotting the phase-plane diagram. Neglect jump. What is the amplitude of the vibration during the dwell period for each motion?

21-7. A dwell-rise-dwell cam rotates at 1,800 rpm and drives a follower through a rise of $1\frac{1}{2}$ in. (nominal) with uniform motion in 90° of cam rotation. The follower retaining spring has a scale of 240 lb/in. and is compressed $\frac{1}{2}$ in. in assembly. Preliminary analysis of the follower system indicates a weight of the follower mass of 0.886 lb and a spring constant of 1,800 lb/in. Neglect damping and jump, and analyze the response of the follower system using the phase-plane method.

21-8. A cam drives a follower through a rise of 2 in. with harmonic motion in 120° of cam rotation, dwells for 60°, and then returns with harmonic motion in 120°. A retaining spring is used having a scale of 200 lb/in. and compressed $\frac{1}{4}$ in. in assembly. The follower mass weighs 39.4 lb and has a spring constant of 3,600 lb/in. Using a cam speed of 900 rpm and assuming no damping, determine the response through the dwell using the phase-plane method.

21-9. Solve Prob. 21-6 using a damping of 15 per cent of critical.

21-10. A follower system is driven through a nominal rise of $1\frac{1}{4}$ in. by a dwell-rise-dwell cam rotating at 1,900 rpm. The rise occurs in 150° of cam rotation with cycloidal motion. A 350 lb/in. retaining spring is used with a preload of 100 lb. The follower system has a spring constant of 4,800 lb/in. and weighs 1.39 lb. Employing a damping ratio of 0.15, analyze the response of the follower system using the phase-plane method. Construct a line on the response diagram to define the beginning of jump.

21-11. The equation for the rise of the 2-3 polynomial cam is

$$y = 3d \left(\frac{\theta}{\beta}\right)^2 - 2d \left(\frac{\theta}{\beta}\right)^3$$

Substitute this motion for the cam mechanism of Prob. 21-10, and analyze the response.

21-12. A dwell-rise-dwell plate cam has a speed of 400 rpm and drives a translating roller follower through a rise of $2\frac{1}{2}$ in. with parabolic motion in 60° of cam rotation. The follower is retained in contact with the cam by a retaining spring having a scale of 120 lb/in. and assembled with a preload of 80 lb. The follower system is estimated to have a spring constant of 3,200 lb/in. and a concentrated mass weighing 13 lb. Determine the response of the system using Johnson's method. Does jump occur?

21-13. A follower system is driven through a rise of 2 in. with simple harmonic motion in 150° of cam rotation. The cam rotates at a speed of 1,350 rpm and has a dwell on each side of the rise. The follower is loaded against the cam by a 300 lb/in. compression spring initially compressed to a distance of $\frac{1}{2}$ in. Taking the follower mass as weighing 25.8 lb and the stiffness as 4,000 lb/in., calculate the response of the system using Johnson's numerical procedure. Does jump occur?

21-14. Illustrated in the figure is a positive-drive flat-face follower system driven by a constant-breadth cam. The follower system consists of the follower rod 3, the lever 4, the connecting rod 5, and a carriage 6 whose weight is 15 lb. In finding the natural frequency of such a system one might, at first, neglect the weight of the links and treat them as series springs. All the mass, then, would be due to the weight of the carriage. Using these assumptions, calculate the equivalent spring constant and find the natural frequency of vibration of the system using the methods of Chap. 18. The links are made of steel having a modulus of elasticity of 28,500,000 psi.

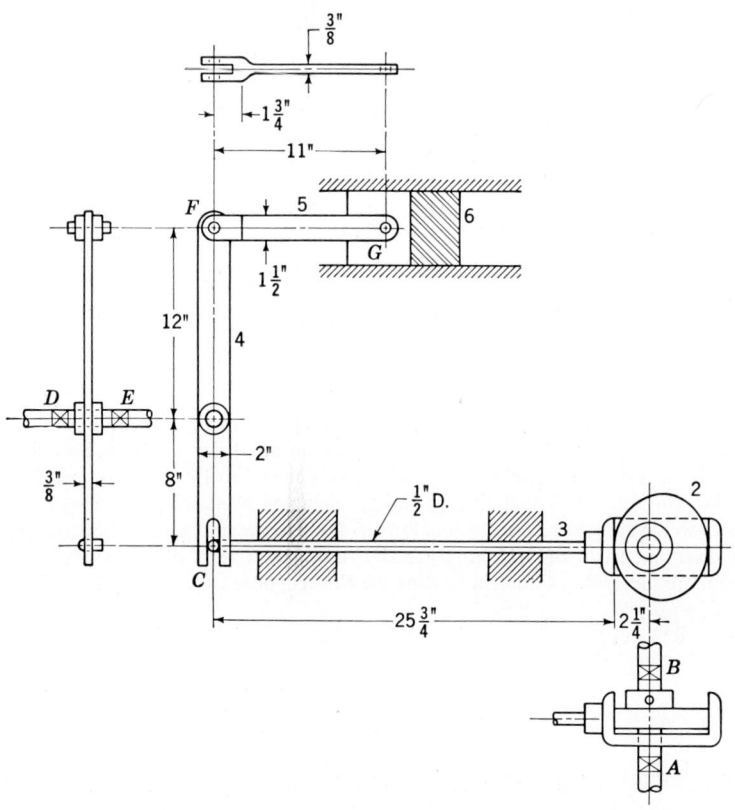

Prob. 21-14

21-15. The cam of Prob. 21-14 moves the flat-face follower a distance of 1½ in. to the left, from the position shown in the figure, in 150° of cam rotation and dwells for 30°. Because the cam has a constant breadth, it is only necessary to specify the motion for 180° of rotation. The cam is cut so as to produce, nominally, a cycloidal motion of the carriage. Using a damping ratio of 0.20 and a cam speed of 900 rpm, determine the response of the carriage.

21-16. Bearings A and B in Prob. 21-15 are spaced the same distance from the center of the cam face. The lever is also mounted on a shaft midway between bearings D and E. Make a complete dynamic-force analysis of the cam mechanism, and plot the results for 360° of cam rotation. The results should include the bearing forces at A, B, D, and E; the pin reactions at C, F, and G; and the camshaft torque. If vibration is present, note that the cam-motion equations of Chap. 11 will not give the true value of the acceleration of the carriage. Neglect the effect of friction upon the bearing forces but do include the damping in analyzing the inertia force change due to the vibration.

CHAPTER 22

DYNAMICS OF FEEDBACK CONTROL SYSTEMS

The trend toward the automatic control of machines and other mechanical systems makes it necessary for the design engineer to have some basic knowledge in the general area of control engineering. Fortunately, a knowledge of the fundamentals of mechanical vibration is all that is necessary in order to understand some of the important principles and basic concepts. The purpose of this chapter, therefore, is to apply our new knowledge of vibrations to a study of automatic control systems.

Naturally the space available here places a severe limitation upon the range of topics which can be explored. For this reason this chapter might be better titled a primer of automatic control. Thus we shall learn something of the nomenclature of the subject, study some of the important concepts, and look into some very interesting problems.[1]

22-1. Examples of Automatic Control Systems. For our first example of a feedback control system we shall select a machine-tool positioning control which is rather widely used. The mechanism, shown in Fig. 22-1, consists of an input shaft driven by gears, an output shaft driven by a motor through another set of gears, a bevel-gear differential as an error detector, and a rotary potentiometer. The input shaft may be turned by hand, or it may be driven by a motor in accordance with commands received from a programmed tape. The ring gear of the differential is the planet carrier. If the input and output shafts are turned through exactly the same angle (though in opposite directions), the ring gear remains stationary, but if the input shaft turns and the output shaft is stationary, then the ring gear will also rotate. Thus rotation of the ring gear indicates relative angular displacement between the two shafts. As shown in the figure, the ring gear is connected through another gear or gears to a potentiometer. This potentiometer is connected in the electrical circuit of a motor so that the motor receives no voltage if the output

---

[1] The Laplace transformation is a method of solving differential equations using operational calculus, and it is used throughout the literature of automatic controls. The transformation makes it possible to solve very difficult differential equations by simple algebraic methods. This method will not be introduced here, however, because it is important for the reader to obtain a good grounding in the classical methods of analysis before advancing to these more sophisticated procedures.

shaft is in the position dictated by the input shaft. If the output shaft is not in the correct position, then a voltage appears across its terminals whose magnitude is proportional to the angular error and whose sign is such as to cause the motor to rotate in the proper direction to decrease the error. The figure is quite schematic because more gears than are shown are required in practice. An electrical amplifier is usually employed, too, between the potentiometer and the motor.

We note particularly, in Fig. 22-1, that the ring gear measures an *error* between the desired position and the actual position of the output shaft. This error is then suitably amplified, because of the gears and input voltage, and fed into the motor. If the error is zero, the motor receives no command. If the error is not zero, the motor is caused to rotate in the direction that will cause the error to decrease.

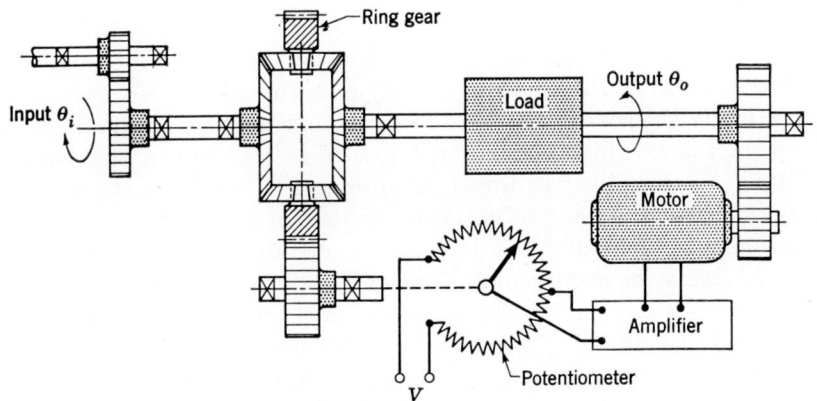

Fig. 22-1. A mechanism to control the angular position of a load using mechanical feedback.

As we examine the system, we can see many interesting possibilities to study. For example, the part to be positioned may be a high-inertia load, and the motor may require an appreciable time to bring the load to the correct position, but because of the inertia, the load overshoots, and now the motor is commanded to exert torque in the opposite direction. Thus the load may oscillate with large amplitudes about the desired position. Some of the factors which influence this oscillation are the amount of backlash in the gears, the mass or inertia of the various parts, the friction and damping, and the amount of amplification in the system.

A similar type of oscillation, due to the control system, is sometimes found in experimental airplanes. A plane in level flight may begin to nose up, causing the pilot to make a correction. Because of the correction, the plane noses down, and the pilot makes another correction.

The second correction causes the plane to nose up again, but more than before. This process goes on until finally the airplane is going through violent oscillations. Much of this is due to the time lags in the control system, and it includes the pilot's reflexes and reaction times. This is a good example of *instability* in a control system, an interesting subject which we shall later investigate in some detail.

Another example of a position control system is the hydraulically operated system illustrated schematically in Fig. 22-2. The valve is a standard spool type with three lands. When the input rod is pushed to signal a change in position, the lever AC is caused to pivot about A. This pushes the valve to the right, admits fluid pressure to the left of the piston, and at the same time opens the exhaust port so that the fluid on the right side of the piston can flow to the sump. When the load is

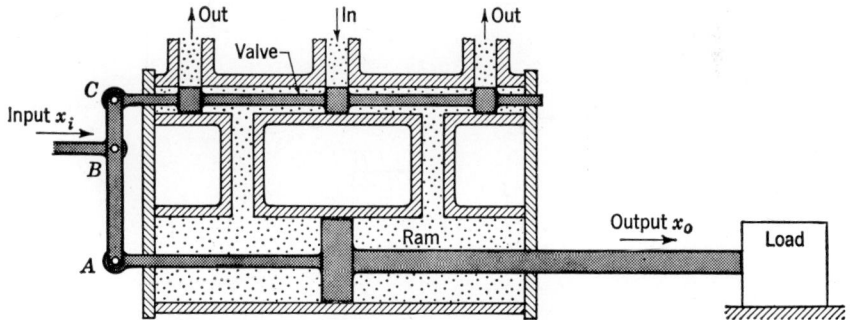

Fig. 22-2. Hydraulic mechanism controls the linear position of the load.

forced to the correct position, the lever AC is vertical again and all ports are closed. Thus for the vertical position of the lever, the output and input correspond and the error is zero. In this example, then, the valve opening is proportional to the error and consequently the volume of fluid flow through the ports is also roughly proportional to the error.

In each of the examples described above we have noted the presence of an input, an output, and an error. The error, each time, has been used to control a source of power directed in such a manner as to make the output approach a certain value. Each of these systems is called a *feedback system* because the output and input were compared and, if an error existed, a signal was *fed back* to the power source as a command to correct the error. The angular positioning mechanism employed the differential as an error detector. Motion of the ring gear represented the magnitude of the error, and this motion was used to actuate a potentiometer which "told" a motor what to do by impressing a voltage of a certain magnitude and polarity across its terminals. In an airplane the pilot is the error detector, and he manipulates certain instruments which direct the power sources and control the flight of the airplane. The hydraulic positioner detected the error between input and output by the angular displacement of a lever which, in turn, displaced a valve and actuated a power source.

It is customary to designate control systems as *open-loop* or *closed-loop*. An open-loop system is not automatically controlled at all. In Fig. 22-3a is shown a crank A, which is turned to position the load B. The crank is shown connected to the load by a long elastic shaft mounted in bearings; this is not essential to the example, however. Elasticity and friction, either coulomb or viscous, are present in practically all control

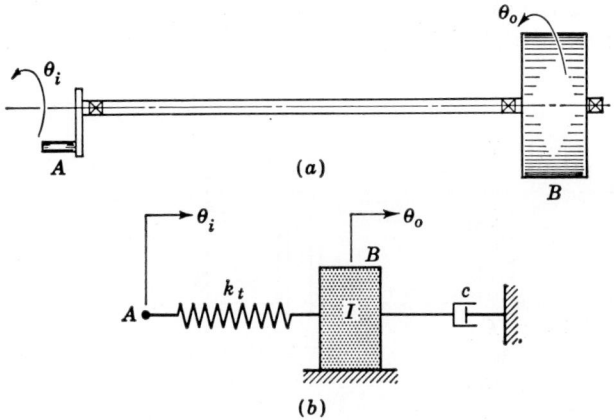

FIG. 22-3. (a) Example of an open-loop control system. (b) Schematic diagram. Input and output functions are θ_i and θ_o, respectively.

systems, and this can exist because of a variety of other elements which might exist between the input and the output. For example, a gear train between the crank and load would also result in an elastic system having friction and damping and elasticity. Consequently, it is convenient to diagram open-loop systems as in Fig. 22-3b. Here I is the moment of inertia of the load, k_t the torsional spring constant due to the elasticity of the shaft, and c the damping factor resulting from the viscous friction. With this form of diagram the problem is simple to formulate in mathematical terms.

We note particularly that the input function θ_i of the open-loop system of Fig. 22-3 is arbitrary and does not depend upon the value of θ_o. If we place a human operator at A and permit him to observe the load B, then the system is closed-loop. The reason for this is that the operator serves as an error-detecting device and automatically applies corrections to the crank in order to place the load in the desired position. Thus the terms *closed-loop* and *feedback*, when applied to control systems, have identical meanings.

The word "servo" appears many times in the literature of automatic controls and requires defining. A *servo system* is generally identified as a feedback control system with *power amplification*. A *servomechanism* would then be a servo system utilizing mechanical elements. Thus the angular positioner of Fig. 22-1 and the hydraulic positioner of Fig. 22-2 are servomechanisms because they utilize feedback with power amplification. If a human operator is used to operate the mechanism of Fig. 22-3, feedback is present but the word servo cannot be used to describe it because of the lack of power amplification.[1]

22-2. Writing the Equations. Mechanical control systems, in general, may contain a great variety of elements, such as gears, cams, levers, clutches, brakes, motors, actuators, etc. The elements may have linear motion or angular motion or both within any given control system. In

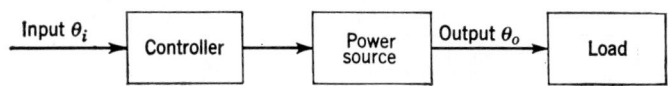

Fig. 22-4. Block diagram of an open-loop system.

addition, these elements possess certain elastic and frictional properties causing them to behave as springs and dampers; the inertia effects must be accounted for, too, because of the mass of the elements. Components such as gears, levers, and motors can be used to amplify or magnify linear displacement, torque, or force, as well as the power used in the system. It is very desirable to study control systems in general without worrying about the restrictions imposed by the selection of a particular group of elements. Consequently, we choose to represent control systems by a *block*, or *black-box*, notation. Such a notation for an open-loop controller is shown in Fig. 22-4. Here, the input and output quantities θ_i and θ_o are nominally used to represent angular displacement, but we shall interpret them quite broadly. Thus in a particular system they might also be used to represent linear displacement. Each block shown in the figure may contain a single element, or it may contain a great number of them.

The differential equation of a control system is always written as an expression of the dynamic equilibrium of the elements. Thus, for the system of Fig. 22-4, we express mathematically that the inertia torque

---

[1] Many of the terms of this section are defined very precisely in J. C. Gille, M. J. Pélegrin, and P. Decaulne, "Feedback Control Systems," pp. 1–24, McGraw-Hill Book Company, Inc., New York, 1959.

is equal to the sum of all the other torques acting upon the system. If the load has both inertia and damping, then the expression is

$$I\ddot{\theta}_o = -c\dot{\theta}_o + K'f(\theta_i) \tag{a}$$

where the input θ_i is understood to be some function of time still to be defined. The function $f(\theta_i)$ depends upon the characteristics of the controller and the power source. A very simple system would result if the components were selected so that the relation between $f(\theta_i)$ and θ_i is linear. For this condition we can make the substitution $K\theta_i = K'f(\theta_i)$; Eq. (a) can then be arranged to give

$$I\ddot{\theta}_o + c\dot{\theta}_o = K\theta_i \tag{22-1}$$

This is the differential equation for an open-loop system in which the output is proportional to the controller input. The quantity K is the constant of proportionality for this example.

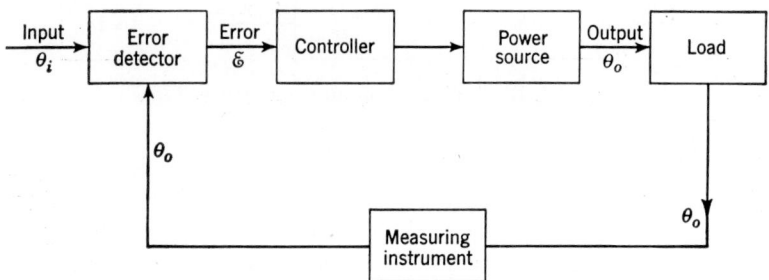

FIG. 22-5. Block diagram of a closed-loop system.

The addition of a feedback loop to the system of Fig. 22-4 results in the closed-loop controller of Fig. 22-5. This requires two additional items of equipment, the measuring instrument which measures the value of the output θ_o and the error detector. The error detector takes the input θ_i and the output θ_o, subtracts them, and sends an error signal ε to the controller of magnitude

$$\varepsilon = \theta_i - \theta_o \tag{22-2}$$

Thus, the difference between the open-loop and the closed-loop systems is that in the open-loop system the input function θ_i commands the controller and in the closed-loop system the controller is commanded by the error function ε.

A closed-loop control system in which the output is directly proportional to the error is called a *proportional-error system*. The differential

DYNAMICS OF FEEDBACK CONTROL SYSTEMS 605

equation for this system is the same as Eq. (22-1) if ε is substituted for θ_i. This gives

$$I\ddot{\theta}_o + c\dot{\theta}_o = K\varepsilon \qquad (b)$$

If we substitute the value of ε from Eq. (22-2), the equation can be rearranged to produce

$$I\ddot{\theta}_o + c\dot{\theta}_o + K\theta_o = K\theta_i \qquad (22\text{-}3)$$

Equation (22-3) can be simplified by the following substitutions:

$$\omega_n = \sqrt{\frac{K}{I}} \qquad (22\text{-}4)$$

$$2\zeta\omega_n = \frac{c}{I} \qquad (22\text{-}5)$$

$$\zeta = \frac{c}{2\sqrt{KI}} \qquad (22\text{-}6)$$

where I = mass moment of inertia, lb-sec²-in.
c = torsional damping factor, lb-in.-sec/rad
K = gain constant, lb-in./rad
ω_n = natural undamped circular frequency, rad/sec
ζ = damping ratio, c/c_c, dimensionless

It will be recalled that the natural frequency ω_n and the damping ratio ζ have been previously defined in Chap. 18 for linear systems. Equation (22-3) can now be written in the form

$$\ddot{\theta}_o + 2\zeta\omega_n\dot{\theta}_o + \omega_n^2\theta_o = \omega_n^2\theta_i \qquad (22\text{-}7)$$

22-3. Standard Input Functions. There is a great deal of useful information which can be gained from a mathematical analysis as well as a laboratory analysis of feedback control systems when standard input functions are used to determine the performance. Standard input functions result in differential equations which are much easier to analyze mathematically than they would be if the actual operating conditions were used as input. Furthermore, the use of standard inputs makes it possible to compare the performance of various proposed control systems.

One of the most useful of the standard input functions is the *unit-step function* shown in Fig. 22-6a. The function is not continuous, and consequently we cannot define initial conditions at $t = 0$. It is customary to specify the conditions at $t = 0+$ and $t = 0-$, where the signs indicate the conditions for values of time greater or less than zero. Thus, for the unit-step function

$$\theta_i = 0 \quad \dot{\theta}_i = 0 \quad \text{when } t = 0-$$
$$\theta_i = 1 \quad \dot{\theta}_i = 0 \quad \text{when } t = 0+$$

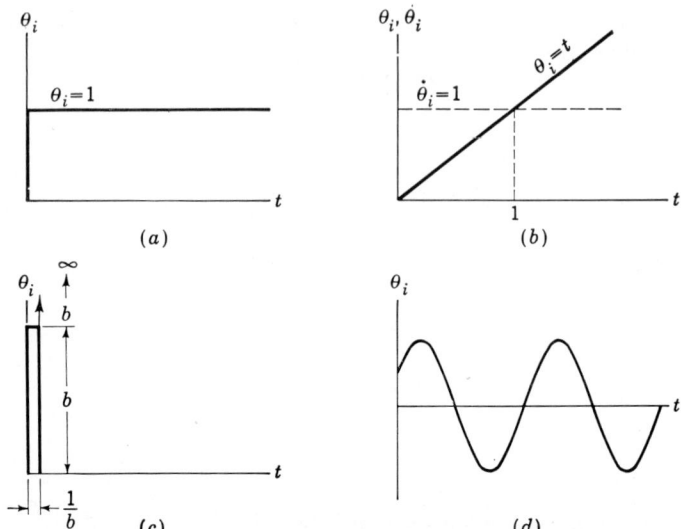

FIG. 22-6. Standard input functions. (a) Unit-step function. (b) Unit-step velocity function. (c) Unit-impulse function. (d) Harmonic input function.

The performance resulting from the application of these input conditions is called the *unit-step response*.

Another standard input function is the *unit-step velocity function*, as shown in Fig. 22-6b. This function is sometimes called the *unit-slope ramp function* because of the form of the displacement diagram. This function is defined as follows:

$$\theta_i = 0 \qquad \dot{\theta}_i = 0 \qquad \text{when } t = 0-$$
$$\theta_i = t \qquad \dot{\theta}_i = 1 \qquad \text{when } t = 0+$$

Also shown in Fig. 22-6 are the *unit-impulse function* and the *harmonic input function*, both of which are useful in determining the system response. We shall find the harmonic function to be especially useful when the response of the system is to be found under steady-state conditions for various input frequencies.

It happens in some control systems that the input function is relatively constant but the output is subjected to suddenly varying forces or torques. For example, an automatically controlled machine tool may run into a deeper cut, causing a greater load torque to be exerted on the machine, thus tending to slow it down. The differential equation for this condition is written

$$I\ddot{\theta}_o + c\dot{\theta}_o + K\theta_o = K\theta_i - T \qquad (a)$$

When this equation is simplified by substituting the values from Eqs. (22-4) to (22-6) and rearranged, it becomes

$$\ddot{\theta}_o + 2\zeta\omega_n\dot{\theta}_o + \omega_n^2\theta_o = \omega_n^2\theta_i - \frac{T}{I} \qquad (22\text{-}8)$$

When the solution to Eq. (22-7) is obtained for a given function θ_i, the solution to Eq. (22-8) can be obtained by superposition.

22-4. Solution of Linear Differential Equations. In the analysis of feedback control systems nth-order linear differential equations are encountered very frequently. The general form of these equations is

$$a_n \frac{d^n\theta}{dt^n} + a_{n-1} \frac{d^{n-1}\theta}{dt^{n-1}} + \cdots + a_1 \frac{d\theta}{dt} + a_0\theta = f(t) \qquad (22\text{-}9)$$

The function $f(t)$ is the driving or forcing function, and so t and θ are the independent and dependent variables, respectively. The coefficients $a_0, a_1, a_2, \ldots$ are constants and are independent of t and θ.

The solution to equations of the form of Eq. (22-9) is composed of two parts. The first part is called the *complementary function*, and it is the solution to the equation

$$a_n \frac{d^n\theta}{dt^n} + a_{n-1} \frac{d^{n-1}\theta}{dt^{n-1}} + \cdots + a_1 \frac{d\theta}{dt} + a_0\theta = 0 \qquad (22\text{-}10)$$

The complementary function is also called the *transient solution* in the literature of automatic controls because this part of the solution dies out quickly for a damped stable system. If the control system should happen to be unstable, then, we shall find, the controlled quantity quickly increases without limit. Note that the transient solution is obtained by making the forcing function zero.

The other part of the solution is called the *particular integral* by mathematicians and the *steady-state solution* by control engineers. It is any particular solution of Eq. (22-9).

The complete solution is the sum of the transient and steady-state components. It will contain in the transient part n arbitrary constants which are evaluated from the boundary conditions. The amplitude of the transient component depends upon the forcing function and the starting conditions, but all the other characteristics of the transient term are completely independent of the forcing function. We shall find, for linear systems, that neither the forcing function nor the amplitude has any effect upon the system stability, this being dependent only on the transient part.

The Transient Component. The following steps are used to obtain the transient solution:

1. Set the forcing function equal to zero, and arrange the equation in the form of Eq. (22-10).
2. Assume a solution in the form

$$\theta = Ae^{st} \tag{22-11}$$

3. Substitute Eq. (22-11) and its derivatives into the differential equation, and solve for the characteristic equation.
4. Solve the characteristic equation for the roots.
5. Obtain the transient solution by substituting the roots back into the assumed solution.

As an example of this procedure let us solve Eq. (22-7) for the transient term. For step 1 we write

$$\ddot{\theta}_o + 2\zeta\omega_n \dot{\theta}_o + \omega_n^2 \theta_o = 0 \tag{a}$$

and for step 2

$$\theta_o = Ae^{st} \qquad \dot{\theta}_o = Ase^{st} \qquad \ddot{\theta}_o = As^2 e^{st} \tag{b}$$

For step 3 we substitute all Eqs. (b) into (a). This gives

$$As^2 e^{st} + 2\zeta\omega_n As e^{st} + \omega_n^2 A e^{st} = 0 \tag{c}$$

The characteristic equation is then obtained by canceling terms. Thus

$$s^2 + 2\zeta\omega_n s + \omega_n^2 = 0 \tag{d}$$

is the characteristic equation. Step 4 is to solve Eq. (d) for the roots. This produces

$$\begin{aligned} s_1 &= -\zeta\omega_n + \omega_n \sqrt{\zeta^2 - 1} \\ s_2 &= -\zeta\omega_n - \omega_n \sqrt{\zeta^2 - 1} \end{aligned} \tag{e}$$

In this book we shall not consider the situation in which $\zeta > 1$. Consequently, for $\zeta < 1$, the radical in Eqs. (e) is imaginary, and we prefer to write the roots in the form

$$\begin{aligned} s_1 &= -\zeta\omega_n + j\omega_n \sqrt{1 - \zeta^2} \\ s_2 &= -\zeta\omega_n - j\omega_n \sqrt{1 - \zeta^2} \end{aligned} \tag{f}$$

where $j = \sqrt{-1}$. Finally, for step 5, we substitute the roots back into the assumed solution. The transient solution then becomes, after factoring,

$$\theta_{o,t} = e^{-\zeta\omega_n t}(Ae^{j\omega_n \sqrt{1-\zeta^2}\,t} + Be^{-j\omega_n \sqrt{1-\zeta^2}\,t}) \tag{g}$$

where A and B are the constants of integration and are two in number, the same as the order of the highest derivative in the differential equation. These cannot be evaluated until the complete solution is found. With the help of Eqs. (2-10) and (2-11) we can transform Eq. (g) into the trigonometric form. The result of this transformation is

$$\theta_{o,t} = e^{-\zeta\omega_n t}(C_1 \cos \omega_n \sqrt{1-\zeta^2}\, t + C_2 \sin \omega_n \sqrt{1-\zeta^2}\, t) \quad (22\text{-}12)$$

The coefficient $e^{-\zeta\omega_n t}$ of Eq. (22-12) indicates its transient nature, since this term becomes approximately zero for large values of t.

It sometimes happens that two or more roots of the characteristic equation are identical. For the case of two identical roots $s_1 = s_2$, the transient solution is

$$\theta_{o,t} = Ae^{s_1 t} + Bte^{s_1 t} \quad (h)$$

For other special cases the reader should refer to a text on the solution of linear differential equations.

The Steady-state Component. Here we shall solve Eq. (22-7) for various input conditions. The equation is repeated for convenience:

$$\ddot{\theta}_o + 2\zeta\omega_n \dot{\theta}_o + \omega_n^2 \theta_o = \omega_n^2 \theta_i \quad (22\text{-}7)$$

For the unit-step function $\theta_i = 1$, and the equation to be solved for the particular integral is

$$\ddot{\theta}_o + 2\zeta\omega_n \dot{\theta}_o + \omega_n^2 \theta_o = \omega_n^2 \quad (i)$$

The solution is

$$\theta_{o,ss} = 1 \quad (22\text{-}13)$$

which can be verified by substituting Eq. (22-13) and its derivatives into Eq. (i).

For the velocity-step function $\theta_i = t$, and the differential equation becomes

$$\ddot{\theta}_o + 2\zeta\omega_n \dot{\theta}_o + \omega_n^2 \theta_o = \omega_n^2 t \quad (j)$$

We assume a solution in the form

$$\theta_{o,ss} = At + B$$

The successive derivatives are

$$\dot{\theta}_{o,ss} = A \qquad \ddot{\theta}_{o,ss} = 0$$

Substituting the assumed solution and its derivatives into Eq. (j) gives

$$2\zeta\omega_n A + \omega_n^2(At + B) = \omega_n^2 t$$

We now arrange the equation in the form

$$(2\zeta\omega_n A + \omega_n^2 B) + (\omega_n^2 A)t = \omega_n^2 t$$

then solve for A and B. Thus

$$A = 1 \qquad B = -\frac{2\zeta}{\omega_n}$$

So the steady-state solution for a velocity-step input is

$$\theta_{o,ss} = t - \frac{2\zeta}{\omega_n} \quad (22\text{-}14)$$

The Complete Solution. The complete solution is the sum of the transient and steady-state terms. Thus

$$\theta_o = \theta_{o,t} + \theta_{o,ss}$$

For the unit-step input this is formed by summing Eqs. (22-12) and (22-13). The result is

$$\theta_o = e^{-\zeta\omega_n t}(C_1 \cos \omega_n \sqrt{1-\zeta^2}\, t + C_2 \sin \omega_n \sqrt{1-\zeta^2}\, t) + 1 \quad (22\text{-}15)$$

The conditions for $t = 0+$ can now be applied in order to evaluate the constants C_1 and C_2. Imposing the condition that $\theta_o = 0$ at the beginning of the step gives

$$0 = 1(C_1 \cos 0 + C_2 \sin 0) + 1$$

or

$$C_1 = -1$$

The second condition to be imposed is that $\dot{\theta}_o = 0$. To apply this condition it is necessary to take the derivative of Eq. (22-15). It is

$$\dot{\theta}_o = -\zeta\omega_n e^{-\zeta\omega_n t}(C_1 \cos \omega_n \sqrt{1-\zeta^2}\, t + C_2 \sin \omega_n \sqrt{1-\zeta^2}\, t)$$
$$+ e^{-\zeta\omega_n t}(-C_1 \omega_n \sqrt{1-\zeta^2} \sin \omega_n \sqrt{1-\zeta^2}\, t$$
$$+ C_2 \omega_n \sqrt{1-\zeta^2} \cos \omega_n \sqrt{1-\zeta^2}\, t)$$

Substituting $t = 0-$ and $\dot{\theta}_o = 0$ gives

$$0 = -\zeta\omega_n(C_1 \cos 0 + C_2 \sin 0)$$
$$+ 1(-C_1 \omega_n \sqrt{1-\zeta^2} \sin 0 + C_2 \omega_n \sqrt{1-\zeta^2} \cos 0)$$

Substituting the value of C_1 and solving yield

$$C_2 = -\frac{\zeta}{\sqrt{1-\zeta^2}}$$

The values of C_1 and C_2 are now replaced in Eq. (22-15), giving the complete solution

$$\theta_o = 1 - e^{-\zeta\omega_n t}\left(\cos \omega_n \sqrt{1-\zeta^2}\, t + \frac{\zeta}{\sqrt{1-\zeta^2}} \sin \omega_n \sqrt{1-\zeta^2}\, t\right)$$
$$(22\text{-}16)$$

This solution will be discussed in detail in the sections to follow. If transformed to a single trigonometric term and a phase angle, as demonstrated in Chap. 18, it is

$$\theta_o = 1 - \frac{e^{-\zeta\omega_n t}}{\sqrt{1-\zeta^2}} \cos(\sqrt{1-\zeta^2}\,\omega_n t - \phi) \tag{22-17}$$

where
$$\phi = \tan^{-1} \frac{\zeta}{\sqrt{1-\zeta^2}}$$

The complete solution for the velocity-step input is obtained in a similar manner. The conditions to be applied in order to evaluate the arbitrary constants are

$$\theta_o = 0 \qquad \dot{\theta}_o = 0 \qquad \text{when } t = 0-$$

The equation to be solved for the arbitrary constants is obtained by adding Eqs. (22-12) and (22-14) and is

$$\theta_o = e^{-\zeta\omega_n t}(C_1 \cos \omega_n \sqrt{1-\zeta^2}\,t + C_2 \sin \omega_n \sqrt{1-\zeta^2}\,t) + t - \frac{2\zeta}{\omega_n} \tag{22-18}$$

The constants are found to be

$$C_1 = \frac{2\zeta}{\omega_n} \qquad C_2 = -\frac{1-2\zeta^2}{\omega_n \sqrt{1-\zeta^2}}$$

and the complete solution is

$$\theta_o = t - \frac{2\zeta}{\omega_n} + e^{-\zeta\omega_n t}\left(\frac{2\zeta}{\omega_n} \cos \omega_n \sqrt{1-\zeta^2}\,t - \frac{1-2\zeta^2}{\omega_n\sqrt{1-\zeta^2}} \sin \omega_n \sqrt{1-\zeta^2}\,t\right) \tag{22-19}$$

or
$$\theta_o = t - \frac{2\zeta}{\omega_n} + \frac{e^{-\zeta\omega_n t}}{\omega_n \sqrt{1-\zeta^2}} \cos(\omega_n\sqrt{1-\zeta^2}\,t - \phi)$$
$$\phi = \tan^{-1} \frac{2\zeta^2 - 1}{2\zeta\sqrt{1-\zeta^2}} \tag{22-20}$$

22-5. Analysis of the Proportional-error Feedback System. We have seen that a proportional-error feedback system of control is one that operates by applying a correction which is directly proportional to the error which exists. Such a system must be analyzed in order to determine the following:

1. The stability of the system.
2. The response time, or the time required to reach steady-state operation after the application of a disturbance to the system. Since a system with viscous damping never reaches a steady-state condition, this is usually defined as the time required for the transient to decay to 5 per cent of its final value.

3. The natural frequency.

4. The steady-state error, that is, the difference between the input and output during steady-state operation.

5. The maximum overshoot, or the maximum deviation between output and input during transient conditions.

In this section we shall discuss only the last four of these performance criteria, reserving a study of stability for a later section.

Unit-step Input. Output response of the system for unit-step input is given by Eq. (22-17), and this equation has been plotted in Fig. 22-7

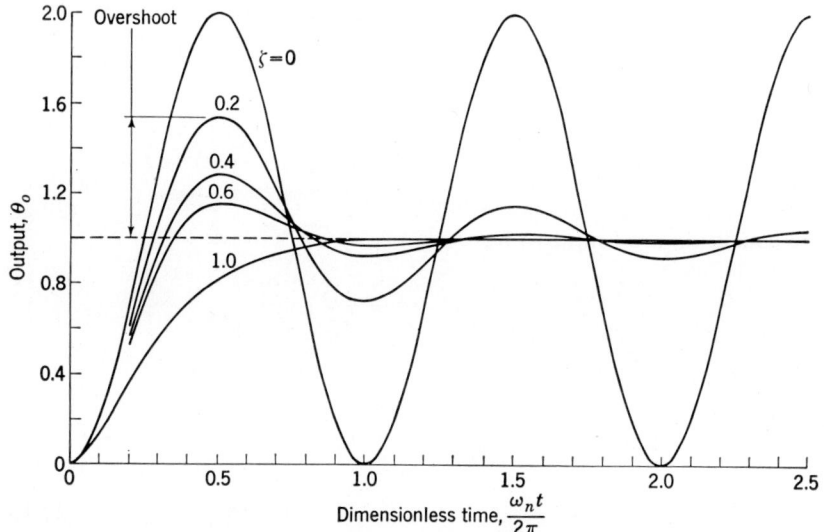

FIG. 22-7. Response to a unit-step input.

for various values of the damping ratio. The abscissa is measured in dimensionless time $\omega_n t/2\pi$, and so the curves can be applied to any physical system. Thus, if the natural undamped frequency ω_n is known for a given system, then the response time can be calculated simply by multiplying dimensionless time by the quantity $2\pi/\omega_n$. The response for zero damping is included for its academic interest; an automatic control system with no damping at all would give completely unsatisfactory performance.

The response for critical damping $\zeta = 1$ shows no overshoot, and the steady-state condition is reached at about $\omega_n t/2\pi = 1.0$.

It is convenient to define an *overshoot ratio* as the ratio of overshoot with damping to the overshoot that would exist if no damping were present. Figure 22-8 shows a plot of this ratio vs. the damping ratio. For $\zeta = 0.40$, the overshoot ratio is approximately 0.25; this quantity can also be read from Fig. 22-7. Notice that the steady-state condition is not reached for $\zeta = 0.40$ until $\omega_n t/2\pi = 2.0$. The response time, therefore, is twice that for critical damping.

Since the response time is proportional to $\omega_n t/2\pi$, we can make this time short simply by making the natural undamped frequency ω_n large.

In practice ζ is usually between 0.4 and 0.7 in physical systems. If a certain deviation from steady state is permitted, say 5 per cent, then a system with $\zeta = 0.7$ will reach steady state before one having $\zeta = 1.0$ because the early portion of its response curve is steeper.

Figure 22-7 shows that the steady-state value of the output is unity. Consequently, there is no steady-state error for step-input functions when applied to proportional-error feedback systems.

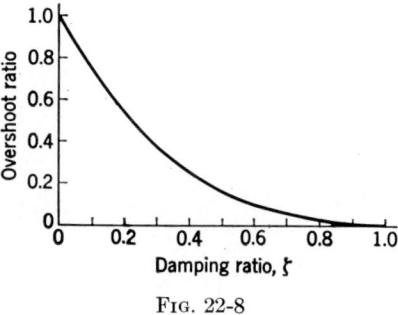

Fig. 22-8

Load Disturbance. Another look at the behavior of the system can be obtained by considering that the input signal is constant or zero and that the load is subjected to a disturbance. Selecting the zero input condition, for simplicity, Eq. (22-8) then becomes

$$\ddot{\theta}_o + 2\zeta\omega_n\dot{\theta}_o + \omega_n^2\theta_o = -\frac{T}{I} \qquad (22\text{-}21)$$

where T represents a constant torque suddenly applied to the output. The steady-state component of the solution is

$$\theta_{o,ss} = -\frac{T}{I\omega_n^2} \qquad (a)$$

so that the complete solution must be

$$\theta_o = e^{-\zeta\omega_n t}(C_1 \cos \omega_n \sqrt{1-\zeta^2}\, t + C_2 \sin \omega_n \sqrt{1-\zeta^2}\, t) - \frac{T}{I\omega_n^2} \qquad (b)$$

The boundary conditions are $\theta_o = 0$, $\dot{\theta}_o = 0$ for $t = 0+$. Solving Eq. (b) for the arbitrary constants gives

$$C_1 = \frac{T}{I\omega_n^2} \qquad C_2 = \frac{\zeta T}{I\omega_n^2 \sqrt{1-\zeta^2}}$$

When these values are substituted into Eq. (b) and the equation transformed to a single trigonometric term and a phase angle, the result is

$$\theta_o = \frac{T}{I\omega_n{}^2}\left[\frac{e^{-\zeta\omega_n t}}{\sqrt{1-\zeta^2}}\cos\left(\omega_n\sqrt{1-\zeta^2}\,t - \phi\right) - 1\right] \quad (22\text{-}22)$$

$$\phi = \tan^{-1}\frac{\zeta}{\sqrt{1-\zeta^2}}$$

This equation is plotted in Fig. 22-9 for three values of the damping ratio to show what happens. As shown, the disturbance dies out at a rate

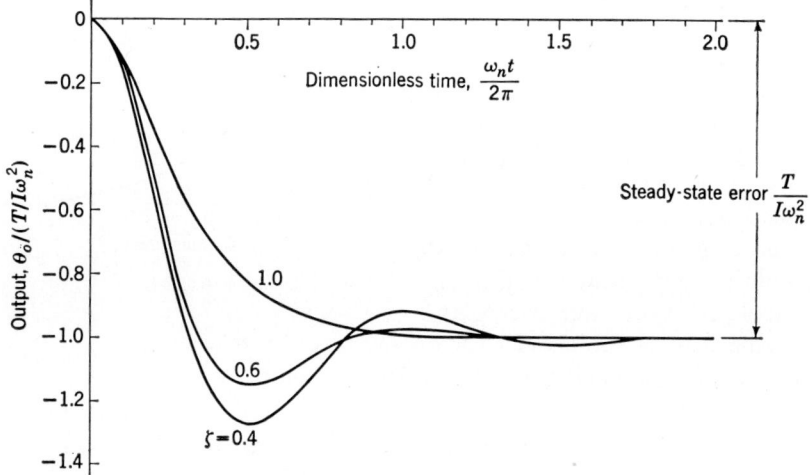

Fig. 22-9. Response of proportional-error feedback control due to a load disturbance.

which is dependent on the amount of damping. It finally reaches a steady-state condition which is not zero but of an amount

$$\varepsilon_{ss} = \frac{T}{I\omega_n{}^2} \quad (22\text{-}23)$$

This is the steady-state error. If we substitute $\omega_n{}^2 = K/I$, the equation becomes

$$\varepsilon_{ss} = \frac{T}{K} \quad (22\text{-}24)$$

so the only method of reducing the magnitude of this error is to increase the gain K of the system.

We have seen that the control system is defined by the parameters I, c, and K. Of these three, the gain is usually the easiest to change. Some variation in the damping is usually possible, but the employment of dashpots or friction dampers is not often a good solution. Variation in the inertia I is the most difficult change to make because this is fixed by the design of the driven element and, furthermore, improvement always requires a decrease in the inertia.

Velocity-step Input. For the unit-velocity-step input $\theta_i = t$, and from Eq. (22-14)

$$\theta_{o,ss} = t - \frac{2\zeta}{\omega_n}$$

after the transient dies out. Therefore the steady-state error is

$$\varepsilon_{ss} = \frac{2\zeta}{\omega_n} = \frac{c}{K} \qquad (22\text{-}25)$$

which is obtained by the substitution of the value of ζ from Eq. (22-6). This error occurs only as long as the input function signals for a constant velocity. Again we see that its magnitude can be reduced by increasing the gain K.

22-6. Response of the Proportional-error System to Harmonic Input. The response of a feedback system to harmonically varying input functions gives very useful results to the control engineer. In fact by employing inputs of varying frequencies he can study the response over the entire frequency range. For this reason such a study is termed a *frequency-response* study, or a study of the *frequency characteristics*. In addition to the amplitude response for various frequencies, such an investigation also reveals information on the phase shift, or the time lag between the input and output vectors.

The input to a feedback system frequently contains spurious signals termed noise or roughness which occur at a relatively high frequency. Obviously no control engineer wants to design a system that will respond to the noise coming to the input. A study of the frequency response will reveal exactly how much noise will get through.

It will be convenient to express the input and output quantities in terms of a rotating vector in the complex polar notation (Sec. 2-13). Thus the harmonic input vector is

$$\theta_i = A(\cos \omega t + j \sin \omega t) = A e^{j\omega t} \qquad (a)$$

where A is the amplitude and ω is the circular frequency in radians per second. The system will respond to this input at the same frequency ω but with a phase lag. Thus the equation of the output is

$$\theta_o = B[\cos (\omega t - \gamma) + j \sin (\omega t - \gamma)] = B e^{j(\omega t - \gamma)} \qquad (22\text{-}26)$$

where γ is the angle by which the input leads the output. A perfect or ideal system is one in which the output follows the input command

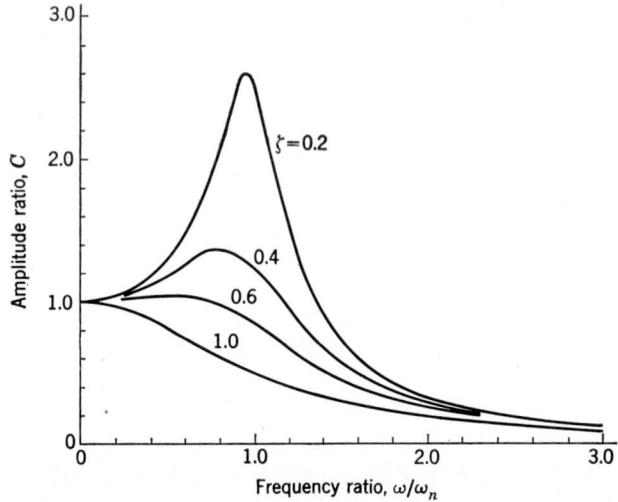

Fig. 22-10. Amplitude response of a proportional-error feedback system to a harmonic input of varying frequency.

without deviation, and consequently such a system would have $B = A$ and $\gamma = 0$ for all input frequencies from zero to infinity.

In studying the frequency response it is desirable to relate the output to the input; this produces dimensionless quantities which are applicable to all systems. If we divide Eq. (22-26) by Eq. (a) we find

$$\frac{\theta_o}{\theta_i} = \frac{Be^{j(\omega t - \gamma)}}{Ae^{j\omega t}} = Ce^{-j\gamma} \qquad (22\text{-}27)$$

Thus C is the amplitude ratio and γ the phase shift. The frequency response is known after these values are obtained for all ω's from zero to infinity.

For convenience we repeat Eq. (22-7) for the proportional-error system here:

$$\ddot{\theta}_o + 2\zeta\omega_n\dot{\theta}_o + \omega_n^2\theta_o = \omega_n^2\theta_i \qquad (22\text{-}7)$$

Substituting for θ_i from Eq. (a) gives

$$\ddot{\theta}_o + 2\zeta\omega_n\dot{\theta}_o + \omega_n^2\theta_o = \omega_n^2 A e^{j\omega t} \qquad (22\text{-}28)$$

To solve this equation using complex polar notation we assume that Eq. (22-26) is the solution. The first and second derivatives are then

$$\dot{\theta}_o = Bj\omega e^{j(\omega t - \gamma)} \qquad (b)$$
$$\ddot{\theta}_o = -B\omega^2 e^{j(\omega t - \gamma)} \qquad (c)$$

Substituting the assumed solution and its two derivatives into Eq. (22-28) produces

$$-B\omega^2 e^{j(\omega t-\gamma)} + 2\zeta\omega_n Bj\omega e^{j(\omega t-\gamma)} + \omega_n{}^2 B e^{j(\omega t-\gamma)} = \omega_n{}^2 A e^{j\omega t}$$

Canceling and rearranging,

$$B[(\omega_n{}^2 - \omega^2) + j(2\zeta\omega_n\omega)]e^{-j\gamma} = A\omega_n{}^2$$

so that

$$\frac{B}{A} e^{-j\gamma} = \frac{\omega_n{}^2}{\omega_n{}^2 - \omega^2 + j(2\zeta\omega_n\omega)}$$

Next, dividing numerator and denominator by $\omega_n{}^2$ yields

$$\frac{B}{A} e^{-j\gamma} = \frac{1}{1 - \omega^2/\omega_n{}^2 + j(2\zeta\omega/\omega_n)} = Ce^{-j\gamma} \quad (22\text{-}29)$$

which is the dimensionless form in terms of the frequency ratio ω/ω_n. Solving for C and γ gives

$$C = \frac{1}{\sqrt{(1 - \omega^2/\omega_n{}^2)^2 + (2\zeta\omega/\omega_n)^2}}$$
$$\gamma = \tan^{-1} \frac{2\zeta\omega/\omega_n}{1 - \omega^2/\omega_n{}^2} \quad (22\text{-}30)$$

The reader should compare these results with Eqs. (18-35) and (18-36). Rectangular plots of Eqs. (22-30) are shown in Figs. 22-10 and 22-11 for various damping ratios. The plots show the reasons for employing moderate values of damping in control systems and why damping ratios

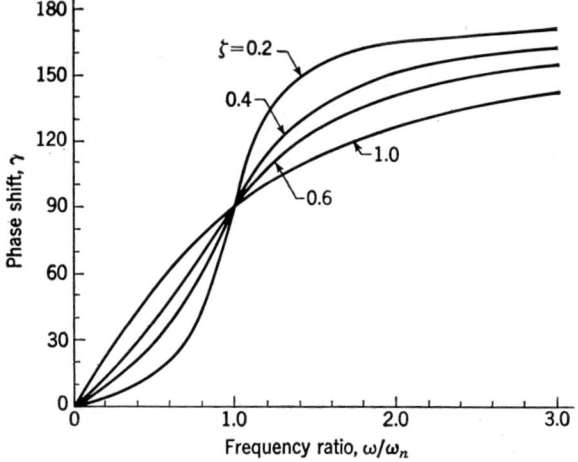

Fig. 22-11. Phase characteristics of a proportional-error feedback system with harmonic input.

less than $\zeta = 0.4$ should not be used. In the region $0.6 < \omega/\omega_n < 1.2$, the output will never exceed 140 per cent of the input as long as $\zeta = 0.4$ or more. All the information contained in the two plots can be plotted on a single chart if the polar form of the vector expressed by Eq. (22-29) is plotted. Figure 22-12 is such a graph.

22-7. Transfer Functions. The path of the error signal in a feedback control system frequently contains a number of separate components. It is often desirable to evaluate the output-input relationships separately,

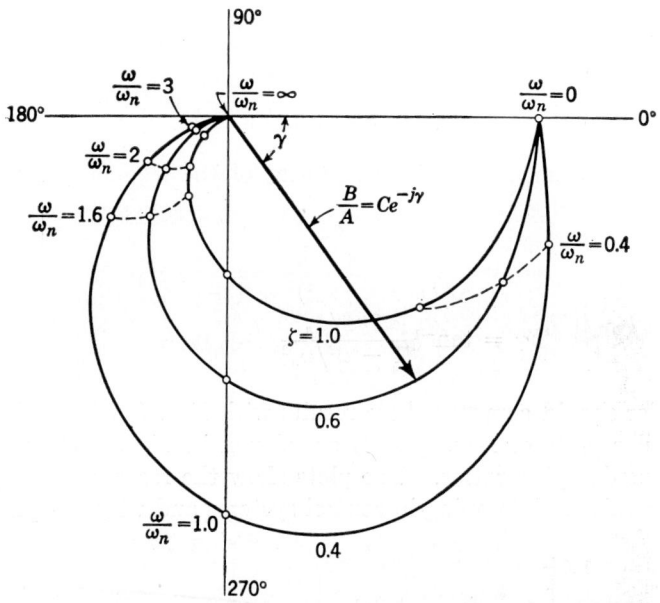

FIG. 22-12. Polar plot of frequency response of proportional-error feedback system.

by analytical or experimental means, and then to combine the relationships in order to obtain a single output-input expression for the total error path. The advantages of such a method are apparent when one considers that the error path may contain many elements each having certain gain, damping, and mass characteristics. Furthermore, since the resulting over-all output-input relation will contain the characteristics of the separate components, it is possible to make changes in the elements of the error path and to observe the effect of these upon the response of the control system.

The *transfer function* is a function which defines the ratio of the output to the input of a single control element or of an entire group of elements. It may or may not be expressed in complex form. In control literature it is often expressed in terms of the Laplace transformation variable s.

When determining the response of a system to various input frequencies, it is convenient to use a sinusoidal input. Under these conditions the Laplace variable can be replaced by the complex variable $j\omega$.

To show how the transfer function can be determined for an element, let us consider that the spring-mass-damper system of Fig. 22-13 is a component in the error path of a closed-loop system. Here ε is the error signal and is the input to this component. The output is designated as θ_o, but this may well be the input to the next component in the path. The differential equation for this system is

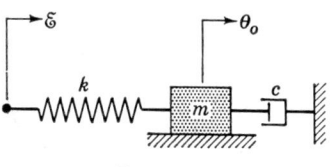

FIG. 22-13

$$m\ddot{\theta}_o = -k(\theta_o - \varepsilon) - c\dot{\theta}_o \qquad (a)$$

Rearranging,

$$\ddot{\theta}_o + \frac{c}{m}\dot{\theta}_o + \frac{k}{m}\theta_o = \frac{k}{m}\varepsilon \qquad (b)$$

Equation (b) can also be written

$$\ddot{\theta}_o + 2\zeta\omega_n\dot{\theta}_o + \omega_n^2\theta_o = \omega_n^2\varepsilon \qquad (22\text{-}31)$$

If harmonic input is used, the error and output signals will be in the form

$$\varepsilon = \varepsilon(j\omega)e^{j\omega t} \qquad \theta_o = \theta_o(j\omega)e^{j\omega t}$$

where $\varepsilon(j\omega)$ is some function of $j\omega$ which describes how the amplitude of the error signal varies with time and $\theta_o(j\omega)$ is the output function. Substituting these quantities and their successive derivatives into Eq. (22-31) gives

$$-\theta_o(j\omega)\omega^2 e^{j\omega t} + 2\zeta\omega_n\theta_o(j\omega)j\omega e^{j\omega t} + \omega_n^2\theta_o(j\omega)e^{j\omega t} = \omega_n^2\varepsilon(j\omega)e^{j\omega t} \qquad (c)$$

Canceling $e^{j\omega t}$ and rearranging,

$$[\omega_n^2 + 2\zeta\omega_n(j\omega) - \omega^2]\theta_o(j\omega) = \omega_n^2\varepsilon(j\omega) \qquad (d)$$

The transfer function KG is

$$KG(j\omega) = \frac{\theta_o(j\omega)}{\varepsilon(j\omega)} = \frac{\omega_n^2}{\omega_n^2 + 2\zeta\omega_n(j\omega) - \omega^2} \qquad (e)$$

or dividing numerator and denominator by ω_n^2,

$$KG(j\omega) = \frac{1}{1 + (2\zeta/\omega_n)(j\omega) - (\omega/\omega_n)^2} \qquad (22\text{-}32)$$

which is then independent of ω_n. The quantity K in the above expression is the constant factor, whereas $G(j\omega)$ is the complex factor.

620   DYNAMIC ANALYSIS OF MACHINES

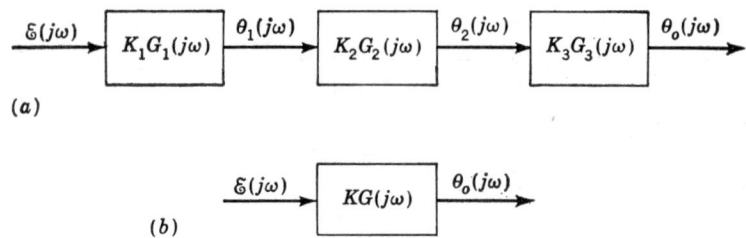

FIG. 22-14. (a) Chain of elements. (b) Equivalent element.

The transfer function for a group of elements can be obtained now. Figure 22-14a shows a chain of three elements in the error path and the transfer functions of each element which we assume have been previously determined. The equivalent or over-all transfer function shown in Fig. 22-14b is found as follows:

$$KG(j\omega) = \frac{\theta_1(j\omega)}{\mathcal{E}(j\omega)} \frac{\theta_2(j\omega)}{\theta_1(j\omega)} \frac{\theta_o(j\omega)}{\theta_2(j\omega)} = \frac{\theta_o(j\omega)}{\mathcal{E}(j\omega)} \quad (f)$$

Since

$$\frac{\theta_1(j\omega)}{\mathcal{E}(j\omega)} = K_1 G_1(j\omega)$$

$$\frac{\theta_2(j\omega)}{\theta_1(j\omega)} = K_2 G_2(j\omega)$$

$$\frac{\theta_o(j\omega)}{\theta_2(j\omega)} = K_3 G_3(j\omega)$$

Then, from Eq. (f)

$$KG(j\omega) = K_1 K_2 K_3 G_1(j\omega) G_2(j\omega) G_3(j\omega) \quad (22\text{-}33)$$

Thus the equivalent transfer function for a chain of components is simply the product of the separate transfer functions.

The transfer function is a vector in complex form which can be plotted in polar form, as in Fig. 22-12, to yield important and useful information concerning the response of the system.

The transfer function provides a relationship between the output and the input of an element or a group of elements in the control path. Since this quantity does not include the feedback path, it is frequently described as the *open-loop equation*. The open-loop equation is especially useful for determining the stability of an element or group of elements in a control system.

As its name indicates, the *closed-loop equation* is a relation between the output and input of a system which includes the feedback path. For the system of Fig. 22-15 the open-loop equation for the control elements is given by $\theta_o/\mathcal{E}$. The closed-loop equation is expressed by θ_o/θ_i. A simple relation between these equations is obtained from the relation

$$\mathcal{E} = \theta_i - \theta_o \quad (g)$$

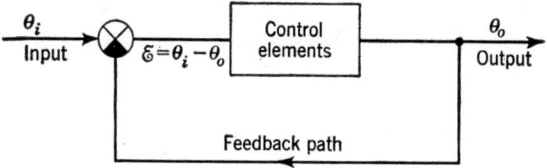

FIG. 22-15. The symbol at the junction of the input, error, and feedback paths is used by control engineers to indicate the addition of signals. In this case one of the branches is blacked in, which indicates that the negative of the quantity is to be used in the addition.

First, divide both sides by θ_o:

$$\frac{\varepsilon}{\theta_o} = \frac{\theta_i}{\theta_o} - 1 \qquad (h)$$

After rearranging, we obtain

$$\frac{\theta_o}{\theta_i} = \frac{\theta_o/\varepsilon}{1 + \theta_o/\varepsilon} \qquad (22\text{-}34)$$

EXAMPLE 22-1. Write the transfer function for the spring-mass-damper system shown in Fig. 22-16a.

Solution. The equivalent system in black-box notation is shown in Fig. 22-16b. The differential equation is

$$m\ddot{\theta}_o = -c\dot{\theta}_o - k(\theta_o - \varepsilon) \qquad (1)$$

which is rearranged to

$$\ddot{\theta}_o + \frac{c}{m}\dot{\theta}_o + \frac{k}{m}\theta_o = \frac{k}{m}\varepsilon \qquad (2)$$

We have already seen that equations of the form of Eq. (2) can be transformed to

$$\ddot{\theta}_o + 2\zeta\omega_n\dot{\theta}_o + \omega_n^2\theta_o = \omega_n^2\varepsilon \qquad (3)$$

where the following substitutions have been made:

$$\omega_n = \sqrt{\frac{k}{m}} \qquad 2\zeta\omega_n = \frac{c}{m} \qquad \zeta = \frac{c}{2\sqrt{km}}$$

Next we assume sinusoidal driving and output functions. In complex polar notation, these are

$$\varepsilon = Ae^{j\omega t} \qquad (4)$$
$$\theta_o = Be^{j(\omega t - \gamma)} \qquad (5)$$

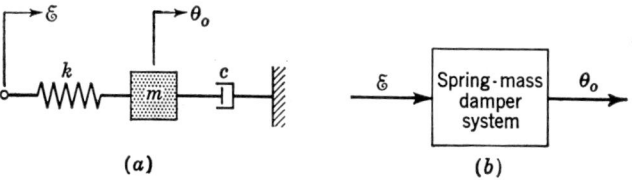

FIG. 22-16

where ω is the frequency of the driving function and γ is the phase angle between input and output. Substituting Eqs. (4) and (5) into Eq. (3) yields, after some manipulation,

$$\frac{\theta_o}{\varepsilon} = \frac{B}{A} e^{-j\gamma} = \frac{1}{1 - \omega^2/\omega_n^2 + j(2\zeta\omega/\omega_n)} \qquad (6)$$

This completes the solution; Eq. (6) is the required transfer function in complex form. Note that it is an open-loop equation.

EXAMPLE 22-2. The linkage attached to the spring-mass-damper system of Fig. 22-17 utilizes the adder of Fig. 13-1a to compute the error ε. Link 2 serves to feed

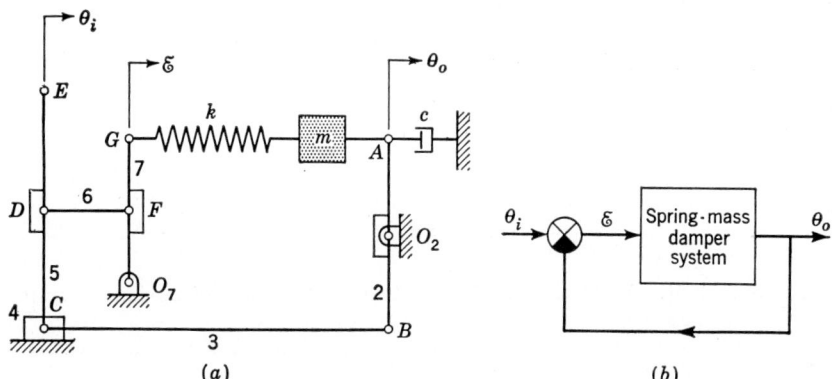

FIG. 22-17. $AO_2 = O_2B$; $CD = DE$; $O_7F = FG$. Links 2 to 7 inclusive are assumed weightless and absolutely rigid, and all pin joints are assumed to be without friction.

a negative value of θ_o into the system such that the signal received at point F is $[\theta_i + (-\theta_o)]/2$. Link 7 doubles the value of this signal so that the input to the spring is $\varepsilon = \theta_i - \theta_o$. Thus the system of this example is the same as in the previous example except that it has a feedback path added to the system. The block diagram is shown in Fig. 22-17b. This is a position control with an amplification of unity. Find the closed-loop equation of the system.

Solution. The open-loop equation is given by Eq. (6) of Example 22-1. Substituting this into Eq. (22-34) gives, for the closed-loop equation,

$$\frac{\theta_o}{\theta_i} = \frac{1}{1 + (1 - \omega^2/\omega_n^2) + j(2\zeta\omega/\omega_n)} = Ce^{-j\gamma} \qquad (1)$$

Then the amplitude and phase angle are

$$C = \frac{1}{\sqrt{(2 - \omega^2/\omega_n^2)^2 + (2\zeta\omega/\omega_n)^2}} \qquad (2)$$

$$\gamma = \tan^{-1} \frac{2\zeta\omega/\omega_n}{2 - \omega^2/\omega_n^2} \qquad (3)$$

22-8. Stability. A control system is *stable* if oscillations created by a disturbance die out or decay with time. A system is *unstable* if the oscillations continue or increase with time. The cause of instability in many control systems is the time delays in the various elements of the system. Time delays in a low-gain system may have little effect on the system stability, but if the gain is increased, these delays may become significant. For example, if a correction is applied to a high-gain system with appreciable time delays, the need for the correction may stop considerably in advance of the time required for all the corrective effort to be applied. Thus the system will overshoot in the direction of the correction more than the error which initiated the action in the first place. This process of continued correction together with the time delays will then eventually produce violent oscillations and, finally, destruction. Of course, some mechanical systems are self-limiting; others become nonlinear when subjected to high-amplitude motions. In any case, however, the presence of sustained high-amplitude oscillations in a control system is to be avoided at all costs.

It should be emphasized that it is important to obtain not only stability but good stability in a control system. The engineer should recall that pronounced overshoots and short bursts of oscillation cause noise, wear, and eventual fatigue failure in the members of machines. Even though a machine will continue to function under oscillation, its performance cannot be described as good if deflections and noise are present because of the existence of transient or steady-state vibrations. It is therefore necessary that the system be *very* stable in the sense that it should recover rapidly from any disturbance or corrective action which may be applied.

Some of the tests devised by control engineers to determine the stability or instability of a system are:

1. Inspection of the differential equation. If a derivative is missing or if the coefficients do not have the same signs, then the system is unstable. While the latter is a necessary condition, it is not a sufficient one.

2. Routh's criterion. In this method the characteristic equation is examined for the existence of positive real roots which would then indicate instability.

3. The Nyquist criterion. This test makes use of a polar plot of the transfer-function equations. An important advantage of the test is that the differential equations need not be written.[1]

[1] This theory was originally developed to define the stability of feedback amplifiers. Before a study of this test is undertaken, a thorough background in control analysis should be established, and for this reason we shall not discuss the Nyquist criterion in this book. The following books develop the theory in considerable detail:

Floyd E. Nixon, "Principles of Automatic Controls," Prentice-Hall, Inc., Englewood Cliffs, N.J., 1953.

C. J. Savant, Jr., "Basic Feedback Control System Design," McGraw-Hill Book Company, Inc., New York, 1958.

George J. Thaler and Robert G. Brown, "Servomechanism Analysis," McGraw-Hill Book Company, Inc., New York, 1953.

The stability of any linear control system depends only upon the parameters of the system. Thus it is completely independent of the driving function. This means that in considering the stability of a system we can always make the portion of the equation containing the driving or forcing function zero. Note that this is equivalent to a consideration of only the transient part of the solution.

Inspection of the Differential Equation. The general form of the differential equation with zero forcing function for a feedback control system is

$$a_n \frac{d^n\theta}{dt^n} + a_{n-1} \frac{d^{n-1}\theta}{dt^{n-1}} + \cdots + a_1 \frac{d\theta}{dt} + a_0\theta = 0 \qquad (22\text{-}35)$$

A necessary, but not sufficient, condition for stability is that all the coefficients a_i have the same sign. Thus, if a_n is positive and one or more of the a's are negative, the system is *unstable*. The system may or may not be stable if all the a's have the same sign.

Routh's Criterion. To apply this test arrange the coefficients of Eq. (22-35) in the form

$$\begin{array}{cccc} a_n & a_{n-2} & a_{n-4} & \cdots \\ a_{n-1} & a_{n-3} & a_{n-5} & \cdots \\ b_1 & b_2 & b_3 & \cdots \\ c_1 & c_2 & \cdots \\ d_1 & \cdots \end{array} \qquad (22\text{-}36)$$

where
$$b_1 = a_{n-2} - \frac{a_n a_{n-3}}{a_{n-1}}$$

$$b_2 = a_{n-4} - \frac{a_n a_{n-5}}{a_{n-1}}$$

$$b_3 = a_{n-6} - \frac{a_n a_{n-7}}{a_{n-1}} \qquad \text{etc.}$$

$$c_1 = a_{n-3} - \frac{a_{n-1} b_2}{b_1}$$

$$c_2 = a_{n-5} - \frac{a_{n-1} b_3}{b_1}$$

$$c_3 = a_{n-7} - \frac{a_{n-1} b_4}{b_1} \qquad \text{etc.}$$

$$d_1 = b_2 - \frac{b_1 c_2}{c_1} \qquad \text{etc.}$$

If all the terms in the first column of the array thus formed are positive, then the system is stable, but if one or more of the terms are negative, then a positive real root exists to the equation and the system is unstable. If there are n changes in sign, then there are n positive real roots.

EXAMPLE 22-3. Determine the stability of a system whose characteristic equation is

$$s^3 + 2s^2 + s + 3 = 0$$

Solution. The array is

$$\begin{array}{cc} 1 & 1 \\ 2 & 3 \\ -0.5 & \\ 3 & \end{array}$$

There are two changes in sign, and consequently there are two positive real roots. The system is unstable. Notice, in this example, that the coefficients of the characteristic equation are all positive. This example illustrates the insufficiency of the condition that the coefficients must all be of the same sign in order that stability exist.

22-9. Types of Controls. Our study of feedback systems has been limited to the simple proportional-error or positioning type of control. With this type of system it was found that the corrective action is directly proportional to the error. Sometimes, because of the magnitude of the steady-state error, the proportional-error system is not a satisfactory one. In this case a control system which anticipates an error and begins its corrective action before the error actually occurs may give better performance.

A complex system may be composed of one or more differentiators or integrators in the control path, the purpose of these elements being to increase response time or decrease the steady-state errors in the system.

Other refinements to control systems are obtained by *compensation.* Thus, certain undesirable features of a control system can be "compensated for" by introducing elements to alter the transfer function of the system. Both stability and response speed can be improved by compensation, and it is frequently used to cancel the effect of external disturbances to the system.

22-10. Nonlinear Systems. Throughout this chapter and, indeed, in many of the previous chapters, the discussions have been confined to so-called "linear" systems. Yet nowhere has the word been precisely defined. Of course, when applied to the relationship between two quantities, "linear" means a straight-line relationship, that is, a relationship always with the same constant of proportionality. Thus, in the study of mechanical vibrations, the inertia forces were considered proportional to the accelerations because the masses were constant and the damping forces were considered proportional to the velocity. Similarly, the constant of proportionality between the deflection of a spring and the force which produced it was assumed to remain the same. All these are *linear* relationships; so *linear* means that the *effect* is directly proportional to the *cause.*

It is doubtful if anything in nature is truly linear. Thus a spring obeys Hooke's law only approximately. Damping is *not* directly proportional to velocity because of the existence of Coulomb friction. Backlash exists in most mechanical systems. The assumption of linearity produces equations which yield beautifully to our mathematical tools. This pleases the theoretician, and so he is prone to turn his head on the realities of the problem and see only the part which is capable of mathematical manipulation. However, an engineer must get a job done. He must see and appreciate the theoretical principles involved in the problem and be able to make a "beautiful" mathematical analysis *if he decides it is necessary.* In order to make such a decision he must understand the magnitude and the contribution that the nonlinearities make in the problem and assess these on the basis of experience and judgment. It may even be that the economics of the problem forbids a complicated analysis of the effect of variations from linearity.

When dealing with linear systems the principle of superposition holds. Thus we were able to sum the transient and steady-state parts of a solution to obtain the total solution to a "linear" differential equation, but superposition is *not* valid for a nonlinear vibrating system. Thus, in one case, we may find that the amplitude and frequency both depend upon the boundary conditions; yet in the next problem it may turn out that amplitude depends upon the system, whereas the frequency is related to the initial conditions.

Consider the differential equation

$$\ddot{x} + \frac{c}{m}\dot{x} + \frac{k}{m}x^2 = 0 \qquad (a)$$

Notice that the spring force is proportional to the square of the deflection. This means that the spring force is not *linearly* related to the variable x. If we attempt to solve this equation in the usual manner by assuming that $x = Ae^{st}$, we find at once that the characteristic equation cannot be solved for the roots.

While there is no general method for solving nonlinear systems, yet a great deal of knowledge has been accumulated in recent years, especially in the control-engineering field. The nonlinearities which occur are usually of two kinds: the small nonlinearities which cannot be avoided because they are inherent in the elements of the system and the larger nonlinearities which are purposefully introduced to improve the performance of the system.

The reader who is interested in this subject will find the literature of nonlinear systems to be growing rapidly. Yet there remains a wealth of unsolved problems to tempt him.

DYNAMICS OF FEEDBACK CONTROL SYSTEMS

PROBLEMS[1]

22-1. It is desired to maintain a constant temperature in a room by means of a thermostat which positions a water valve on a radiator. Considering the heating system as a feedback control system, discuss the following proposals: (a) The temperature-sensitive element of the thermostat is located in the air of the room to be heated; (b) the temperature-sensitive element of the thermostat is located in the radiator discharge water; (c) the temperature-sensitive element of the thermostat is located in the radiator supply water, and the "set point" of the thermostat is determined by another thermostat which senses the outside air temperature. This means that as the outside air temperature increases, for example, the outside air thermostat causes the indoor thermostat to control the radiator supply water at lower temperatures.

22-2. The block diagrams in the figure represent three different feedback control systems. The Y's represent the transfer functions of the separate components in each system. For each system find an expression for the closed-loop transfer function. In c find also the transfer function θ_1/θ_i.

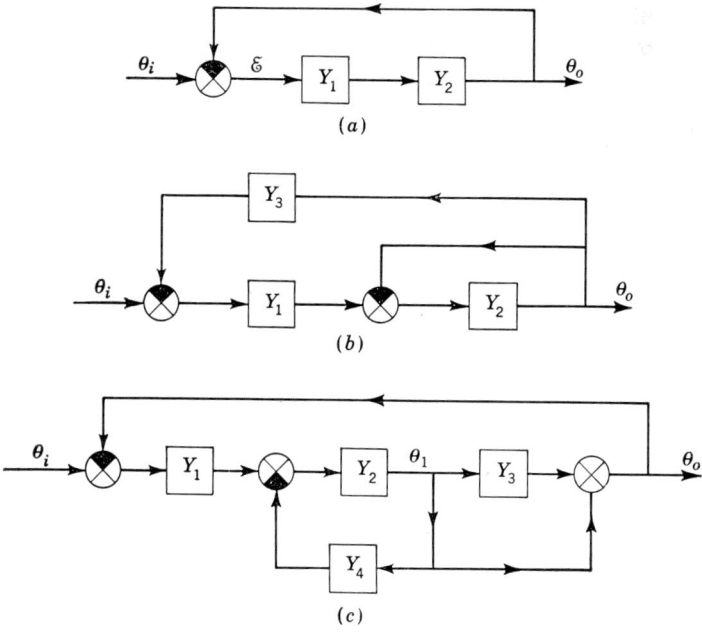

Prob. 22-2

22-3. The spring-mass-damper system shown in the figure is one of the elements in the error path of a closed-loop system. The error signal ε is the input to this particular component. (a) If θ_1 is the output of the component, determine the open-loop transfer function $\theta_1(j\omega)/\varepsilon(j\omega)$. (b) If θ_2 is the output of this component, determine the open-loop transfer function $\theta_2(j\omega)/\varepsilon(j\omega)$. (c) For each case write the differential equation which relates the output to the input for the closed-loop system.

[1] By Robert F. Timm, Teaching Fellow, College of Engineering, University of Michigan, Ann Arbor, Michigan.

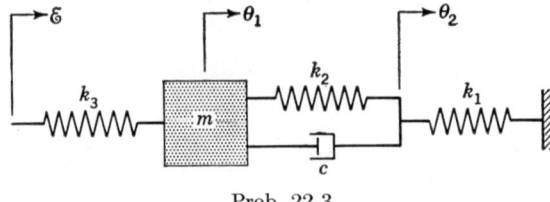

Prob. 22-3

22-4. The differential equation for an underdamped second-order system is

$$I\ddot{\theta}_o + c\dot{\theta}_o + k\theta_o = k\theta_i$$

where $I = 2$ lb-ft-sec^2 and $c = 20$ lb-ft-sec. (a) Determine the gain K of the system if the natural frequency of the damped oscillation is 8.66 rad/sec. (b) What is the damping ratio? (c) If $\theta_i = 100t$ rad, by what angle does θ_o lag θ_i?

22-5. Noting that the open-loop transfer function is

$$KG(j\omega) = \frac{\theta_o(j\omega)}{\mathcal{E}(j\omega)}$$

where
$$\mathcal{E}(j\omega) = \theta_i(j\omega) - \theta_o(j\omega)$$

determine the differential equation for the closed-loop system shown in the figure. What is the steady-state response when $\theta_i = 10 \cos 2t$?

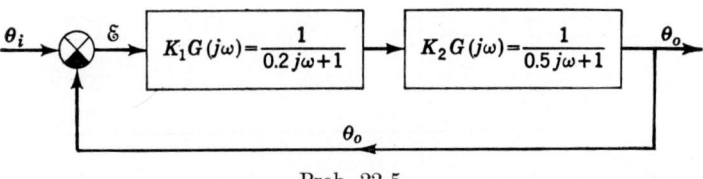

Prob. 22-5

22-6. The differential equation for a control system has the characteristic equation

$$s^5 + 2s^4 + 60s^3 + 100s^2 + 100Bs + 100B = 0$$

The factor B represents a gain constant in the system. Using the Routh criterion, determine the permissible range of B for stable operation of the system.

22-7. In some control systems the viscous damping is not sufficient to provide satisfactory transient response. If viscous damping is added directly, the system response can be improved, but this will result in a further dissipation of energy at the system output. This problem is intended to illustrate the principle of derivative feedback, which is one method of effectively increasing damping without involving an undesirable energy loss. A shaft-positioning servo system is represented schematically in the figure. The over-all gain and the damping and inertia effects have been lumped as indicated in the diagram, and an additional signal $C_v\dot{\theta}_o$, proportional to the time derivative of the output, is fed back and subtracted from the error signal. A torque amplifier with gain K then applies a correcting torque $K(\theta_i - \theta_o - C_v\dot{\theta}_o)$ to the output shaft. (a) By summing torques on the output shaft determine the differential equation of the control system. (b) Find an expression for the damping ratio of the system, and compare it with the damping ratio of the same system with the derivative feedback path disconnected. (c) With no external load on the system ($T_L = 0$), determine the error in the steady-state response for a unit-step input $\theta_i = 1$ rad. Repeat for a velocity input $\theta_i = \omega_i t$. (d) Compare the error obtained with velocity input in (c) with that obtained with the same system when the derivative feedback path is disconnected.

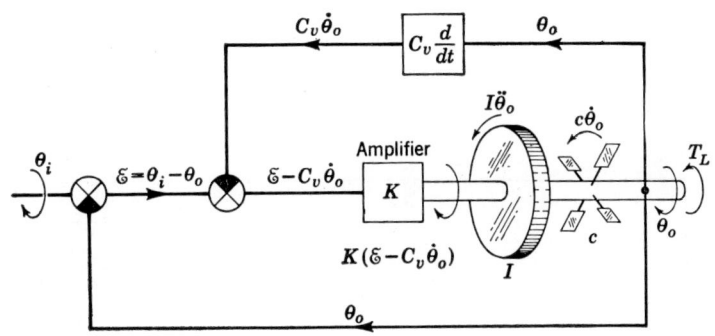

Prob. 22-7

22-8. In the second-order servo system shown in the figure the error signal is added to its time derivative. (a) Write the differential equation for the system. (b) Compare the damping ratio and the steady-state error for a velocity input $\theta_i = \omega_i t$ with the case where there is no error-rate signal. (c) How is this system of error-rate derivative control superior to the derivative feedback system of Prob. 22-7?

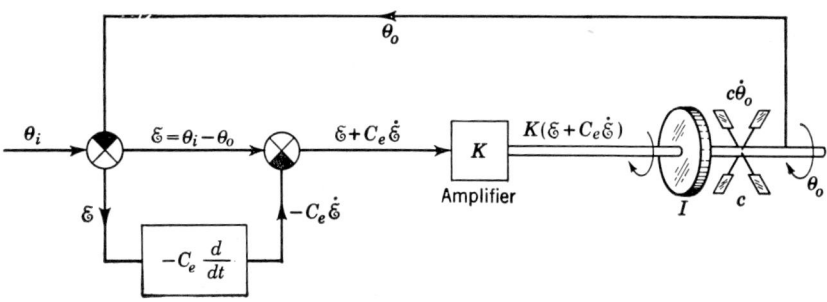

Prob. 22-8

22-9. The figure shows a control system in which an error signal together with its time integral is fed into the amplifier. This type of control is employed in the chemical industry, where it is called *reset control*. A more descriptive name for it, however,

630 DYNAMIC ANALYSIS OF MACHINES

is *error-integral control*. The purpose of this problem is to illustrate its advantages. The reader should first verify that summing the torques on the output shaft gives the following integrodifferential equation:

$$I\ddot{\theta}_o + c\dot{\theta}_o + K\theta_o + KC_r\int\theta_o\,dt = k\theta_i + KC_r\int\theta_i\,dt - T_L$$

and that term-by-term differentiation then yields the third-order differential equation

$$I\dddot{\theta}_o + c\ddot{\theta}_o + K\dot{\theta}_o + KC_r\theta_o = K\dot{\theta}_i + KC_r\theta_i - \dot{T}_L$$

(a) With no external torque load ($T_L = 0$), show that the steady-state error in the system response is zero for a constant velocity input $\theta_i = \omega_i t$. (b) If T_L varies in accordance with the expression $T_L = At + B$, find the steady-state error in the output if $\theta_i = \omega_i t$.

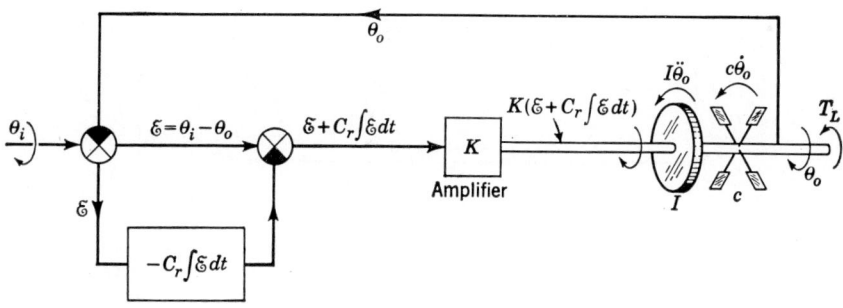

Prob. 22-9

22-10. Both error-rate and integral control, as discussed in the previous problems, can be combined with position control in a closed-loop system. For a system with gain K, inertia I, viscous damping c, and a load torque T_L, draw a block diagram indicating the error signal and feedback paths when incorporating both error-rate and integral control. Write the differential equation for the closed-loop system.

22-11. The same as Prob. 22-10 but use derivative feedback instead of error-rate control.

22-12. The figure shows a hydraulic valve amplifier together with the block diagram for the system. As shown, the amplifier is coupled to a load where the mass, spring stiffness, and viscous damping for the entire system are assumed to be lumped. As shown in the block diagram, the gain of the valve amplifier is to be taken as K_a for small motions. The transfer function for the mass-spring-damper system is

$$\frac{1}{m(j\omega)^2 + c(j\omega) + k}$$

The transfer function for the feedback path is $d_1/(d_1 + d_2)$. Show that the differ-

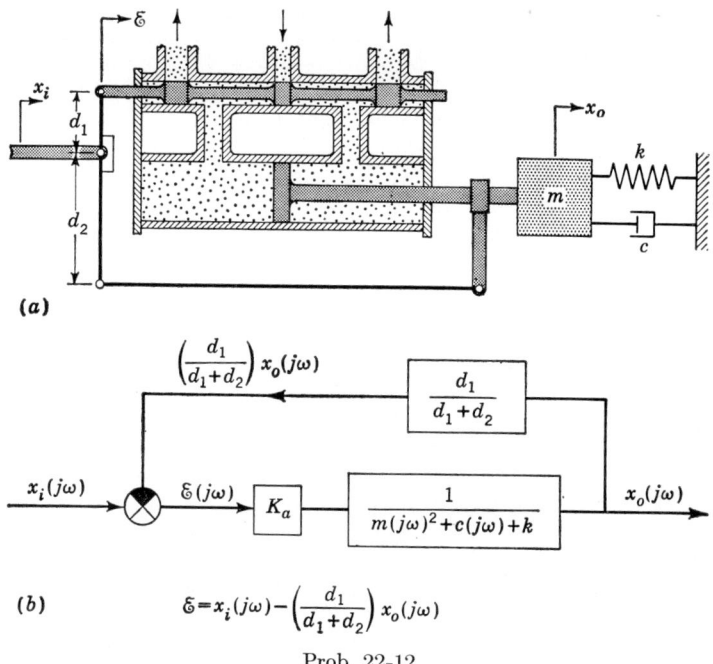

Prob. 22-12

ential equation for the system is

$$m\ddot{x}_o + c\dot{x}_o + \left(k + K_a \frac{d_1}{d_1 + d_2}\right) x_o = K_a x_i$$

Determine the steady-state response of the system if $x_i = A \cos \omega t$

APPENDIX III

MOMENTS OF INERTIA

Rod	$i_y = i_z = \dfrac{ml^2}{12}$
Round disk	$i_x = \dfrac{mr^2}{2}$ $i_y = i_z = \dfrac{mr^2}{4}$
Rectangular prism	$i_x = \dfrac{m(a^2 + b^2)}{12}$ $i_y = \dfrac{m(a^2 + c^2)}{12}$ $i_z = \dfrac{m(b^2 + c^2)}{12}$
Cylinder	$i_x = \dfrac{mr^2}{2}$ $i_y = i_z = \dfrac{m(3r^2 + l^2)}{12}$
Hollow cylinder	$i_x = \dfrac{m(a^2 + b^2)}{2}$ $i_y = i_z = \dfrac{m(3a^2 + 3b^2 + l^2)}{12}$

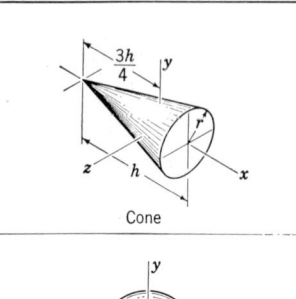
Cone

$$i_x = \frac{3mr^2}{10}$$
$$i_y = i_z = \frac{m(12r^2 + 3h^2)}{80}$$

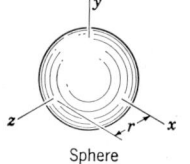
Sphere

$$i_x = i_y = i_z = \frac{2mr^2}{5}$$

APPENDIX IV

INDICATOR DIAGRAMS

Experimentally, an instrument called an *engine indicator* is used to measure the variation in pressure within a cylinder. The instrument constructs a graph, during operation of the engine, which is known as an *indicator diagram*. Known constants of the indicator make it possible to study the diagram and determine the relationship between the gas pressure and the crank angle for the particular set of running conditions in existence at the time the diagram was taken.

When an engine is in the design stage, it is necessary to estimate a diagram from theoretical considerations. Fortunately we have the recorded experiences of many designers to guide us in making this estimate[1] and the approximation will not be too rough. From such an approximation a pilot model of the proposed engine can be designed and built and the actual indicator diagram taken and compared with the theoretically devised one. This will provide much useful information for the design of the production model.

An indicator diagram for the ideal air-standard cycle is shown in Fig. A-3 for a four-stroke-cycle engine. During compression the cylinder volume changes from v_1 to v_2 and the cylinder pressure from p_1 to p_2. The relationship at any point of the stroke is given by the polytropic gas law as

$$p_x v_x^k = p_1 v_1^k = \text{const} \qquad (A\text{-}7)$$

In an actual indicator card the corners at points 2 and 3 are rounded and the line joining these points is curved. This is explained by the fact that combustion is not instantaneous and that ignition occurs before

---

[1] H. Caminez and C. W. Iseler, Standard Method of Engine Calculations, *Air Service Inform. Circ.*, Material Div., Army Air Corps, vol. 5, no. 421, 1923.

R. L. Hersey, J. E. Eberhardt, and H. C. Hottel, Thermodynamic Properties of the Working Fluid in Internal Combustion Engines, *SAE J.*, vol. 39, p. 409, 1936.

Joseph Liston, "Aircraft Engine Design," pp. 26–34, McGraw-Hill Book Company, Inc., New York, 1942.

Milton C. Shaw and Fred Macks, "Analysis and Lubrication of Bearings," pp. 11–19, McGraw-Hill Book Company, Inc., New York, 1949.

James B. Hartman, "Dynamics of Machinery," pp. 172–180, McGraw-Hill Book Company, Inc., New York, 1956.

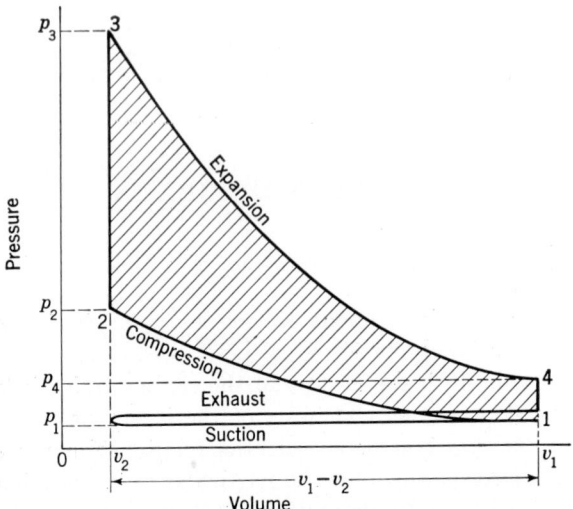

Fig. A-3. Ideal indicator diagram for a four-cycle engine.

the end of the compression stroke. An actual card is also rounded at points 4 and 1 because of the fact that the valves do not operate instantaneously.

The polytropic exponent k in Eq. (A-7) is often taken to be 1.30 for both compression and expansion, although differences probably do exist.

The relationship between the horsepower developed and the dimensions of the engine is given by the equation

$$\text{bhp} = \frac{p_b l a n}{(33{,}000)(12)} \qquad \text{(A-8)}$$

where bhp = brake horsepower per cylinder
p_b = brake mean effective pressure, psi
l = length of stroke, in.
a = piston area, in.^{2}
n = number of working strokes per minute

The amount of horsepower that can be obtained from a cubic inch of piston displacement varies considerably, depending upon the engine type. For automotive engines it ranges from about 0.55 up to 0.90 hp/in.^{3}, with an average of perhaps 0.70 at the present time. On the other hand many marine diesel engines have ratios varying from 0.10 to 0.20 hp/in.^{3} About the best that can be done in designing a new engine is to use standard references to discover what others have done with the same type of engines and then to choose a value which seems to be reasonably attainable.

INDICATOR DIAGRAMS   637

For many engines the ratio of bore to stroke varies from about 0.75 to 1.00. The tendency in automotive-engine design seems to be toward shorter-stroke engines in order to reduce engine height.

Decisions on the bore-stroke ratio and horsepower per unit displacement volume will be helpful in solving Eq. (A-8) to obtain suitable dimensions when the horsepower, speed, and number of cylinders have been decided upon.

The ratio of the brake mean effective pressure p_b to the indicated mean effective pressure p_i, obtained experimentally from an indicator card, is the mechanical efficiency e_m:

$$e_m = \frac{p_b}{p_i} \qquad \text{(A-9)}$$

Differences between a theoretical and an experimentally determined indicator diagram can be accounted for by applying a correction called a *card factor*. The card factor is defined by the equation

$$f_c = \frac{p_i}{p_i'} \qquad \text{(A-10)}$$

where p_i' is the theoretical indicated mean effective pressure and f_c is the card factor, usually about 0.90 to 0.95.

Defining compression ratio (Fig. A-3) as

$$r = \frac{v_1}{v_2} \qquad \text{(A-11)}$$

then the work done during compression is

$$U_c = \int_{v_2}^{v_1} p \, dv = p_1 v_1^k \int_{v_2}^{v_1} \frac{dv}{v^k} = \frac{p_1 v_1}{k-1}(r^{k-1} - 1) \qquad (a)$$

The displacement volume can be written

$$v_1 - v_2 = v_1 - \frac{v_1}{r} = \frac{v_1(r-1)}{r} \qquad (b)$$

Substituting v_1 from Eq. (b) into Eq. (a) yields

$$U_c = \frac{p_1(v_1 - v_2)}{k-1} \frac{r^k - r}{r-1} \qquad (c)$$

The work done during expansion is the area under the curve between points 3 and 4 of Fig. A-3. This is found in the same manner; the result is

$$U_e = \frac{p_4(v_1 - v_2)}{k-1} \frac{r^k - r}{r-1} \qquad (d)$$

638   DYNAMIC ANALYSIS OF MACHINES

The net work accomplished in a cycle is the difference in the amounts given by Eqs. (c) and (d), and it must be equal to the product of the indicated mean effective pressure and the displacement volume. Thus

$$U = U_e - U_c = p_i'(v_1 - v_2)$$
$$= \frac{p_4(v_1 - v_2)}{k-1} \frac{r^k - r}{r-1} - \frac{p_1(v_1 - v_2)}{k-1} \frac{r^k - r}{r-1} \quad (A\text{-}12)$$

If the exponent is the same for expansion as for compression, Eq. (A-12) can be solved to give

$$p_4 = p_i'(k-1) \frac{r-1}{r^k - r} + p_1 \quad (e)$$

Substituting p_i from Eq. (A-10) produces

$$p_4 = (k-1) \frac{r-1}{r^k - r} \frac{p_i}{f_c} + p_1 \quad (A\text{-}13)$$

Equations (A-7) and (A-13) can be employed to create the theoretical indicator diagram. The corners are then rounded off so that the pressure at point 3 is made about 75 per cent of that given by Eq. (A-7). As a check, the area of the diagram can be measured and divided by the displacement volume. The result should equal the indicated mean effective pressure.

EXAMPLE A-1. Construct an indicator diagram for a V8 four-cycle gasoline engine having a 4-in. bore, $3\frac{11}{16}$-in. stroke, and a compression ratio of 9.75. The operating conditions are 270 bhp at 4,600 rpm. Use a mechanical efficiency of 77 per cent and a card factor of 0.90.

Solution. Rearranging Eq. (A-8) we find the brake mean effective pressure as follows:

$$p_b = \frac{(33{,}000)(12) \text{ bhp}}{lan} = \frac{(33{,}000)(12)(270/8)}{(3.6875)[\pi(4)^2/4](4{,}600/2)} = 125 \text{ psi}$$

The indicated mean effective pressure is calculated from Eq. (A-9):

$$p_i = \frac{p_b}{e_m} = \frac{125}{0.77} = 162 \text{ psi}$$

Taking the exponent $k = 1.3$ and atmospheric pressure for p_1, Eq. (A-13) is solved next:

$$p_4 = (k-1) \frac{r-1}{r^k - r} \frac{p_i}{f_c} + p_1$$
$$= (1.3 - 1) \frac{9.75 - 1}{(9.75)^{1.3} - 9.75} \frac{162}{0.90} + 14.7$$
$$= 64.2 \text{ psi}$$

The piston displacement volume is

$$v_1 - v_2 = la = (3.6875) \left[\frac{\pi(4)^2}{4} \right] = 46.34 \text{ in.}^3$$

INDICATOR DIAGRAMS

Since
$$v_1 - v_2 = \frac{v_1(r-1)}{r}$$

we have
$$v_1 = \frac{(v_1 - v_2)r}{r-1} = \frac{(46.4)(9.75)}{9.75-1} = 51.63 \text{ in.}^3$$

and
$$v_2 = 51.63 - 46.34 = 5.29 \text{ in.}^3$$

Therefore the clearance volume is

$$\left(\frac{5.29}{46.34}\right)(100) = 11.43 \text{ per cent}$$

of the displacement volume. Expressing the volumes as percentages of the displacement volume enables us to write Eq. (A-7) in the form

$$p_x(X + 11.4)^{1.3} = p_1(100 + 11.4)^{1.3}$$

where X is the percentage of piston travel measured from the head end of the stroke. Thus, the formula

$$p_{xc} = p_1\left(\frac{100+11.4}{X+11.4}\right)^{1.3} = 14.7\left(\frac{100+11.4}{X+11.4}\right)^{1.3} \tag{1}$$

is used to calculate the pressure during the compression stroke for any piston position between $X = 0$ and $X = 100$ per cent. For the expansion stroke Eq. (1) is written

$$p_{xe} = p_4\left(\frac{100+11.4}{X+11.4}\right)^{1.3} = 64.2\left(\frac{100+11.4}{X+11.4}\right)^{1.3} \tag{2}$$

Equations (1) and (2) are now solved to give the results shown in Table A-1. As previously stated, the indicator diagram is plotted by reducing the maximum pressure 75 per cent and by rounding off the low-pressure end of the expansion curve. When

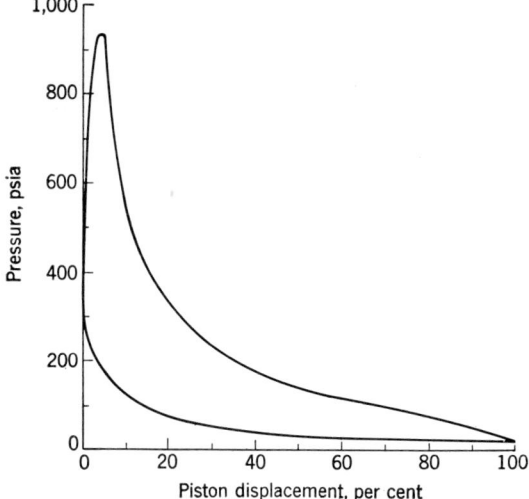

Fig. A-4

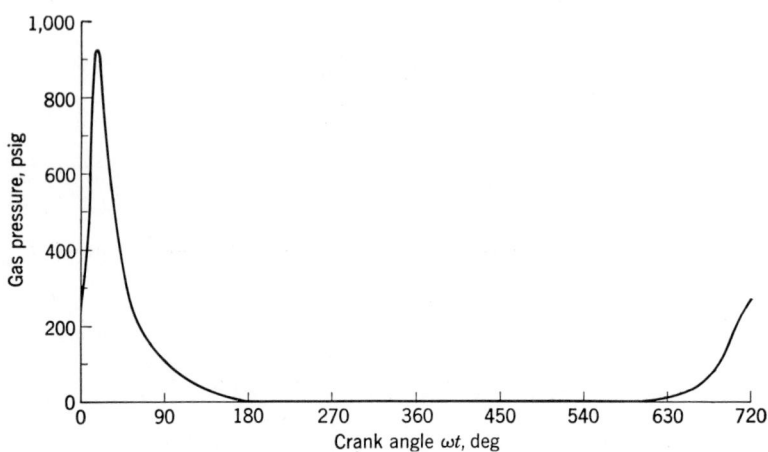

Fig. A-5. Variation of gas pressure in a four-cycle engine.

this is done, the indicator diagram of Fig. A-4 is obtained. With such a diagram the absolute pressure corresponding to any crank angle in the cycle can be obtained. The procedure is to determine the piston position corresponding to the given crank angle. This defines the abscissa of the indicator diagram, and the pressure is then read directly from the chart. Repeating this for all angles between 0 and 720° enables the graph of gas pressure of Fig. A-5 to be constructed. Note that the pressures on this plot are gauge pressures, obtained by subtracting 14.7 psi from each value.

TABLE A-1

Piston travel X, per cent	Compression pressure p_{xc}, psia	Expansion pressure p_{xe}, psia
0	284	1,240
10	125	546
20	76	333
30	52	234
40	40	176
50	32	140
60	26	115
70	22	97
80	19	83
90	17	73
100	14.7	64

INDEX

Acceleration diagrams, 447
Air-standard cycle, 635
Amplitude, dimensionless, 470
  meaning, 444
Amplitude ratio, 616
Angular momentum, 411
Axes, of inertia, 417
  principal, 416
Axis, centroidal, 392
  precession, 431
  of spin, 430
Ayre, Robert S., vii, 438n., 464

Balancers, single-plane, 503
Balancing, by compensation, 517
  of engines, 519, 547, 554
  field, 521
  flexible parts, 520
  nodal-point method, 514
  pivoted-cradle method, 511
  reason for, 500
  (*See also* Unbalance)
Balancing machines, dynamic, 511
  static, 503
Beams, natural frequency of, 479
Bearing force, 363, 426, 501, 506
Beer, Ferdinand P., 405n., 422n.
Block diagrams, 604
  notation, 621
Bore-stroke ratio, 637
Brake horsepower, 636
Brenner, Bernard, 422n.
Brown, Robert G., 623n.

Cam mechanisms, differential equation of, 577
  natural frequency of, 577
Caminez, H., 635n.
Cams, elastic, 576

Cams, forces in, 569
  models of, 574, 576
Card, indicator, 636
Card factor, 637
Center of percussion, 393, 515, 539
Centroidal axis, 392
Characteristic equation, 608
Church, Austin H., 438n., 540
Clearance volume, 639
Closed-loop equation, 620
Closed-loop system, 603, 604
Coefficient of static friction, 365
Cole, E. B., 502n.
Compensation of control systems, 625
Compression ratio, 637
Connecting rod, mass distribution, 539
  types, 527
Correction planes, 514
Couple, arm of, 348
  components of, 348
  inertial, 386
  moment of, 348
  plane of, 348
  shaking, 548
Crabtree, Harold, 430n.
Crank pin, 541
Crankshaft, 558
Crede, C. E., 422n., 472n.
Critical damping, 461
Critical speed, 479, 487, 502
Curreri, John R., 438n., 523n., 558n.
Cycles of engines, 527

D'Alembert's principle, 384
Damping, 366
  in cam mechanisms, 576
  critical value, 461
  phase angle, 464
  viscous, 366
  (*See also* Friction)

Damping constant, 366
Damping factor, 366, 438
  torsional, 441
Damping ratio, 461
  torsional, 605
Damping template, 465
Decaulne, P., 603n.
Decrement, 463
Deflection, static, 477
Deflection curve, 482
Deimel, Richard F., 430n.
Den Hartog, J. P., 438n.
Derivative feedback, 628
Diagrams, circle, 551
  displacement (*see* Displacement diagram)
  free-body, 350
  indicator, 532, 636
  loading, 481
  phase-plane, 459, 460, 467, 584
  shear, 481
  vector, 449, 451
  velocity, 446, 450, 451
Differential equations, complex solution, 616
  of control systems, 603
  nonlinear, 452
  solution of, 607
    steady-state, 609
    transient, 608
Displacement, of cam followers, 596
  phase of, 469
  virtual, 403
Displacement diagram, 446, 451
  of flexible cam system, 580
  by phase plane, 450
Distributed mass, 491

Eberhardt, J. E., 635n.
Efficiency, 371
Elonka, Steve, 516n.
Engines, bearing loads in, 544
  crank arrangements, 527, 535
  diesel, 526
  gas forces, 535
  horsepower, 531, 636
  in-line, 527
  internal-combustion, 526
  multicylinder, 528
    balancing, 551
  specifications of, 530–535

Engines, V, 527
Equipment, isolation of, 473
Equivalent mass, 539
Equivalent systems, 488
Error detector, 604
Error-integral control, 630
Error-rate control, 629
Error signal, 619
Euler's equations, 422, 427

Factors, safety, 373
  service, 373
Feedback, meaning, 601
Firing order, 527
Flywheels, 555
Follower command, 581
Follower response, by Johnson's method, 589
  by phase plane, 581
Force, application of, 347
  bearing, 363, 426, 501, 506
  cam, 569
  collinear, 410
  damping, 469
  direction of, 347
  exciting, 472
  external, 350
  friction, 364–367
  gyroscopic, 432
  harmonic, 467
  inertia, 386, 392, 469, 540, 542
    primary, 542, 551
    secondary, 542, 551
  internal, 350
  isolation of, 473
  magnitude of, 347
  moment of, 348
  normal, 365
  periodic, 439
  phase of, 469
  on piston wall, 537
  shaking, 401, 548
  shear, 481
  spring, 469
  static friction, 365
  step, 445
  transmitted, 357, 474
  unbalanced, 506
  viscous damping, 366
  worm gear, 373
  wrist pin, 538

INDEX 643

Forcing functions, 457
Fourier series, 467
Free-body diagrams, 350
Frequency, 439
  natural, 443, 477
Frequency ratio, 579
Frequency response, 615
  polar, 618
Friction, 364
  Coulomb, 366
  internal, 366
  sliding, coefficient of, 366
  solid, 366
  static, coefficient of, 365
  viscous, 366, 440
  (*See also* Damping)
Friction angle, 368
Friction circle, 368

Gas forces, 640
Gas law, 635
Gear teeth, friction in, 372
Gears, bevel, forces, 359, 361
  helical, forces, 357, 358
  worm, 373
Gille, J. C., 603n.
Green, W. G., 405n.
Gruber, Warren M., 513n.
Gyroscopes, 428

Harmonic function, 606
Hartman, James B., vii, 523n., 635n.
Helix angle, 358
Hersey, R. L., 635n.
Holowenko, A. R., 528n., 549n.
Holzer's method, 484, 558
Hooke's law, 626
Horsepower formula, 636
Hottel, H. C., 635n.
Hydraulic control, 601

Impulse, angular, 411
  linear, 410
Indicator diagrams, 635
Inertia, axes of, 423
  equivalent moments of, 490
  moment of (*see* Moment, of inertia)
  pendulum for measuring, 418
  product of, 413

Inertia force (*see* Force)
Inertia torque, 386, 392, 486, 542
Instability, 601
  (*See also* Stability)
Integration, constants of, 442
Iseler, C. W., 635n.

Jacobsen, Lydik S., vii, 438n., 464
Johnson, Ray C., 575n., 589
Johnston, E. Russell, Jr., 405n., 422n.
Jump, 585
  criteria, 587

Laplace transformation, 599, 619
Law of inertia, 342
Linear, definition, 625
Linear momentum, 410
Liston, Joseph, 635n.
Logarithmic decrement, 463

Mabie, Hamilton H., 422n.
Macduff, John N., 438n., 523n., 558n.
Machines, shock testing, 393
Macks, Fred, 635n.
Mass, of connecting rod, 539
  distributed, 491
  equivalent, 539
  units, 389n.
Mechanical compensation, 518
Mechanical efficiency, 637
Mechanisms, plane motion, 384
  slider-crank, analysis of, 537, 541
Modulus of elasticity, 481
Moment, of couple, 348
  of force, 348
  of inertia, 413, 633
    axis of, 416
    measurement, 418–422, 435
    (*See also* Inertia)
  of momentum, 411
Moment vectors, 348
Moments due to unbalance, 508
Momentum, angular, 411
  components of, 412
  conservation of, 412
  linear, 410
  moment of, 411
Myklestad, N. O., 523n.

Newton's laws, 342
Nixon, Floyd E., 623n.
Node, 484, 492
Nonlinear, definition, 626
Notation, block, 603
  dynamics, 346
Null point, 515
Nyquist criterion, 623

Ocvirk, Fred W., 422n.
Open-loop equation, 620
Open-loop system, 602
Otto cycle, 526
Overshoot, 612

Page, Leigh, 430n.
Particular integral, 607
Pélegin, M. J., 603n.
Pendulum, for measuring inertia, 418
  torsional, 420
Performance, 439
Period, 439
Phase angle, 444, 469, 471, 513, 616
Phase plane, 450
Phelan, Richard M., vii
Piston, 528, 538
Piston position, 537
Polygon, moment, 508
Position control, 600, 601
Position error, 581
Precession, 431
Pressure angle, transverse, 358
Product of inertia, 413
Proportional-error system, analysis of, 611
  definition, 604

Radius of gyration, 393, 416
Ramp function (unit-step velocity function), 606
Rayleigh, Lord (John William Strutt), 478n.
Rayleigh's method, 478
Reset control, 629
Resonance, 439, 471
Response, 439
  amplitude, 470, 472
  of cam followers, 576

Rotating vectors, 446–448
Rotation, 391
Rothbart, Harold A., vii, 569n., 595
Routh's criterion, 623, 624

Safety factors, 373
Savant, C. J., Jr., 623n.
Scarborough, James B., 430n.
Servo, 603
Servomechanism, definition, 603
Shaft, critical speed of, 479
Shaking couple, 548
Shaking forces, 401, 548
  modification of, 549
Shaw, Milton C., 635n.
Speed, critical, 479, 487, 502
Speed fluctuation, coefficient of, 556
Spiral template, 465
Spring constant, 440
  torsional, 441
Spring surge, 594
Springs, energy of, 493
  force of, 469
  helical, 488
  parallel, 490
  precompression of, 586
  retainer, 571
  scale of, 440, 477, 488
    equivalent, 490
  series, 490
  torsional, 489
Spur gears, forces, 357
  friction, 372
  torque, 357
Stability, 622
Stability tests, 623
Standard functions, 606
Static deflection, 477
Static equilibrium, 350
Steady-state operation, 345, 439
Steady-state solution, 607
Step disturbance, 454
Step functions, 606
Stiffness, 440
Strutt, John William, 478n.
Systems, models of, 488

Tabulation method, 484
Thaler, George J., 623n.

Timm, Robert F., vii, 627*n*.
Torque, camshaft, 571
　crankshaft, 538, 547
　inertia, 386, 392, 486, 542
Transfer formula, 392, 417
Transfer function, definition, 618
Transient conditions, 345
Transient operation, 439
Transient solution, 607
Transients, 456
Translation, 391
Transmissibility, 476
Transmitted force, 357, 474

Unbalance, analysis of, 506, 507
　cams, 594
　correction of, 504
　direction, 503
　　measurement of, 518
　dynamic, 504, 506
　effect of, 501
　graphical analysis, 508
　location, measurement of, 518
　magnitude, 503
　moments due to, 508
　phase of, 513
　reciprocating, 548
　rotating, 471
　static, 500, 506
　trouble shooting, 522
　(*See also* Balancing)
Unit-impulse function, 606
Unit-step function, 606
Unit-step input, solution, 609, 610
Unit-step velocity function, 606
Unit-step velocity input, solution, 611
Unit vectors, 347

Vector diagram, 449, 451
Vector triad, 347
Vectors, force, 347
　free, 349
　moment, 348
　phase of, 447
　rotating, 446–448
　unit, 347
Velocity, amplitude, 445
Velocity diagrams, 446, 451
　by phase plane, 450
Velocity-step input, solution, 610
Vibration, 438
　amplitude, 444
　damped, 461
　　amplitude of, 462
　　rate of decay, 462
　force, phase of, 469
　forced, 438
　free, 438, 441
　frequency, 443
　isolation of, 473
　mode, 484
　period, 444
　single degree of freedom, 440
　steady-state, 439, 467
　torsional, 439, 441, 483
　transient, 439, 469
Virtual work, method of, 402
Viscous damping, 366
Viscous friction, 366

Wada, Sanae, 372*n*.
Wilson, W. K., 558*n*.
Windup of camshafts, 595
Worm gears, 373
Worthington, A. M., 430*n*.
Wrist pin, 538, 541